HERITAGE MATTERS

DISPLACED HERITAGE

RESPONSES TO DISASTER, TRAUMA, AND LOSS

Heritage Matters

ISSN 1756–4832

Series Editors
Peter G. Stone
Peter Davis
Chris Whitehead

Heritage Matters is a series of edited and single-authored volumes which addresses the whole range of issues that confront the cultural heritage sector as we face the global challenges of the twenty-first century. The series follows the ethos of the International Centre for Cultural and Heritage Studies (ICCHS) at Newcastle University, where these issues are seen as part of an integrated whole, including both cultural and natural agendas, and thus encompasses challenges faced by all types of museums, art galleries, heritage sites and the organisations and individuals that work with, and are affected by them.

Previous volumes are listed at the back of this book.

Displaced Heritage

Responses to Disaster, Trauma, and Loss

Edited by

IAN CONVERY, GERARD CORSANE
AND PETER DAVIS

THE BOYDELL PRESS

First published 2014
The Boydell Press, Woodbridge
Paperback edition 2019

ISBN 978-1-84383-963-7 hardback
ISBN 978-1-78327-430-7 paperback

The Boydell Press is an imprint of Boydell & Brewer Ltd
PO Box 9, Woodbridge, Suffolk IP12 3DF, UK
and of Boydell & Brewer Inc.
668 Mt Hope Avenue, Rochester, NY 14620–2731, USA
website: www.boydellandbrewer.com

The publisher has no responsibility for the continued existence or accuracy of URLs for external or
third-party internet websites referred to in this book, and does not guarantee that any content on
such websites is, or will remain, accurate or appropriate

A CIP record for this book is available
from the British Library

This publication is printed on acid-free paper

Printed and bound in Great Britain by
TJ International Ltd, Padstow, Cornwall

Contents

Illustrations

COVER IMAGES

(Top) The remains of an Artukid madrassah near the onion-domed tomb of the
  Akkoyunlu prince Zeynel Bey at Hasankeyf.
  Photo: Sarah Elliott
(Middle) Standing next to the Berlin Wall at Checkpoint Charlie, Ellie and Lloyd Land, 1986.
  © *Ellie Land*
(Bottom) *nanoq: flat out and bluesome*, 2004, Snæbjörnsdóttir/Wilson, Spike Island, Bristol.
  Alan Russell

TABLES

Acknowledgments

When preparing the outline proposal for this volume we were anxious to include a discussion of the impact of disasters on natural and cultural heritage, and achieve international coverage. The authors who have contributed to this volume not only accepted the wide-ranging nature of the book but also were committed to exploring both theory and practice; as a result they have delivered a remarkable collection of chapters, and we are grateful to them for their hard work and dedication. They have also accepted and responded to editorial demands and timescales, and we thank them for their patience. We are especially grateful to Kai Erikson, who wrote the Preface, and to Phil O'Keefe, who wrote the Endpiece, for their fascinating and insightful contributions.

Thanks are due to our respective institutions, the University of Cumbria and Newcastle University. The latter has been especially supportive of the *Heritage Matters* initiative, and we are extremely grateful for their continuing involvement in developing a wide-ranging series of publications on varied aspects of heritage; this volume in particular demonstrates the commitment of both institutions to taking new theoretical and multidisciplinary approaches to the concept of heritage, and we are particularly thankful for their support.

We are especially indebted to Catherine Dauncey, Publications Officer in the International Centre for Cultural & Heritage Studies, Newcastle University (ICCHS), who has provided support and encouragement throughout; she has dealt calmly and effectively with the demands of the editors and contributors, ensuring that deadlines were met and queries addressed. She has also ensured that final copy meets the demands of the series style and we are very grateful for her careful and thorough work on the texts. We would also like to thank Caroline Palmer at Boydell & Brewer for her continued support, understanding and good humour during the development of this volume in the *Heritage Matters* series.

Ian Convery, University of Cumbria
Gerard Corsane, Newcastle University
Peter Davis, Newcastle University

Abbreviations

ARD	asbestos-related diseases
ASL	above sea level
BMN	Balkans Museums Network
BSE	bovine spongiform encephalopathy
bTB	bovine tuberculosis
CCP	Chinese Communist Party
CHwB	Cultural Heritage without Borders
COTS	Crown of Thorns starfish
DEFRA	Department for Environment, Food and Rural Affairs
ECGD	Export Credits Guarantee Department
ECLAC	Economic Commission for Latin America and the Caribbean
EU	European Union
FMD	Foot and Mouth Disease
FRELIMO	Frente de Libertação de Moçambique
FUW	Farmers' Union of Wales
GBR	Great Barrier Reef
GBRMPA	Great Barrier Reef Marine Park Authority
GDR	German Democratic Republic
GMB	General, Municipal, Boilermakers and Allied Trade Union
GNP	Gorongosa National Park
HEP	hydro-electric power
HMP	Her Majesty's Prison
HPSEB	Himachal Pradesh State Electricity Board
IAPA	'intensive action' pilot area
ICCHS	International Centre for Cultural & Heritage Studies, Newcastle University
ICJB	International Campaign for Justice in Bhopal
ICOM	International Council of Museums
iDTR	Institute for Dark Tourism Research, University of Central Lancashire
IICT	Indian Institute of Chemical Technology
ISG	Independent Scientific Group
IST	Indian Standard Time
IUCN	International Union for Conservation of Nature
KGB	Komitet gosudarstvennoy bezopasnosti
KMT	Kuomintang
KOCB	Kinabatangan Orang-utan Conservation Programme
MAFF	Ministry of Agriculture, Fisheries and Food
MIC	methyl isocyanate
MLK PDU	Maze/Long Kesh Programme Delivery Unit
MP	Madhya Pradesh
NFU	National Farmers' Union
NGO	non-governmental organisation

NLM	National Lithuanian Museum
PAC	Pembrokeshire Against the Cull
PKK	Partiya Karkarên Kurdistan (Kurdistan Workers' Party)
POST	Parliamentary Office of Science and Technology
POW	prisoner of war
PRC	People's Republic of China
PUMAH	Planning, Urban Management and Heritage
RENAMO	Resistencia Nacional Moçambicana
ROSPA	Royal Society for the Prevention of Accidents
RRC	Regional Restoration Camps
SIB	Strategic Investment Board Limited
SOAS	School of Oriental and African Studies, University of London
SOHC	Scottish Oral History Centre Archive, University of Strathclyde
SQPV	squirrel poxvirus
TGWU	Transport and General Workers' Union
TOI	Times of India
TVBS	Television Broadcasts Satellite
UCC	Union Carbide Corporation
UN	United Nations
UNDP	United Nations Development Programme
UNESCO	United Nations Educational, Scientific and Cultural Organization
WAG	Welsh Assembly Government
WCD	World Commission on Dams
WEIRD	Western, educated, industrialised, rich and democratic
WHO	World Health Organization
WILD	Women's International Leadership Development
ZANU	Zimbabwe African National Union

Preface

Kai Erikson

I have been asked to provide a brief foreword to this collection of chapters: not to introduce the contents of what is to follow but to offer 'a personal reflective account of working with disasters that can help position their place in human experience'. I was offered this honour because I happen to have spent the past 40 years visiting scenes of disaster and writing about them. At the risk of appearing to open with a mindless personal travelogue, then, I propose to provide a brief sampling of those places – in rough chronological order – as I have on other occasions.

A coal mining valley in West Virginia known as Buffalo Creek, where a winter flood, caused by a faulty impoundment, caused a terrifying amount of damage to the people living downstream.

An Ojibwa Indian Reserve in Northwest Ontario, Canada, where a mercury spill that had taken place many miles upstream entered the waterways along which the natives had lived, and from which they had drawn sustenance, since the beginning of their reckoning of time. They viewed those waters as a living thing and called the alien substance that had insinuated its way into them 'pijibowin', Ojibwa for poison. This shook not only the native economy profoundly but also the native way of life, leaving them with what their wise young chief called 'a broken culture'.

A nuclear power plant in Pennsylvania called Three Mile Island, where a near meltdown resulted in unexpected levels of dread on the part of people who lived in its shadow. By the standards of the time, that reaction appeared puzzling and even 'irrational', but it became apparent before long that it was the standards, not the human reaction, that failed the test of reason.

A migrant farm worker camp in South Florida, where a large group of numbingly poor migrants from Haiti discovered that the money they had earned working the fields – money their families back home had counted on for everyday survival – had simply been stolen.

Native villages along the coasts of Alaska where the seas from which people had extracted their living for centuries had been blackened and polluted for a thousand miles by a gigantic oil spill. It began as a threat to the native fishery in ways that go far beyond easy calculation, and ended as a threat to their very culture – already a way of life made more fragile by the Russians and then the Americans who had moved into their homeland.

A town in the Western Slavonia region of Croatia, where the townspeople were almost equally divided between ethnic Serbs and Croats, and where close to half the marriages conducted in the last 50 years had been between a Croat and a Serb. The town became a scene of violent civil war at the time of the collapse of Yugoslavia in the early 1990s. The physical landscape was badly damaged by the tides of war, but that was no more than a surface layer of wreckage. The *human* landscape was even more sharply disrupted, in ways both visible and invisible. Whole neighbourhoods, even families, were split apart as sharply as crystals struck by a hammer.

A remote atoll in Micronesia that had been visited more than half a century earlier by a huge cloud of radioactive fallout, the result of an above-ground nuclear test in Bikini, many miles away. In the years since, an unusual number of ailments have been discovered there of the sort known to be related to radioactivity, although the islanders knew little about that, and a dark sense of uneasiness soon worked its way into the very fabric of local culture, affecting the way people viewed themselves, the way they raised children, and the way they measured their own abilities and strengths.

And, finally, the Gulf region of the United States, including the celebrated city of New Orleans, where a tropical storm with the fetching name of 'Katrina' moved across a broad stretch of terrain, doing harm to everything in its path – and, in a way, reaching out to people all over the world through the attention it managed to attract. As we approach the 10th anniversary of Katrina, it is more than reasonable to declare that the storm is still raging.

This is quite an assortment of horrors: a flood, mercury contamination, a nuclear emergency, an act of larceny, an oil spill, a civil war, radioactive fallout and a fierce hurricane. And they touch the lives of a rich assortment of peoples: coal miners from the mountains of Appalachia, native hunters and trappers from subarctic Canada, townsfolk from central Pennsylvania, migrant farm workers from Haiti, native fishers along the hard coasts of Alaska, South Slavs from what was once Yugoslavia, Micronesian islanders on a remote speck of land out in the South Pacific, and those still reeling persons who live – or once lived – in New Orleans and along the Gulf coast of the United States. This assortment does not reach across as wide a range as this volume does either culturally or geographically, of course, but it will be my warrant for speaking in generalities.

My invitation is to speak of what I think I have learned over those years of study, and I have to begin with what should be obvious – that my 'knowledge' is formed in part by my time in the field, but it owes a far greater debt to the work of colleagues following the same pursuit, including several whose work is represented in this volume. I will be speaking of these conclusions as if they were established *findings*, but they should be understood as *questions*: things to ponder as we continue our inquiry into the nature of catastrophic events.

Most of us have approached disasters over the years taking it more or less for granted that they can be divided into 'natural' and 'human-made' categories. The difference seems evident on its face. It is becoming ever more clear, however, that the closer one looks at the distinction, the more blurred it becomes. Tsunamis and earthquakes and similar disturbances seem to leap into the flow of life out of nowhere as they strike the human settlements in their path, and they are clearly of natural origin. But what makes those events 'disastrous' has less to do with their ferocity than what lies before them. An earthquake registering at the very top of the Richter Scale – a tremendous force of nature – will not be called a 'disaster' in the evening news if it hits an empty island in the South Seas and does no harm to human habitation. It will not even be called 'news'. Something ranking as a disaster, then, is a collision between a natural or a human-made event of some kind and a site shaped by human hands. It is what those hands built and not what nature wrought that makes of that collision a disaster.

Voltaire once noted that the Lisbon earthquake of 1755 was an act of nature, and he received a well-known rebuke from Rousseau: 'Admit … that nature did not construct twenty thousand houses of six to seven stories there and that if the inhabitants of this great city had been more equally spread out and more lightly lodged, the damage would have been much less and perhaps of no account' (Masters and Kelly 1992 [1756], 110). (The argument Rousseau was making took him elsewhere, but the point remains valid.) We are in the habit of naming disasters for the events that brought them to attention. We know the day, the hour, the minute that the tsunami hit the shore or the earthquake made itself felt or the waters began to rise or the dam collapsed. But more often than not, when we draw attention to a disaster by naming it or by locating it in time we are in effect marking the moment it became news, the moment it became widely noticed, and not necessarily the moment that specialists, looking back, would cite as the real starting point of the trouble. Katrina offers a striking example. The city of New Orleans was devastated by an assault known by everyone as 'Katrina', but the irony is that the winds of that hurricane had abated and its surging waters had slipped back meekly into the Gulf before the first hints of damage appeared in the city itself. In a very real sense, the real disaster did not even begin in New Orleans until 'Katrina' had disappeared from the radar.

How then should we date the appearance of that disaster? By the day the levees failed? By the day the decision was taken to build levees as a way of fending off inconvenient natural processes? In some other way?

Once we find ourselves looking backward in time to moments well before the official beginning of a disaster, we sometimes learn, as Anthony Oliver-Smith did in his study of what he aptly identified as 'Peru's Five-Hundred Year Earthquake', that its true origins are to be found far back in the course of history. Katrina, again, offers a good example.

If we were to ask thoughtful persons from New Orleans how that terrible event came to be, we are quite likely to hear about the building of the city's levees; earlier decisions to cut canals through that urban space, making the levees necessary; hidden distinctions of race and class that have cut other lines through social space; and so on. And if we reach far enough into that past, the collapse of the levees at the time of Katrina can be understood as the end point in a continuing sequence as readily as it can the beginning point in a new one.

The same would be the case if we were to confer with persons who lived along the Gulf coast and were exposed to a quite different facet of that complex disaster. They were hit hard by the winds and waters of the hurricane directly, and for them it was a straightforward storm – a natural occurrence of the kind they knew the look and feel of from long experience. But they, too, look back now and realise that those raging waters reached them as quickly and as cruelly as they did because the land they live on and once counted on to cushion them from the storms of the Gulf had not only been entirely reshaped but vastly diminished by human efforts to outwit the forces of nature and to extract its riches quickly.

So it can be difficult to know when to date the beginning of a calamitous event, and it is just as difficult to know how to date its end. The social and behavioural sciences are rich with schemes tracing the 'stages' of a disaster, the final one, almost inevitably, being something called 'recovery'. To use a term like that is to suggest that victims have been restored to their original state and have been able to return to their former lives. I will not go into that in any detail as I bring these brief remarks to a close, but I will share my doubt that this actually happens very often. Most survivors, we have reason to know, find ways to stabilise and carry on and take a place in everyday social life. But they have been changed by the experience, and they look out at the world through eyes that can be sadder and wiser, as the old expression goes. The sadness comes from realising how grim and how unforgiving life can sometimes be. The wisdom comes from knowing things that the rest of us can only guess at, and I have long felt that we should be taking advantage of that wisdom. It is difficult to think of anyone better qualified to serve as the teachers and the healers of fellow human beings who come to find themselves in similar straits. It would be good for patient and healer alike.

But that is an aside. The main point I want to make is that a disaster cannot be understood fully as a discrete event. In order to get a sense of its true dimensions, we have to allow it to settle back into the larger flow of history, and then study what happened before and what happened – or is happening – afterwards.

And that may be as good a place as any for a discussion of the impact of disasters on heritage – natural, cultural and intangible – and on the processes and practices of memorialisation, to begin.

BIBLIOGRAPHY AND REFERENCES

Masters, R D, and Kelly, C (eds), 1992 [1756] *The Collected Writings of Jean-Jacques Rousseau*, vol 3, University Press of New England, Hanover NH

Introduction

IAN CONVERY, GERARD CORSANE AND PETER DAVIS

INTRODUCTION

Disasters, whether they are natural or caused by people, change the environment and 'displace' heritage resources. They can be dramatic natural impacts such as tsunami and volcanic eruptions, or terrible events unleashed by humankind, including holocaust and genocide. Sometimes disasters are more insidious, such as the logging of rainforests for short-term gain, or elevated sea temperatures, possibly linked to global climate change, that result in thermal stress and bleach coral ecosystems; these may be slower events but their impact is still hugely significant. Disasters can be high-impact events or occur on a small, localised scale. Whether natural or human-made, rapid or slow, great or small, the impact is effectively the same; nature, people and cultural heritage are displaced or lost.

What constitutes 'disaster'? At first this might seem a fairly straightforward question, but 'disaster' eludes simple definition, or as Philip Buckle (2007) puts it, defining disaster is never easy and rarely definitive. Indeed, the word disaster is so frequently used in everyday dialogue, from misplaced house keys to major events such as earthquakes and hurricanes, as to be almost meaningless (Convery *et al* 2008). This ubiquity is problematic, and as López-Ibor (2005) notes, in academic disciplines it is almost impossible to find an acceptable definition of a disaster. The term originates from the unfavourable aspect of a star, from the French *désastre* or Italian *disastro*, and suggests that when the stars are poorly aligned, unfortunate things are likely to happen; the implication is that disastrous events are outside human control. Indeed, disasters may still be viewed as 'events from the physical environment... caused by forces which are unfamiliar' (López-Ibor 2005, 2) and frequently unforeseen.

Nesmith (2006, 59) writes that the word disaster has many synonyms that add conceptual significance to the term in communicating misery, death, destruction, helplessness, sudden reversal of what is expected and unhappy resolutions to distressing events. She provides a set of defining characteristics:

- Event that disrupts the health of and occurs to a collective unit of a society or community
- The event overwhelms available resources and requires outside assistance for management and mitigation
- Represents tremendous relative human losses
- Negative impact event of natural, financial, technologic, or human origin, for example, armed conflict
- Represents a breakdown in the relationship between humans and the environment

Disasters can therefore take many forms, but what links together all the above is the notion that, whatever form an event might take, it becomes a disaster when the resulting situation exceeds our human ability to recover (Convery *et al* 2008). This broad definition of disaster has been adopted by many international agencies and NGOs. For example, The World Health Organization (2002) states that a 'disaster is any event that exceeds the capacity of individuals or communities affected to alleviate their suffering or meet their needs without outside assistance'. The United Nations Development Programme (2004, 136) also highlights a serious disruption of the functioning of society, causing widespread 'human, material, economic or environmental losses, which exceed the ability of those affected to cope'. Similarly, Blaikie *et al*'s (1994, 21) influential book *At Risk* highlights that disasters occur when a 'significant number of vulnerable people experience a hazard and suffer severe damage and/or disruption of their livelihood system in such a way that recovery is unlikely without external aid'. Even with external aid, disasters may have significant impacts on the living conditions, economic performance and environmental assets and services of affected areas (UN/ECLAC 2003). Consequences may be long-term and may even irreversibly affect economic and social structures and the environment.

While disasters often tend to have a specific focal point, they are not completely discrete events. Their possibility of occurrence, time, place and severity can often be predicted in advance and is likely to link to events elsewhere. As O'Keefe and Middleton (1998) indicate, there are ambiguities in the ways in which disasters are identified, and while some natural events may act as a trigger for disaster, economic, social and political factors also play an important role. Similarly, Hearns and Deeny (2007) argue that many of the world's disasters are now complex emergencies because they include a multiplicity of problems such as war, ethnic conflict, famine, endemic diseases and political unrest.

Yet at the same time the impact of disasters is highly situated; they occur in locales that have their own very distinctive cultural and natural heritage resources. As McLean and Johnes (2000) indicate, disasters are embedded within specific socio-economic, historical, cultural and chrono-logical contexts. While heritage is at risk from disasters, in time sites of suffering are sometimes reframed as sites of memory; through this different lens these 'difficult' places become heritage sites and even attract tourists. Disasters can also make us better informed about the heritage we have lost, or are losing, leading to attempts to better preserve what remains by creating new mechanisms to recognise and protect natural and cultural heritage resources. Disasters and catastrophic events can be seen as 'happenings' that entangle people, place and their heritage, and disasters and displacement can leave people overcome by a 'loss of self' and a 'loss of place'. Particularly at risk of displacement are the cultural and natural capital assets of communities and their place; both forms of capital can be deeply affected. Material culture – buildings, objects, possessions – is a vital element in the framing of people's identities; similarly nature makes a deep psychological contribution to our construction of identity and place (Clayton and Opotow 2003). While the intangible cultural heritage of a community might be considered as less at risk from catastrophe, in extreme cases the loss of culture bearers, or dramatic shifts in society, can result in the loss of these heritage assets. For example, Onciul's chapter in this volume (Chapter 16) deals with events which changed forever the lives of Native American peoples and impacted on their intangible cultural heritage.

Harada's (2000) view that 'individuality makes sense only in a social context ... we construct ourselves and we are constructed both "socially" or "individually" in space and with relations with the materials' chimes with these reflections on the intangible nature of place, society and heritage. When familiar spaces, situations and materials are changed, we are forced to reconsider our social and individual relations. Harada argues that sociality is made up out of the material objects which people use and which surround them in daily life. Disasters can reorder meaning, and 'things' can either become out of place, out of proportion or lost altogether during disasters. As Convery *et al* (2008) note, everyday objects, materials, tools and spaces take on heightened significance, become transformed or enlarged into resonant materials which carry traumatic associations, even agency. Objects collected during times of trauma can evoke strong responses and trigger reminiscence, a fact that museum educators use not just to educate but to encourage healing and closure. Museums, through their object collections and educational activities, can bring an important lens to the significance and power of 'things': materials, artefacts, places, buildings and events can come together to demonstrate the relationship between material order and social meaning and to reflect upon and manage the memories of crisis. Most museums in the UK offer a 'reminiscence service', allowing local people to recall, value, share and preserve their personal memories; while not all of them attempt to deal with traumatic events, 'reminiscence boxes' – particularly those relating to periods such as World War II – often contain objects that trigger emotional responses. Northern Ireland's 'Reminiscence Network' (www.rnni.org.uk), a consortium of Education and Library Services, Museums, Health and Social Services, Sheltered Housing, Churches and Community Groups, has very specific programmes (and helpful bibliographies for reminiscence providers), which deal with Northern Ireland's 'Troubles', truth-telling and reconciliation. Museums' active collecting of artefacts associated with traumatic events is now regarded as a vital aspect of curatorial practice, as discussed by Besley and Were (Chapter 4, this volume).

Hetherington and Munro (1997, 197) note that places – in the discussions in this book these include museums, galleries and heritage sites – have the effect of 'folding of spaces, times and materials together into complex topographical arrangements that perform a multitude of differences'. We are aware that objects can act as important emotional triggers; similarly photography can provide a 'cultural shock' that links the local to wider society, and extends some understanding of the trauma to others 'outside' the disaster (Ashmore 2013; see also Chapter 23, this volume). Writing in relation to an exhibition of images detailing the 2001 Foot and Mouth epidemic in the UK, Ashmore writes that when these landscape images have been exhibited, comments reveal that they have been effective vehicles for conveying the trauma of Foot and Mouth to those people affected and an outside audience alike. Landscape imagery can thus communicate the traumatic human experience of a changed 'lifescape' and is actually essential to understanding the traumatic experience of a crisis which happened within, and to, the landscape.

The management of *place* post-disaster is frequently difficult and complex. Convery *et al* (2008) argue that the ability to recognise the new identities and resonances taken on by objects and places in disasters is vitally important. Commemorating and recording disasters, *in situ*, is important. The work of museums and heritage agencies can provide a different lens to view disasters; it enables us to see beyond the initial event and view them as happenings that entangled the local and the global, the historic and the future, continuity and dramatic change; however, it is important that specificity of place is regarded as paramount. In some instances, it is important for *place* to be left intact. This is evidenced in holocaust sites such as Auschwitz (Poland) and the Tuol Sleng Genocide Museum (Cambodia); disaster sites such as the Union Carbide plant in Bhopal (India); or places of internment such as Robben Island (South Africa) or the Maze Prison (Northern Ireland). By maintaining these sites they can function not just as reminders of dreadful events and difficult pasts but also as a means of deterrence (Stone 2011).

At other times there are compelling reasons to remove the evidence of disasters and consequently change physical places. One such example is the site of the 1966 Aberfan disaster in which 144 people were killed, 116 of them children, buried when a coal waste-tip engulfed the village school. Following extensive site restoration, Portland and Nabresina stone memorials were erected in Aberfan cemetery, accompanied by archways and a memorial garden. These original memorials were replaced in 2007 by the Aberfan Memorial Charity, using polished Pearl White granite; re-engraving of all inscriptions and the erection of additional archways created a site of continuing remembrance. The concept of gardens as memorials has been taken a step further in the National Memorial Arboretum (Staffordshire, England); planting began in 1997 and the site now acts as a special place to honour men and women who have served in the armed forces, police, fire and rescue and ambulance services. It is not a cemetery but a peaceful and beautiful place to remember and reflect.

Whether retaining traumatic sites intact or creating new forms of memorial (and hence changing places), the notion of creating a place to reflect on hardship, disaster and loss is one of the most difficult areas for heritage, museum and tourism professionals to manage successfully. Their actions have the potential to result in understanding, commemoration and deterrence, but care must be taken to consult, listen and respond to local communities – those who have suffered the trauma – at every stage of the process. In any disaster situation, affected communities need the opportunity to regain some control of their own lives, personal spaces and collective space; they must be regarded as essential contributors to recording, remembering and commemorating disaster events. The importance of community response is also noted by Bakker *et al* (2005, 808), who state that the 'lives of citizens are enhanced by, and indeed inseparable from, the construction of collectivities' (consisting of humans, materials and agency). The development of such relationships is clearly important in terms of empowering people to act during a disaster, but also, we would argue, in assisting remembrance and commemoration, post disaster.

Keiffer (2013) notes that the process of remembering highlights the progress made, but also how much we still have to learn about the impact of disasters. This book is therefore timely because of the rapidly developing interest in disasters and their connections to place and heritage. While a number of books have been written about heritage and sense of place, few have considered these issues in relation to disasters, trauma and suffering. The contributors selected for this collection all work within the arena of heritage, sense of place, disasters and sites of suffering, and cover a wide range of disciplines. All are keen to be involved in exploring and theorising the links between disasters, trauma, displacement, human suffering, heritage and place. Most importantly, they all see the need for an edited collection which draws together these issues – exploring both theory and practice – into a single volume. The book includes the views of academics, practitioners and writers from heritage and museum studies, sociology, history, geography, disaster studies, health and the arts.

BIBLIOGRAPHY AND REFERENCES

Ashmore, R, 2013 'Suddenly There Was Nothing': Foot and Mouth, communal trauma and landscape photography, *Photographies* 6 (2), 289–306, DOI: 10.1080/17540763.2013.766632

Bakker, K, Braun, B, and McCarthy, J, 2005 Hurricane Katrina and abandoned being, Guest Editorial, *Environment and Planning D: Society and Space* 23, 795–809

Blaikie, P, Cannon, T, Davis, I, and Wisner, B, 1994 *At Risk: Natural hazards, people's vulnerability, and disasters*, Routledge, London

Buckle, P, 2007 *Building Partnerships for Disaster Risk Reduction and Natural Hazard Risk Management*, World Bank and the United Nations International Strategy for Disaster Reduction

Clayton, S, and Opotow, S, 2003 *Identity and the Natural Environment*, MIT Press, Cambridge MA and London

Convery, I, Mort, M, Baxter, J, and Bailey, C, 2008 *Animal Disease and Human Trauma: Emotional Geographies of Disaster*, Palgrave, Basingstoke

Harada, T, 2000 Space, materials, and the 'social': in the aftermath of a disaster, *Environment and Planning D: Society and Space* 18, 205–12

Hearns, A, and Deeny, P, 2007 The Value of Support for Aid Workers in Complex Emergencies: A Phenomenological Study, *Disaster Management and Response* 5, 28–35

Hetherington, K, and Munro, R, 1997 *Ideas of Difference: Social Spaces and the Labour of Division*, Blackwell, Oxford

Keiffer, S, 2013 *The Dynamics of Disaster*, W W Norton, New York

López-Ibor, J, 2005 What is a disaster?, in *Disasters and Mental Health* (ed J López-Ibor), Wiley, New York

McLean, I, and Johnes, M, 2000 *Aberfan: Government and Disasters*, Welsh Academic Press, Cardiff

Nesmith, E, 2006 Defining 'Disasters' with Implications for Nursing Scholarship and Practice, *Disaster Management and Response* 2, 59–62

O'Keefe, P, and Middleton, N, 1998 *Disaster and Development (The Politics of Humanitarian Aid)*, Pluto Press, London

Stone, P R, 2011 Dark Tourism Experiences: mediating between life and death, in *Tourist Experience: Contemporary Perspectives* (eds R Sharpley and P R Stone), Routledge, Abingdon, 21–7

United Nations Development Programme, 2004 *Reducing Disaster Risk: A Challenge for Development*, United Nations Development Programme Bureau for Crisis Prevention and Recovery, New York

United Nations Economic Commission for Latin America and the Caribbean, 2003 *Handbook for Estimating the Socio-economic and Environmental Effects of Disasters* [online], available from: http://www.eclac.cl/publicaciones/xml/4/12774/lcmexg5i_VOLUME_Ia.pdf [21 May 2014]

World Health Organization (WHO), 2002 *Disasters & Emergencies*, Panafrican Emergency Training Centre, Addis Ababa

Displaced Heritage:
Histories and Tourism

1

Dark Tourism and Dark Heritage: Emergent Themes, Issues and Consequences

Catherine Roberts and Philip R Stone

Introduction

The ways in which societies (re)present death, dying and their dead has long been symbiotic with particular cultural representations of mortality. These representations are often bound up with heritage and tourism, whereby travelling to meet with the dead has long been a feature of the touristic landscape. Examples of early travel to sites of death and the dead can be found in medieval pilgrimages and their reliquary associations, or in Grand Tour visitations to tombs and petrified ruins of the ancient world, or in touristic visits to deceased authors' homes, haunts and graves during the Romantic period of the 18th and 19th centuries. The historical precedent of how travel (and tourism) provided compelling techniques for imaginatively contacting the dead is well founded (Westover 2012). Thus, despite an increasing academic and media focus on contemporary 'dark tourism' – that is, travel to sites associated with death, disaster or the seemingly macabre – the act of travel to such sites is not a new phenomenon (Stone 2011). Nonetheless, the practice of present-day dark tourism has the capacity to expand boundaries of the imagination and to provide the contemporary visitor with potentially life-changing points of shock. Consequently, sites of dark tourism are vernacular spaces that are continuously negotiated, constructed and reconstructed into meaningful places (Sather-Wagstaff 2011). Furthermore, dark tourism can represent inherent political dichotomies of a 'heritage that hurts' and, in so doing, offer a socially sanctioned, if not contested, environment in which difficult or displaced heritage is consumed. Given its transitional elements and potential to influence the psychology and perception of individuals, dark tourism as a rite of social passage occurs within constructivist realms of meaning and meaning making. Arguably, dark tourism as part of a broader (dark) heritage context provides a contemporary lens through which the commodification of death may be glimpsed, thus revealing relationships and consequences of the processes involved that mediate between individuals and the societal frameworks in which we reside. The purpose of this chapter, therefore, is to offer an overview of key themes, issues and consequences of how dark tourism can construct and disperse knowledge through touristic consumption of traumascapes that, in turn, can help make contested heritage places salient and meaningful, both individually and collectively. Firstly, however, a review of dark tourism and the tourist experience provides a context for subsequent discussions.

DARK TOURISM AND THE 'DARK TOURIST' (EXPERIENCE)

Dark tourism as a field of academic scrutiny is where death education and heritage tourism studies collide. Consequently, the scholarly attention on dark tourism and the inherent visitor experience it entails has generated a wealth of typologies, including a surge in descriptive additions to heritage and tourism vocabularies, including thanatourism (Seaton 1996), black spots (Rojek 1993), grief tourism (West 2004) and morbid tourism (Blom 2000). However, despite often-protracted debates over what is and what is not 'dark tourism', the contested term of *dark tourism* has been increasingly applied to a diverse range of global heritage sites, attractions and exhibitions that showcase death. Developing the idea that particular touristic sites of death can either be subjectively lighter or darker (Miles 2002), Stone (2006) offered a dark tourism classification or 'spectrum' that outlined a qualitative set of site-related factors, including political ideologies, educational orientations and interpretation authenticity, that influence 'shades' of touristic experience. Subsequently, there have been concurrent tendencies towards an expansion of the dark tourism typological base, as new locations are brought into the body of research. Correspondingly, there has been a distillation of research within specific subsets of dark tourism, particularly toward the 'darker' poles of positional spectrums: graveyards and cemeteries (Seaton 2002), Holocaust sites (Beech 2009), places of atrocity (Ashworth and Hartmann 2005), prisons and crime sites (Wilson 2008; Dalton 2013) and slavery-heritage attractions (Dann and Seaton 2001; Rice 2009). These subsets of dark tourism are frequently symbolised by iconic landscapes that are often instantly recognisable as well as being recurrent in the academic literature as case studies. For example, Holocaust sites such as Auschwitz-Birkenau in Poland, or Ground Zero in New York – site of the 9/11 attacks – or the Killing Fields in Cambodia where the former Khmer Rouge leader Pol Pot committed genocide against his own people, carry extraordinary semiotic weight. Hence, this uncanny significance may influence not only public perception and visitor behaviours, but also research approaches and processes. While discussion of such influences is beyond the scope of this chapter, impacts and consequences of consuming dark tourism may relate to deep-rooted psychosocial concerns about appropriateness, deviance and the taboo (Stone and Sharpley 2013).

The juxtaposition of sites where historic and human significance is of particular magnitude (for example, death camps of the Holocaust) with less socially consequential sites (such as dungeon visitor attractions in the UK and elsewhere) further problematises the dark tourism 'brand', especially within broader heritage terms. Concern about seemingly arbitrary correlation of remarkably different experiences leads some commentators to highlight the risk of dark tourism research findings becoming ambiguous (Stone 2011). Continuing efforts to finesse dark tourism definitions find resonance in Crick's (1989, 313) comment that touristic taxonomies 'separate phenomena that are clearly fuzzy or overlapping'. Meanwhile, Stone's (2006, 146) reservation – 'whether it is actually possible or justifiable to collectively categorise a diverse range of sites, attractions and exhibitions that are associated with death and the macabre as "dark tourism"' – highlights the inherent vulnerability of conceptual frameworks founded on positional spectrums with a limited set of potentially deeply subjective axes – these being, firstly, place/product and, secondly, light/dark qualities. Nevertheless, this vulnerability can be reduced when focus on place attributes is matched by scholarly interrogation of the tourist experience, to inform a more holistic and, crucially, a consequential societal approach, rather than simply a tourist motivation research perspective.

Even so, while tourist motivation has connotations of impetus or attraction that may reinforce a reductive supply/demand paradigm, related research offers useful hypotheses around experiential, contemplative and/or psychological motivations, and corresponding mediating devices. Seaton (1996), for instance, proposes dark tourism as the desire to 'experience' a kind of death as a motivating factor, while later research by Stone (2012a; 2012b) theorises the consequences of visiting some dark tourism sites as a means by which individuals might contemplate their own life and mortality through the tourist gaze on death. Moreover, while Lennon and Foley (2000, 11) suggest dark tourism is an 'intimation of post-modernity', Seaton (2010) traces manifestations of what he terms 'thanatourism' throughout the history of Western civilisation, and its subsequent traditions of thanatopsis – that is, the contemplation of death. Whether seen as a linear consequence to or a distinct postmodern divergence from thanatopic traditions, contemporary dark tourism has some relevance to present-day thanatopic behaviours – especially when located within a thesis of death sequestration and mediating mortality within contemporary society (Stone and Sharpley 2008; Stone 2012a). Dark tourism and the sequestration of death proposition offer a significant context to lines of scholarly enquiry which, subsequently, proposes dark tourism as a contemporary mediating medium by which societies may negotiate notions of mortality. Unsurprisingly, however, given numerous variables and diverse factors influencing the socio-cultural framing of death and dying, the role of dark tourism as a contemporary mediating institution of mortality is not absolute, nor can it ever be. Moreover, while the treatment of death and dying rites and rituals have been used as a means of shielding society from a public consciousness of mortality, such processes have been medicalised and privatised which, in turn, suggests a collective drive to conceal or deny death in the public domain. Yet, robust critiques of the death-denial thesis challenge its discriminative qualities whereby antithetical increases in public (re)presentations of death within societal domains have been proposed (Kellehear 2001). Such arguments problematise research that suggests a supposed sequestration of death and a consequent dichotomy that death is publicly absent but privately present (Giddens 1991; Mellor 1993; Mellor and Shilling 1993). Howarth (2007, 35) goes on to argue that 'it may be that in their quest to uncover hidden death, social theorists have neglected to acknowledge the more public face of death'. Subsequently, dark tourism as a context to scrutinise and acknowledge a more public face of death takes its thanatological research cue.

Less absolute treatments of this absent/present death paradox acknowledge these ambiguities and suggest more nuanced mediations of mortality. Consequently, dark tourism research may be seen as directing traditional thanatoptic discourse away from a schismatic argument in which death is *either* concealed *or* revealed, toward different mediations and even metamorphoses of death depending on multifarious societal needs – for example, via different behaviours, institutions and transactions. The proposal by Stone (2012a) that *certain kinds* of death are de-sequestered back into the public domain for contemporary consumption raises complex questions about the public presentation of death, and why and how certain kinds of death may be de-sequestered. Indeed, dark touristic praxis may itself function as a means by which certain kinds of death are de-sequestered and mediated and consumed in specific public domains (Stone 2009a). Moreover, dark tourism can provide transitory moments of mortality in which significant Other death is confronted and where death is rendered into *something else* that is comfortable and safe to deal with and to contemplate (Sharpley and Stone 2009).

Conversely, the motivations of so-called 'dark tourists' may correlate so closely with those of heritage, pilgrimage and special interest tourists (Hyde and Harman 2011) that to infer a

particular interest in death and/or mortality is merely speculative – though post-visit consequences of visiting sites that present the Significant Other Dead may indeed raise broader issues
of mortality. Arguably, therefore, closed supply/demand paradigms represent the tourist experience as more culturally reactive to, than directive of, heritage institutions. Yet the designation and
emergence of dark tourism locations is often influenced by a combination of public visitation,
media scrutiny and political discourse. Indeed, Seaton (1996) privileges touristic demand over its
sources and supply, thereby placing dark tourism in the context of behavioural phenomenology.
Meanwhile, Sharpley (2009) conceptualises dark tourism as interplay between the characteristics
of a site, with all the concomitant variables, and its touristic reception – including consideration
of touristic drivers, expectations and perceptions. This invites a nuanced consideration of the
tourist and their destination as collaborative agents, engaged in a range of transactional encounters that influence and are influenced by external meanings systems and cultural representations
of death and dying.

Despite the diverse range of socio-cultural factors that affect points of access to, engagement
with and exit from dark tourism experiences, political, logistical, materialistic and other causal
factors help describe and comprehend the fundamental nature of (dark) touristic behaviour.
Analysis of the so-called dark tourist experience can be critically validated only when such experiences are understood to exist beyond Seaton's proposed phenomenological 'vacuum', and instead
located in a broader context of socio-cultural identities and roles. This is particularly so when
researching dark tourism and concomitant visitor experiences and whether it makes sense at all to
divide people into different types without taking into account their full life spans. Thus, within
current dark tourism scholarship there is an obligation to, and indeed calls for, a more rigorous
attention to wider socio-cultural and psychosocial contexts (Biran *et al* 2011; Stone 2013). Moreover, such scholarship might usefully be informed by consideration of broader heritage concepts
and it is to these relationships that this chapter now turns.

Dark Tourism and Dark Heritage: Towards a Common Ground

Lennon and Foley (2000) locate the concept of dark tourism within postmodernist contexts,
highlighting its key characteristics and mapping them against postmodernist philosophical
frameworks. While frequently challenged, this premise represents an openness to, and engagement with, new conceptual dimensions and philosophical underpinnings in tourism studies.
These changes include the evolution of cultural tourism, and its associated agendas, into heritage
tourism, allowing a theoretical convergence with heritage studies that offers useful perspectives
on dark tourism frameworks, experiences and transactions.

Of particular resonance to dark tourism concerns are theories relating to built and/or inhabited environments and the way in which they obtain socio-cultural significance. The typology
of place offered by Williams (2009) distinguishes qualified environments, for example *built-
scapes*, *workscapes*, *technoscapes* and *peoplescapes*. More specifically, Jansen-Verbeke and George
(2012) observe changing identifications of 'war landscapes' over the past century or so as *memoryscapes*, *heritage landscapes* and *tourism landscapes*. The dark tourism lexicon adds '*deathscapes*' and
'*traumascapes*' to this taxonomy and, as such, designation of (death) space according to social use
and the making of meaning suggest psychologised processes that inform treatment of communal
landmarks and landscapes. Where such landscapes relate to significant conflict, violence or tragedy,
intense controversies may arise around their use and development. Such intensity is perhaps

proportionate to the various kinds of investments (socio-cultural, political and emotional) that are perceived to have been made by various stakeholders, both individually and collectively. The examination of the developmental processes – that is, convergences of people, place and time – by which dark tourism sites come into being is, therefore, vital to an enhanced understanding of those sites' functionality and identity within collective heritage contexts.

A useful template by which these developmental processes might be modelled is offered by Foote (1997) in which he examined sites associated with tragic events and, subsequently, suggested a prevailing set of conceptual outcomes. Foote's proposed continuum incorporates stages of *rectification*, *designation* and *sanctification*, through which the historical/cultural identity of sites is created or amended. He also proposes a state of *obliteration*, whereby the locus of violent or tragic events is forgotten in time; obliteration may occur for different reasons (and at various levels of deliberateness and consciousness), but they can all, arguably, be traced to a failure to rectify, sanctify and/or designate the site. Foote (2009, 38–9) maintains that 'no one outcome is ever final. Sanctification, designation, rectification, and obliteration are not static outcomes, but only steps in a process.' The flexibility of this model reflects the case histories of several iconic sites, such as at particular battlefields, where, for example, designation as a public memorial site may take many years; or redesignation may take place depending on cultural or political shifts (Chronis 2005). It also allows for the rapid creation of temporary memorials (or spontaneous shrines) and their potential, eventual permanence or obliteration. Of course, these processes will be informed and influenced by a diverse set of stakeholders as well as a range of other cultural, historical and ideological factors.

In some cases, authorities may avert spontaneous and non-authorised designation of particular sites through preventative obliteration, especially where the 'attraction of death' for visitors might be met with perceptions of deviance and the taboo (for example, the demolition by authorities of the house inhabited by, and witness to the violent crimes of, Fred and Rosemary West in the UK). Conversely, the Whitehall Cenotaph in London, originally intended by the authorities as a temporary monument (to be *obliterated*), was *designated* a permanent site due to the pressure of public opinion, which *sanctified* the site through mass visitation. This exemplifies, in displaced heritage contexts at least, 'more or less spontaneous gestures of public emotion, as often occurs after wars or public disasters, and the needs they create' (Benton 2010, 1). The question that of course arises is that where such social and emotional needs are met, by and within physical space – the obtaining nexus that may be defined as cultural heritage – might it also, under certain circumstances, be specified as dark tourism?

Arguably, there is no remarkable leap between the impact of prevailing public opinion and a comparable agency within the touristic community. Future research around the lifecycles (and designatory stages) of dark tourism sites may evaluate the impact of the tourist experience on institutional authenticity. Indeed, an examination of the agency of the tourist is of particular value with regard to cross-cultural participation and narrative congruence, and expressions of socio-cultural need. The function of, and challenge to, dark or displaced heritage is 'presenting or constructing monuments and ceremonies that attempt to meet these needs, and to match the inevitable differences in a "collective" memory of the event in question' (Benton 2010, 1). At sites of trauma of international and historic significance, physical and moral spaces may be required to enclose and represent diverse narratives and needs. Here, the iconic tourism site is challenged by what Stone (2009b, 63) describes as 'a post-conventional society' and its need for 'an open identity capable of conversation with people of other perspectives in a relatively egalitarian and open communicative space'. The issue remains, of course, of how sites located in the often over-whelming and appalling historical contexts of dark tourism can manage such conversation and create such space. Even so, the role of participating communities, including tourist communities, in the development and designation of such sites is critical. Indeed, it may ensure the success of such sites and, thus, prevent obliteration.

Hence, the nature of dark touristic transactions with dark heritage sites invites close study. Of course, while 'dark tourism does not need "dark tourists" – just people who are interested in learning about this life and this world' (Philip Stone quoted in Coldwell 2013, 1) – the tourist experience may have a powerful capacity to direct and influence the landmarks of cultural heritage and its narratives. Interdisciplinary research approaches may include issues of social change, social action and cultural orientation, and the agency of individuals and of groups in influencing significant institutions. Other research avenues for examining the dark tourism/ displaced heritage nexus may focus on post-materialist theory and values systems, and on cultural theories of the post-museum and consumer authority in public contexts. In turn, these interdisciplinary discourses may offer germane, complex contexts in which to explore the social significance of dark touristic transactions and, ultimately, their convergence with broader cultural heritage concerns.

Dark Tourism vs Dark Heritage: A Narrative Dissonance?

In the latter half of the 20th century, heritage studies increasingly privileged the role of memory in identifying what is important in society. In turn, the development of heritage systems were built on and around memory and meaning, rather than, necessarily, on fact and artefact. Benton (2010, 1) reveals a heritage/tourism convergence that emphasised 'the power of collective memory, where large or small groups within society share an idea of what happened in the past and why it was important [which] translates into patterns of tourism'. Clearly, where such groups hold ideas, and perceptions of importance, which are not shared (either with other groups, or with others within a group), their translation is likely to be problematic and even dissonant. Where memories relate to events of trauma, violence and/or conflict, the likelihood of difference in perception of the past is increased. Moreover, where diverse cultures and faith systems are factors, narrative discord may be further exacerbated. For this reason, the memorialisation of extraordinary events and efforts to acknowledge multiple memorial narratives may be fundamentally problematised in modern cultural heritage contexts and, particularly, in contexts in which dark touristic transactions occur. Here, we encounter situations where memory and its translation – or put another way, heritage and tourism – becomes discordant, and we find reflection of those situations in developing conceptual discourse relating to difficult, displaced and/or dissonant heritage.

With regard to touristic concerns, (dark) heritage scholarship allows a focus on the real-world functioning of heritage sites, and specific contemporary dilemmas encountered in their management. Perceptions and interpretations of heritage in modern multicultural societies, and in visitation to Other cultures, are ambiguous; they necessitate consideration of justifiable contestation of heritage and perceived dissonance between 'closed' heritage narratives and 'open' experience and memory. Such considerations are the nucleus of much of the recent literature on heritage messaging and meaning-making systems which may provoke heritage dissonance or even displacement (Ashworth and Hartmann 2005; Poria and Ashworth 2009). Ashworth's (2008) examination of historic trauma and violence and its implications for heritage tourism resonates with, although its agenda clearly differs from, dark tourism research.

Similarly, the authors in the edited volume by Logan and Reeves (2009) introduce the term 'difficult heritage' in their consideration of sites dealing with genocide, political imprisonment and conflict. However, the term and contextualising case studies used by contributing authors suggest a potential and relevant convergence with dark tourism research; yet only one specific reference is made to dark tourism concepts – that is, the examination of Auschwitz-Birkenau by Young (2009). Arguably, therefore, dark tourism has yet to be fully recognised as a mutually relevant cross-referential field in heritage studies contexts. However, White and Frew (2013), in their examination of sites of dark heritage, suggest an emergent tendency in broader heritage research to evoke dark tourism tropes, where given sites and their associations relate to profound and historic human experience.

CONCLUSION

This chapter set out to outline key parameters of dark tourism and its fundamental interrelationships with dark heritage. In so doing, the chapter has revealed that dark tourism, while a contested term, is an academic brand that can shine critical light on the touristic consumption of 'heritage that hurts'. Consequently, discourses of both cultural heritage and dark tourism converge and cluster readily when themes of war, disaster, atrocity or social conflict, and memory and identity are in question. However, interpretations of these themes are understandably prone to concerns about dissonance, inclusion, exploitation, sensitivity and appropriateness, and are vulnerable to ideological shifts. There may also be a perceived responsibility, or indeed political direction to support or engage on some level with conflict resolution processes, including rehabilitation and reintegration, especially in pedagogic and interpretation activities. Therefore, developing touristic opportunities at particular dark heritage sites is an increasing, perhaps inevitable, feature of creating contemporary traumascapes in shifting political and socio-cultural contexts. Of course, the practical possibility of travelling to landscapes of conflict and atrocity is one influencing factor in their evolution as tourism destinations, as is their historic and human significance. Therefore, it is likely that dark tourism scholarship will continue to find significant, even growing, mutuality with those of cultural heritage studies and indeed other associated fields. Ultimately, as heritage concerns and systems are further globalised and integrated by political institutions and processes, dark tourism will provide a heritage mechanism in which death is democratised and shared and narrated for the contemporary visitor economy.

BIBLIOGRAPHY AND REFERENCES

Ashworth, G J, 2008 The Memorialization of Violence and Tragedy: Human Trauma as Heritage, in *The Ashgate Research Companion to Heritage and Identity* (eds B J Graham and P Howard), Ashgate, Aldershot and Burlington VT, 231–44

Ashworth, G, and Hartmann, R, 2005 *Horror and Human Tragedy Revisited: The Management of Sites of Atrocities for Tourism*, Cognizant Communications Corporation, New York

Beech, J, 2009 'Genocide' Tourism, in *The Darker Side of Travel: The Theory and Practice of Dark Tourism* (eds R Sharpley and P R Stone), Aspects of Tourism Series, Channel View Publications, Bristol, 207–23

Benton, T (ed), 2010 *Understanding Heritage and Memory*, Manchester University Press, Manchester

Biran, A, Poria, Y, and Oren, G, 2011 Sought Experiences at (Dark) Heritage Sites, *Annals of Tourism Research* 38 (3), 820–41

Blom, T, 2000 Morbid tourism: a postmodern market niche with an example from Althorpe, *Norwegian Journal of Geography* 54 (1), 29–36

Chronis, A, 2005 Coconstructing Heritage at the Gettysburg Storyscape, *Annals of Tourism Research* 32 (2), 386–406

Coldwell, W, 2013 Dark Tourism: why murder sites and disaster zones are proving popular, *The Guardian*, 31 October, available from: http://www.theguardian.com/travel/2013/oct/31/dark-tourism-murder-sites-disaster-zones [31 October 2013]

Crick, M, 1989 Representation of International Tourism in the Social Sciences: Sun, Sex, Sights, Savings, and Servility, *Annual Review of Anthropology* 18, 307–400

Dalton, D, 2013 *Dark Tourism and Crime*, Routledge, Abingdon

Dann, G M S, and Seaton, A V (eds), 2001 *Contested Heritage and Thanatourism*, Haworth Press, New York

Foote, K, 1997 *Shadowed ground: America's landscapes of violence and tragedy*, University of Texas Press, Austin TX

— 2009 Heritage tourism, the geography of memory, and the politics of place in Southeastern Colorado, in *The Southeast Colorado heritage tourism project report* (ed R Hartmann), Wash Park Media, Denver CO, 37–50

Giddens, A, 1991 *Modernity and Self-Identity: Self and Society in the Late Modern Age*, Polity Press, Cambridge

Howarth, G, 2007 *Death and Dying: A Sociological Introduction*, Polity Press, Cambridge

Hyde, K, and Harman, S, 2011 Motives for a secular pilgrimage to the Gallipoli battlefields, *Tourism Management* 32 (6), 1343–51

Jansen-Verbeke, M, and George, E W, 2012 Reflections on the Great War Centenary: From warscapes to memoryscapes in 100 years, in *Tourism and War: A Complex Relationship* (eds R Butler and W Suntikul), Routledge, Abingdon

Kellehear, A, 2001 Denial, criticisms of, in *The Encyclopedia of Death and Dying* (eds G Howarth and O Leaman), Routledge, London

Lennon, J, and Foley, M, 2000 *Dark Tourism: The Attraction of Death and Disaster*, Continuum, London

Logan, W, and Reeves, K (eds), 2009 *Places of Pain and Shame: Dealing with 'Difficult Heritage'*, Routledge, Abingdon

Mellor, P, 1993 Death in high modernity: the contemporary presence and absence of death, in *The Sociology of Death* (ed D Clark), Blackwell, Oxford, 11–31

Mellor, P, and Shilling, C, 1993 Modernity, self-identity and the sequestration of death, *Sociology* 27, 411–32

Miles, W, 2002 Auschwitz: Museum Interpretation and Darker Tourism, *Annals of Tourism Research* 29 (4), 1175–8

Poria, Y, and Ashworth, G, 2009 Heritage tourism – current resource for conflict, *Annals of Tourism Research* 36 (3), 522–5

Rice, A, 2009 Revealing Histories, Dialogising Collections: Museums and Galleries in North West England Commemorating the Abolition of the Slave Trade, *Slavery and Abolition* 30 (2), 291–309

Rojek, C, 1993 *Ways of Escape: Modern transformations in leisure and travel*, Macmillan Press, London

Sather-Wagstaff, J, 2011 *Heritage that Hurts: Tourists in the Memoryscapes of September 11*, Left Coast Press Inc, Walnut Creek CA

Seaton, A V, 1996 Guided by the dark: From thanatopsis to thanatourism, *International Journal of Heritage Studies* 2 (4), 234–44

— 2002 Thanatourism's final frontiers? Visits to cemeteries, churchyards and funerary sites as sacred and secular pilgrimage, *Tourism Recreation Research* 27 (2), 73–82

— 2010 Thanatourism and Its Discontents: An Appraisal of a Decade's Work with Some Future Issues and Directions, in *The Sage Handbook of Tourism Studies* (eds T Jamal and M Robinson), Sage Publications, London, 521–42

Sharpley, R, 2009 Dark Tourism and Political Ideology: Towards a Governance Model, in *The Darker Side of Travel: The Theory and Practice of Dark Tourism* (eds R Sharpley and P R Stone), Aspects of Tourism Series, Channel View Publications, Bristol, 145–63

Sharpley, R, and Stone, P R, 2009 Representing the Macabre: Interpretation, Kitschification and Authenticity, in *The Darker Side of Travel: The Theory and Practice of Dark Tourism* (eds R Sharpley and P R Stone), Aspects of Tourism Series, Channel View Publications, Bristol, 109–28

Stone, P R, 2006 A Dark Tourism Spectrum: Towards a typology of death and macabre related tourist sites, attractions and exhibitions, *Tourism: An Interdisciplinary International Journal* 54 (2), 145–60

— 2009a Making Absent Death Present: Consuming Dark Tourism in Contemporary Society, in *The Darker Side of Travel: The Theory and Practice of Dark Tourism* (eds R Sharpley and P R Stone), Aspects of Tourism Series, Channel View Publications, Bristol, 23–38

— 2009b Dark Tourism: Morality and New Moral Spaces, in *The Darker Side of Travel: The Theory and Practice of Dark Tourism* (eds R Sharpley and P R Stone), Aspects of Tourism Series, Channel View Publications, Bristol, 56–72

— 2011 Dark Tourism: towards a new post-disciplinary research agenda, *International Journal of Tourism Anthropology* 1 (3/4), 318–32

— 2012a Dark Tourism and Significant Other Death: Towards a model of mortality mediation, *Annals of Tourism Research* 39 (3), 1565–87

— 2012b Dark tourism as 'mortality capital': The case of Ground Zero and the significant other dead, in *The Contemporary Tourist Experience: Concepts and Consequences* (eds R Sharpley and P R Stone), Routledge, Abingdon, 30–45

— 2013 Dark Tourism Scholarship: A Critical Review, *International Journal of Culture, Tourism and Hospitality Research* 7 (2), 307–18

Stone, P R, and Sharpley, R, 2008 Consuming Dark Tourism: A Thanatological Perspective, *Annals of Tourism Research* 35 (2), 574–95

— 2013 Deviance, Dark Tourism and 'Dark Leisure': Towards a (re)configuration of morality and the taboo in secular society, in *Contemporary Perspectives in Leisure: Meanings, Motives and Lifelong Learning* (eds S Elkington and S Gammon), Routledge, Abingdon

West, P, 2004 *Conspicuous Compassion: Why sometimes it really is cruel to be kind*, Civitas, London

Westover, P, 2012 *Necromanticism: travelling to meet the dead, 1750–1860*, Palgrave Macmillan, Basingstoke and New York

White, L, and Frew, E (eds), 2013 *Dark Tourism and Place Identity: Managing and Interpreting Dark Places*, Routledge, Abingdon

Williams, S, 2009 *Tourism Geography: A new synthesis*, Routledge, London

Wilson, J Z, 2008 *Prison: Cultural Memory and Dark Tourism*, Peter Lang, New York

Young, K, 2009 Auschwitz-Birkenau: the challenges of heritage management following the Cold War, in *Places of Pain and Shame: Dealing with 'Difficult Heritage'* (eds W Logan and K Reeves), Routledge, Abingdon, 50–67

2

Anthropogenic Disaster and Sense of Place: Battlefield Sites as Tourist Attractions

STEPHEN MILES

Disasters are commonly associated with large-scale catastrophic events causing material destruction, economic and social hardship, loss of life and suffering. They are often unanticipated and unpredictable with consequences that last long after the immediate crisis itself. In popular understanding they are normally seen as the result of natural phenomena on a large, newsworthy scale. But disasters can also be man-made, as successive oil spills, nuclear incidents, industrial accidents, explosions and insidious environmental degradation attest. Although definitions of disaster tend to support a sense of accidental causation, war and deliberate acts of terrorism can also be included in this anthropogenic category. War is certainly a disaster for those caught up in it, both for combatants and non-combatants alike, and is often accompanied by the attendant suffering of famine, epidemics, homelessness and population displacement, both within and outside of theatres of conflict. Moreover the emotional and mental suffering of war is totally incalculable. The destruction of war is arguably more catastrophic than most natural disasters: World War II resulted in the deaths of between 60 and 78 million people, whereas the most deadly recorded natural disaster, the Bengal Famine of 1770, killed 10 million (Bowen 2012).

The destructive effects of war, however, spread much further than human suffering: the damage and depletion of a nation's buildings; the displacement of populations with their inherent languages, traditions, music and cultural practices; and the damage and dispersion of a nation's art treasures are just some of the examples of how both tangible and intangible heritage can become displaced and distorted by human disaster. This chapter will give examples of this process from a spatial perspective and show how, while heritage can suffer from displacement, there is much refashioning of heritage after warfare. This can be seen in a new interpretation of place and as '(re)-placed heritage'.

Both natural and anthropogenic disasters lead to a loss of national and personal location and place. War destroys cultural heritage either accidentally or deliberately as a 'tactical tool' (as, for example, Adolf Hitler's Baedeker Raids to target the built heritage of British cities in 1942). But, like the phoenix rising from the ashes, an unintended consequence of destruction is the added poignancy and cultural resonance attached to places affected by war in the public consciousness. Sites of destruction can act as foci for national resolve after the event and become important cultural touchstones. With the destruction of the World Trade Center on 11 September 2001, 'Ground Zero' became a rallying point for American national resilience in the face of the 'War on Terror', as well as an important locus for personal and collective grief (Stone 2012). When peace is restored, places of human tragedy develop an important role in becoming centres for commemoration of the fallen, using the highly charged iconography and symbolism of memorialisation. The battlefield sites of the Western Front in France and Belgium are part of such a 'commemorative landscape'. Far from destroying place, war as an anthropogenic disaster can thus create a new sense of place. Instead of displacing heritage resources, war can create them afresh and stimulate interest in seeing the places where disaster happened. New cultural landscapes can develop after the disaster of war, and tourism has an important role to play in re-inscribing and perpetuating memory (Butler and Suntikul 2013).

Battlefield sites have an important role to play in situating war, both spatially and symbolically. Adopting a broad interpretation of the 'battle site', this chapter describes how, instead of destroying our material and conceptual world, such sites are able to create new cultural meaning. As space is turned into place, battlefield sites can adopt new cultural and symbolic identities, and tourism is used as a vehicle for understanding this.

THE CREATION OF A SENSE OF PLACE

To be human is to be located in geographical space and to engage with it materially and perceptually. But space *per se* is an empty concept. It is only turned into place when it is given meaning. Otherwise it is a *location*, the setting for routine and the quotidian. Place is thus a 'meaningful location' (Creswell 2004, 7) and through its personal or collective significance is made distinctive above other areas of space. As Yi-Fu Tuan has commented, 'place is a pause in movement' and 'the pause makes it possible for a locality to become a centre of felt value' (Tuan 1977, 138). A place is somewhere where we stop, rest and become involved. This significance is bestowed through human forces such as organisations, buildings, management, interpretation and signage. It is also defined by the way people react to these and the emotional and cognitive processes which they engender. But the creation of place is never static. As Creswell has commented, 'Place…is an embodied relationship with the world. Places are constructed by people doing things and in this sense are never finished but are constantly being performed' (Creswell 2004, 37). There is a constant and reiterative practice in the creation of place. It is noteworthy that the battlefield site at Hastings (1066) in southern England has been seen successively as a monastic game park, stately home, army camp, tourist attraction, battlefield re-enactment location and municipal bonfire site (English Heritage 2007). Several of these stages have overlapped temporally, providing a palimpsestic definition of space.

'Sense of place' is usually defined in terms of the special qualities that distinguish an area of space from others around it. These places normally have a strong sense of identity, character and authenticity, providing a mark of distinction. Commentators would regard this identity as wholly dependent on human engagement and the creation of meaning through cultural process. Where things happened is a very powerful agent in the creation of a 'sense of place'. Important historical events are inevitably defined by historians, and this determines their existence as heritage tourism attractions.

The major battle sites at Hastings (1066), Bannockburn (1314), Bosworth (1485) and Culloden (1746) hold an important place in British national consciousness as the location for particularly important events which brought about dramatic change. The Battle of Hastings marked the end of the Anglo-Saxon era and, with the death of Harold Godwinson, the last Saxon king, the beginning of Norman rule in England. Bannockburn was where the Scottish king Robert the Bruce defeated the English under Edward II, which led to the granting of Scottish independence from England in 1328. Bosworth was the culmination of the dynastic conflict known as the Wars of the Roses in the late 15th century between the 'houses' of York and Lancaster, which resulted in the death of Richard III, the last Plantagenet monarch, and the start of the Tudor dynasty. Finally, Culloden was the last battle in the Jacobite Rebellion against the Hanoverian dynasty. With the defeat of the rebel forces led by the Stuart 'pretender', Bonnie Prince Charlie, this serious challenge to the British throne was decisively eliminated at Culloden. It was also the last battle to be fought on British mainland soil.

All four of these sites have large visitor centres but their historical importance might not account for their iconic status alone. Naseby (1645) was a key battle in deciding the outcome of the English Civil War (1642–51) and arguably one of the most decisive in English history. It does not, however, have a visitor centre and interpretation panels were only recently erected. The creation of an 'attraction' is believed to be more complicated and the result of a protracted process of 'site-sacralisation' (Seaton 1999).

Places can also be prominent in the imagination. For example, at Mons Graupius (AD 85) the Romans were defeated by a Caledonian host, but the location of this event has never been determined with any certainty. The battle nevertheless remains prominent in public consciousness, particularly in Scotland. Many actual sites are located but remain unknown to all but the expert, with no 'markers' such as signposts, monuments or heritage interpretation. As places they might only exist in the consciousness of the minority or for the local community.

These battlefields have a sense of place, either perceptually or physically, but as heritage sites many are marked by dissonance. There have been a number of conflicting theories as to the location of Bosworth, and it was only in 2009 that a scientifically plausible location for the battle was discovered, two miles away from the visitor centre that had been built in the 1970s (Foard and Curry 2013). As a heritage site, Bosworth was 'placed' when the visitor centre was built, 'displaced' as doubts over the location increased and then 're-placed' as the site was moved to a more archaeologically and historically convincing location. Naseby currently stands as a cultural contradiction and an example of how a nation's heritage can become 'displaced' as its position as an iconic event in history is not matched by any physical 'marking' of location. Mons Graupius is 'displaced' in that its location is a mystery and needs 'placing'. These are examples of 'places that are displaced' in that they are contested and subject to constant reinterpretation. One of the problems in trying to place or 're-place' heritage is that the process is open to tension and contestation as different voices attempt to appropriate the way heritage is presented and interpreted. The irony is that as heritage is put in its proper place, further 'displacement' is experienced.

MEMORY AND SENSE OF PLACE: '*LES VILLAGES DÉTRUITS*' OF VERDUN

Between 21 February and 18 December 1916, during World War I, the French and German armies conducted a savage battle of attrition near the town of Verdun in north-eastern France. This was the longest battle of the war with a total of 714,231 casualties: 377,231 on the French side and 337,000 on the German (Reemtsma 2004, 26). Verdun strikes a deep resonance in French culture to this day and, like the Somme for the British, is shorthand for meaningless slaughter, national sacrifice and the horrors of war. It is estimated that 32 million shells fell on the wooded hills above Verdun. So extensive was the physical destruction along the Western Front that after the war a *Zone Rouge* (Red Zone) was established in France, where agriculture and housing development was forbidden because of the large amount of unexploded ordinance still in the ground. This covered 1200 square kilometres (460 square miles) of north-eastern France

(Niess 2009, 123). Some destroyed villages were never rebuilt, and a decision was made to afford them a unique individual status as places which had '*Mort pour La France*' ('Died for France'). In the Verdun area nine of these are listed as '*villages détruits*' ('*destroyed villages*'): three were rebuilt after the war and repopulated whereas six were left as ruins (see Figs 2.1 and 2.2).[1] The villages were evacuated before being involved in the fighting and there were relatively few civilian casualties in the area. To the French state the six left in ruins do still exist. Douaumont still returns a mayor, voted in by an electorate descended from the original inhabitants and appears on the French national dictionary of *communes* (Lichfield 2006).

At these sites a sense of place is powerfully created by the legacy of destruction, even though as destroyed villages they are 'displaced' heritage. There is something appealing about the presence of ruins and, as with old stately homes, destruction and displacement is often a prelude to appreciation because it provides a stark reminder of something that has been built through effort and then lost (Woodward 2001). The *villages*, even those now in ruins, are tangible memorials to the destruction and loss of war. At Douaumont, among the overgrown rubble and visible shell craters, the layout of the streets has been preserved along with the location of the *mairie* (Town Hall) and school. In 1914 the village had 288 inhabitants living in more than 50 houses. Most poignantly, in many of the villages the homes of the inhabitants are marked by plaques which state the name and profession of the householder, and at Fleury the village café/grocer has been marked. Even water is memorialised, with one marker at Douaumont stating 'To the memory of water. In this place was one of the fountains of Douaumont.'

There is a close relationship between a sense of place and the dynamics of memory. In his *Les lieux de mémoire* (Nora 1984–92), Pierre Nora suggested that the French nation builds its memory around various 'sites' (*lieux*). The sites can be interpreted in their broadest sense and include places, buildings, personages, works of art or texts. They stand at the intersection of official (or civic) and vernacular cultures and are loci around which a social group can create a cultural memory and construct a shared past. The destroyed villages of Verdun are good examples of 'sites of memory', as places where momentous events took place. Even so, Nora believed that *lieux* were not able to constitute a true 'collective memory' since by the late 20th century the environments (*milieux*) conducive to fostering such an identity had been lost. As Erll has commented, 'sites of memory function as a sort of artificial placeholder for the no longer existent, natural collective memory' (Erll 2011, 23).

Nevertheless tourism to sites such as Verdun does provide a collective experience and serves to reify and reinscribe the memory of past events. As Gough has shown for the Newfoundland Memorial Park – another World War I 'memory site' in France – 'without frequent reinscription, the date and place of commemoration simply fades away as memory atrophies' (Gough 2004, 238). The waning of memory can also be an agent of displacement as heritage becomes detached from social memory; this is particularly true if there is no mechanism for re-engaging in heritage through memory. Memories are re-established through ceremonies around memorials

---

1   The villages were Beaumont, Bezonvaux, Cumières, Douaumont, Fleury, Haumont, Louvemont, Ornes and Vaux. The creation of a memorialised sense of place by leaving buildings destroyed by war as they were is also represented by Coventry Cathedral, the Kaiser Wilhelm Memorial Church in Berlin and the Frauenkirche in Dresden (rebuilt 1993–2005). These were all damaged by bombing raids in World War II. The Spanish town of Belchite (Zaragoza province) was destroyed during the Civil War in 1937 but left as a living reminder of the superiority of Nationalist forces against the Republicans.

FIG 2.1. A GENERAL VIEW OF THE 'DESTROYED VILLAGE' OF BEZONVAUX.

FIG 2.2. THE MAIN STREET AT THE 'DESTROYED VILLAGE' OF BEZONVAUX.

which utilise ritual as an aid to remembrance (Connerton 1989). Tourists visiting the destroyed villages take part in such ceremonies, but also by their very presence and interest in the events contribute to the perpetuation of the memory of the victims and places involved. This tourist involvement can also mitigate the effects of displacement.

THE ROLE OF TOURISM AND INTERPRETATION IN CREATING A SENSE OF PLACE

In moving from notions of space to place, distinction is bestowed upon locations through a process of physical 'marking', which can be anything from a simple signpost through to a large building such as a visitor centre (MacCannell 1999). A 'nowhere' becomes a 'somewhere' through the creation of identity. To use Tuan's phrase again, there is a 'pause in movement' as people are persuaded to stop and look. The 'destroyed villages' of Verdun are marked out and separated as worthy of our attention and their story is related to us as a powerful adjunct to their tangibility as places. Tourism is thus important in the creation of place. But not all places are created by tourism; those given private meanings such as the home and the workplace have distinction for the individual alone. Places special to each of us – such as where we proposed to our spouses, the graves of loved ones, locations for special but personal memories, and our own favourite places – are outside of the public realm. Some might have been turned into 'attractions' subject to the spectatorial tourist 'gaze' (Urry 2011) for other reasons, but to us they are still private.

Tourism does create place, but how far does it create a 'sense of place'? The antithesis of the special attributes of place is the idea that in post-modern society there is 'placelessness' (Relph 1976). Shopping malls, petrol stations and department stores are often cited as places with no sense of distinction in a mass culture where standardisation is normal. The 'non-place' is more predictable in a risk-averse society more comfortable with familiarity. Perhaps we are seeing the end of place where places are now '… strategic installations, fixed addresses that capture traffic' (Thrift 1994, quoted in Creswell 2004, 48) with little or no identifiable physical or symbolic characteristics to mark them out as distinctive in any way. Relph (1976) has argued that tourism is the worst culprit in encouraging inauthenticity and an anodyne sense of place. This is because rapid movement is at the very heart of the tourist experience where the tourist 'gaze' is glancing and ephemeral. Airport lounges, railway stations and hotel foyers are empty spaces designed to 'capture traffic' and ease the boredom of waiting, mobility's by-product. It is nevertheless unfair to blame so much placelessness on the industry. Tourism does provide a wide range of highly distinctive places which have become attractions precisely because of their uniqueness. Heritage attractions are prime examples of places with distinct physical presence and unique 'stories'. Instead of a predictable standardisation there is in fact great heterogeneity in the heritage tourism industry.

But the creation of this heterogeneity does come at a price. It is unlikely that anyone would know where a battlefield site was without the marker; many battlefield sites are 'displaced', not necessarily because the location of the place is uncertain or unknown, but because there is nothing to mark them as such on the ground. With the presence of a large visitor centre this marking is prominent and an unequivocal statement that 'something' happened here. Place is thus created but along with it comes the intrusiveness of buildings and an infrastructure of paths, signs and car parks. One could argue that this diminishes a special sense of place especially when accompanied by the negative effects of large numbers of visitors. Place is created and then tarnished by the very fact that it is now a place worthy of human attention. The authenticity of the site is 'displaced' through the need to mark, conserve and interpret the site for greater cultural and heritage purposes. Nevertheless, aside from the management challenges that such a scenario presents, this assumes that infrastructure automatically ruins sense of place. While this can certainly be true, some of the most 'place-worthy' heritage sites do have a tourist infrastructure. The site at Culloden is often regarded as having a special sense of place (McLean *et al* 2007), but at the same time has a visitor centre, which in 2010 attracted over 99,000 visitors (Miles 2013, 223).

The creation of a sense of place is also reliant on the way the 'story' behind the events is communicated to the visitor. Interpretation is a vital ingredient of heritage presentation (Helms and Blockley 2006). As places of morbidity and drama, battle sites have the potential to provide a more 'edgy' form of interpretation than many other types of heritage attraction. But this does pose problems of how to present extreme violence to a wide public. One of the most effective ways of doing this is re-enactment which, although heavily stylised, is able to portray the events in as realistic a manner as possible. At Hastings this takes place on the actual field of battle itself, which does diminish the aura of being at the site of one of the most iconic events in English history. Alternatively the simulation of a real historical event could be a determining factor in the enhancement of a sense of place, and more research needs to be conducted to explore what meanings are associated with these events. One of the most effective ways of creating a sense of place, however, is the highlighting of human stories at battlefield sites, which is brought out most dynamically in the interpretation at Hastings, Bannockburn, Bosworth and Culloden (Miles 2013). People can relate much more closely to human dilemmas, hardships and predicaments than to the political background or the facts and dates from history. The question one is always prompted to ask is: what would I have done if I had been there? This does bring past events down to a personal and immediate level, the import of which is greatly magnified by proximity to where the events took place on the ground (where known). In addition to this, the connection of sites with famous historical personages adds a further mix to the sense of being at the place where history was made: Hastings is Harold; Bannockburn Robert the Bruce; Bosworth Richard III; and Culloden Bonnie Prince Charlie. A further important factor in creating a sense of place is the use of artefacts in any battlefield interpretation. At Culloden this is used very effectively with objects found in archaeological excavations at the site on display as a complement to the narrative and socio-political background to the conflict.

Sense of place is a function of the physical attributes of a place itself and also how it is conceptually 'read' by the viewer. Tourism and interpretation have a significant role to play in governing how places are perceived and understood. But there are important human factors which also need to be considered in that people react to the structures provided, and process 'messages', differently. They choose where to go and what to see and inevitably have varying rates of knowledge absorption; places mean different things to different people. Socio-cultural context can also be crucial, and the creation of sense of heritage place is inevitably influenced by the contemporary societal importance of 'histories' and their sites strongly influenced by the media. This underlines the previous point that places are continually subject to reinterpretation. At the time of writing, a new visitor centre is being built at the Bannockburn site, the opening of which will coincide with the 700th anniversary of the battle in 2014. As a Scottish victory over England, Bannockburn has long been a focal point for Scottish nationalism, and the opening of a new centre, although not overtly nationalistic in tone, is a highly symbolic act. It is no coincidence that 2014 is also the year of Scotland's referendum on independence. Places are never definitively established and left in aspic.

One of the presumptions in this chapter has been the locational accuracy of battlefield sites. To create distinction from other places, battlefield sites exploit their tangibility as places where events happened. This is not always possible, and although this raises questions over authenticity, which are beyond the capacities of this chapter, it can be said that precise accuracy is not important. In the case of Bannockburn there is still uncertainty as to where the battle took place. Nevertheless the idea that the event was 'near here' still provides a level of distinctiveness sufficient to delineate space as place. The site of the battle of Bosworth has only recently been discovered by archaeologists, yet this never diminished the experiences of visitors to the nearby visitor centre. Raivo (1998) has asserted that the locational accuracy of battlefields is unimportant since the sense of place is never created by the site alone. The special qualities of place, the *genius loci*, are created by people and their reactions to the site.

CONCLUSION

This chapter has explored the sense of place present at one particular type of heritage attraction, the battlefield site, which represents anthropogenic disaster, and how these sites can be 'displaced'. It has also shown how places are important in the construction and perpetuation of memory and thus important at sites of disaster. Tourism can create a sense of place, by highlighting and developing places so that they are worthy of attention, and in 're-placing' them. It can provide a context for the presentation of the 'story' and in certain sites be a vehicle for the reinscription of important personal and societal memories.

In a society rapidly losing any sense of place, and where places are threatened and fought over (Creswell 2004), the creation and maintenance of places with identity is ever more important. The flows of postmodern life have disconnected people from the idea of place and led to much heritage 'displacement'. It is important therefore that we do not lose our sense of the past and the places where events 'took place'. For sites of anthropogenic disaster, subject to 'displacement', a new afterlife can be created and tourism can be instrumental in fostering this. But it is important that this is done with a deep sense of respect so that places of disaster can stand as memorials to human events, the meaning of which we are unlikely to truly fathom.

BIBLIOGRAPHY AND REFERENCES

Bowen, H V, 2012 *Revenue and Reform: The Indian Problem in British Politics* 1757–1773, Cambridge University Press, Cambridge

Butler, R, and Suntikul, W, 2013 Tourism and War: an ill wind?, in *Tourism and War* (eds R Butler and W Suntikul), Routledge, London, 1–11

Connerton, P, 1989 *How Societies Remember*, Cambridge University Press, Cambridge

Creswell, T, 2004 *Place: A Short Introduction*, Blackwell, Oxford

English Heritage, 2007 *Battle Abbey and Battlefield Guidebook*, English Heritage, London

Erll, A, 2011 *Memory in Culture*, Palgrave, Basingstoke

Foard, G, and Curry, A, 2013 *Bosworth 1485: A Battlefield Rediscovered*, Oxbow Books, Oxford

Gough, P, 2004 Sites in the imagination: the Beaumont Hamel Newfoundland Memorial on the Somme, *Cultural Geographies* 11, 235–58

Helms, A, and Blockley, M (eds), 2006 *Heritage Interpretation*, Routledge, London

Lichfield, P, 2006 Verdun: myths and memories of the 'lost villages' of France, *The Independent* [online], 21 February, available from: http://www.independent.co.uk/news/world/europe/verdun-myths-and-memories-of-the-lost-villages-of-france-467285.html [30 April 2013]

MacCannell, D, 1999 *The Tourist: A New Theory of the Leisure Class*, 3 edn, University of California Press, Berkeley

McLean, F, Garden, M, and Urquhart, G, 2007 Romanticising Tragedy: Culloden battle site in Scotland, in *Battlefield Tourism: History, Place and Interpretation* (ed C Ryan), Elsevier, Oxford, 221–34

Miles, S, 2013 From Hastings to the Ypres Salient: Battlefield Tourism and the interpretation of fields of conflict, in *Tourism and War* (eds R Butler and W Suntikul), Routledge, London, 221–31

Niess, A, 2009 From the Chemin des Dames to Verdun: The Memory of the First World War in War Memorials in the Red Zone, in *France and its Spaces of War: Experience, Memory, Image* (eds P M E Lorcin and D Brewer), Palgrave Macmillan, New York, 121–32

Nora, P, 1984–92 *Les lieux de mémoire*, 7 vols, Gallimard, Paris

Raivo, P, 1998 Politics of Memory: Historical Battlefields and Sense of Place, in *NGP Yearbook* 27 (1) (eds J Vuolteenaho and T Antti Aikas), Nordia Geographical Publications, Oulu, 59–66

Reemtsma, J P, 2004 The Concept of the War of Annihilation, in *War of Extermination: The German Military in World War II* (eds H Heer and K Naumann), Berghahn Books, London, 13–35

Relph, E C, 1976 *Place and Placelessness*, Pion, London

Seaton, A V, 1999 War and thanatourism: Waterloo 1815–1914, *Annals of Tourism Research* 26 (1), 130–58

Stone, P R, 2012 Dark tourism as 'mortality capital': The case of Ground Zero and the Significant Other Dead, in *Contemporary Tourist Experience: Concepts and consequences* (eds R Sharpley and P R Stone), Routledge, London, 71–94

Tuan, Y-F, 1977 *Space and Place: The Perspective of Experience*, Edward Arnold, London

Urry, J, 2011 *The Tourist Gaze 3.0*, 3 edn, Sage, London

Woodward, C, 2001 *In Ruins*, Vintage, London

3

Memorialisation in Eastern Germany: Displacement, (Re)placement and Integration of Macro- and Micro-Heritage

SUSANNAH ECKERSLEY AND GERARD CORSANE

INTRODUCTION

Histories of disaster, trauma and loss are an integral part of examining Germany's 20th century heritage. In this chapter three heritage sites which represent these aspects of displacement are introduced: firstly, the Frauenkirche in Dresden as a place embodying human-made disaster; secondly, the Zeithain Grove of Honour as a site of trauma; and finally, the *Betonzeitschiene* – Plattenbau Micromuseum as a space of loss. The Dresden Frauenkirche is perhaps Saxony's most well-known general tourist attraction, celebrating the city's Baroque architectural history, but also a place of memory following the disaster of the Allied firebombing of Dresden during World War II. This occurred on 13 and 14 February 1945, reducing the church to heaps of rubble. It was rebuilt during 1992–2005 following reunification, having been left as a ruin during the German Democratic Republic (GDR) period. The Zeithain Grove of Honour and Micromuseum are a commemoration and interpretation of the trauma of a former World War II prisoner of war (POW) camp housing Allied soldiers. The *Betonzeitschiene* – Plattenbau Micromuseum consisted of a spatial, architectural installation commemorating the history of the *Plattenbau* (prefabricated concrete panel apartment blocks) as an architectural form and local industry.

This chapter considers what constitutes heritage against notions of displacement and (re) placement in relation to: disaster and recovery; destruction and regeneration; trauma and reconciliation; loss and renewal; remembering and forgetting; place and time; past, present and future. Underlying these notions are the shifting perceptions of what is, or is not, 'worthy' of preservation or memorialisation at any one moment in time; perceptions which are dependent on numerous political and public factors, themselves always in a state of flux. These issues are heightened in locations such as eastern Germany, which has undergone major political and social upheaval from the end of World War II, through the period of the German Democratic Republic (GDR) 1949–1990, reunification with the Federal Republic of Germany (West Germany), and right up to today. This chapter examines the changing nature of heritage memorialisation over time, using examples from the current eastern German state of Saxony. These examples of heritage memorialisation range from the 'macro' (nationally and internationally recognised symbolic heritage) to the 'micro' (niche interest and/or 'unpopular' heritage).

The Frauenkirche is an example of macro-level heritage, widely recognised for both its architectural significance and as a post-reunification identity marker for Germany. The Zeithain Grove of Honour and *Betonzeitschiene* – Plattenbau Micromuseums are micro-level heritage; places which connect to individual or family histories and experiences as much as to the national and international macro-histories of the 20th century. The interesting juxtapositions of different layers of the past found within all three sites suggest that there may be the potential to integrate micro- and macro-histories, in order to enrich each individual site, but also to expand upon the overall understandings of heritage places as part of a wider network. This will be proposed within the final section of the chapter.

Macro- and Micro-level Heritage

Inevitably, in processes of heritage memorialisation, choices need to be made that may result in contestation, dissonance, displacement and (re)placement through a desire to forget aspects of the past. Often 'official' state-directed, recognised and acknowledged interpretations and (re)presentations of heritage are pushed to the fore. With the dominance that this brings, the representation of other aspects of the past – perhaps those which are deemed less 'desirable' by the current authorities – are displaced within the public sphere. This displacement can be a result of 'selective social amnesia' and/or the desire to present 'positive' and appealing touristic versions of a location's past, where a (re)structured 'sense of place' is considered worthy of promotion for international audiences. This may be in contrast to local people's understandings of the significance of the places which they inhabit, their histories and heritages. The resultant displacement is an example of contestation within 'heritagisation', memorialisation and the difficulty of appealing to differing notions of heritage acceptable to 'insiders' and 'outsiders' at the same time. In some cases a series of political 'displacements' has taken place, within which the heritage associated with or sanctioned by each time period, or political regime, has replaced that of the previous one.

Heritage considered to be on the macro-level is most often that which is perceived to have international and national recognition and significance. This type of heritage is likely to receive large-scale public financial support and is frequently placed at the heart of tourism strategies. Examples of macro-heritage are often the accepted visual symbols and large-scale identity markers used within 'outward' facing representations of sense of place. At the macro-level, the debates over the historical and contemporary approaches to the memorialisation of Dresden's Frauenkirche (the Church of our Lady) part of Dresden's former[1] UNESCO World Heritage Site (UNESCO 2009) will be compared to lesser known and sometimes 'unpopular' cases of micro-heritage memorialisation.

Useful examples in eastern Germany of micro-heritage memorialisation – sometimes linked to 'heritage from below' – include the micromuseums in Saxony, introduced below. The term 'Micromuseum', and its accompanying philosophy, was developed by the Irish architect and designer Ruairí O'Brien in relation to his architectural museum installations. These act as discrete museum-like interventions and 'showcases' or, in the case of Zeithain Grove of Honour, as a 'walk-in vitrine' (glass display case). His micromuseums have been described, in contrast to large-scale traditional museums, as a form of 'museum experience' which:

---

[1]   Dresden Elbe Valley was removed from the UNESCO World Heritage List 'due to the building of a four-lane bridge in the heart of the cultural landscape which meant that the property failed to keep its "outstanding universal value as inscribed"' (UNESCO 2009).

offers a richness of experience which belies its tiny size… keeping the visitor's attention by seeking to stimulate their imagination, rather than by the volume of attractions on offer. It is an approach where personal instinct and aesthetic judgement are prioritised over group discussions about taxonomy, narrative, or the contents of particular zones, and which is efficient in terms of time, resources and space. Cultural institutions and their funders would do well to pay attention. It is also the key to ensuring that, in these post-mega budget days, our museum sector can continue to thrive.   (Allen 2003, 20)

Each micromuseum is unique, yet they have in common: careful consideration to layering space; attention to sense of place; integration of varied heritage evidence; and use of low-cost interpretive interventions creating exhibition environments which encourage individual discovery.

The lasting impacts of GDR politics and history on the contemporary approaches to memorialisation at two different 'micro' heritage sites – the Zeithain Grove of Honour near Riesa and the Plattenbau Micromuseum, known as the *Betonzeitschiene*, in Dresden – will be introduced. Both sites have been subject to micromuseum interventions by O'Brien and yet they each relate to very specific local histories and heritage. These micromuseum inventions have met with varying levels of support within the local area.

Frauenkirche Dresden

The Frauenkirche was set alight during the Dresden firebombing of 13 and 14 February 1945 and reduced to rubble on 15 February, when the remaining structure collapsed under the weight of the dome. There were plans immediately after the war to rebuild the Frauenkirche, with individual stones and fragments being painstakingly removed from the rubble, given inventory numbers and stored for future rebuilding. However, this all came to nothing for a combination of financial and ideological reasons (Angermann 2014). During the GDR period it was then left as a ruin, and planted with roses to both deter souvenir hunters and to stabilise the rubble heap, preventing landslips. The ruin began to take on the symbolism of an anti-war memorial, displacing its original religious heritage with a new political significance more in line with the dominant political ideology, and in 1967 a small plaque documenting the date of the church's building and destruction was placed on the monument. A further development in the displacement and replacement of the Frauenkirche's symbolism took place in 1982, when Dresden city council decided to officially declare the ruin as a memorial (*Mahnmal*)[2] and to add a new plaque stating that: 'the ruin commemorates the tens of thousands of dead and exhorts the living to fight against imperialist barbarism, for the peace and happiness of humanity'.[3] The site became part of the annual commemorative events on 13 February, and then during the latter part of the GDR, it was also adopted as a symbolic site for peaceful protests against the Communist system by the democratic peace movement (Stiftung Frauenkirche Dresden n.d.). This in turn was replaced with the further development of the ideas (originally mooted immediately post-war) to rebuild and restore the Frauenkirche to its former Baroque glory following German reunification

2   A *Mahnmal* is a particular type of memorial; the German word in fact indicates that it is a memorial which has the purpose of warning future generations.
3   'Ihre Ruine erinnert an Zehntausende Tote und mahnt die Lebenden zum Kampf gegen imperialistische Barbarei, für Frieden und Glück der Menschheit' (Angermann 2014) (author's translation).

FIG 3.1.   THE RESTORED FRAUENKIRCHE, DRESDEN, IN 2011.

in 1990. The rebuilding of the Frauenkirche was completed in 2005 (see Fig 3.1), in time for the city of Dresden's 800th anniversary in 2006, with supporters of the project seeing it as 'a symbol of healing, reconciliation, and European Culture' (James 2006, 246).

Since then the Frauenkirche has acted as a key visual symbol within Dresden's tourism marketing and is seen by many as a focal point for local pride, combined with a sense that it allows the city and its population to move on from the traumas of war (James 2006). However, this is by no means the only perspective on the Frauenkirche's new symbolism and position, with many mourning the loss of the ruin as an evocative and unique monument (ibid), unparalleled and irreplaceable. The Frauenkirche's history of destruction, symbolism and restoration link it to the debates and discourses surrounding other significant items of built heritage within German history whose value within current society is contested, such as the Berlin Wall, the Palast der Republik and the Berlin Stadtschloss (see Lisiak 2009 for further discussion of such 'disposable pasts'). The wish to move on from a traumatic past is embodied in the displacements and replacements of these built heritage symbols, but the risk remains that by doing so, aspects of the past are ignored or conveniently 'forgotten' in order to fit in with a contemporary political narrative that potentially negates the memories and experiences of the population.

Zeithain Grove of Honour Micromuseum

The Zeithain Grove of Honour and Micromuseum is a commemoration and interpretation of the former World War II POW camp that housed Allied soldiers (predominantly Russians). During the GDR period, and in line with official propaganda of the time, the interpretation of the site focused disproportionately on the Communist and Soviet prisoners. However, this interpretation and the site as a whole fell into disuse and decline following the collapse of the East German state. Since German reunification, the site has undergone changes to revitalise it and the micromuseum interventions were introduced. These interventions use two spaces within the larger memorial site of the former POW camp.

On arrival at the memorial site visitors are confronted with the imposing red stone arches inscribed with hammers and sickles and stars, leading along a central tree-lined vista closed by a memorial bell tower. To the left of this area is the whitewashed documentation building (which also houses the library, administration offices and visitor facilities), and adjacent to that sits the black wooden (former camp) barrack hut.

One part of the micromuseum is located within the whitewashed documentation building. This intervention consists of two tower-like display units with drawers, shelves and apertures on all sides for artefacts and copies of archival material. A key feature of these display towers is that they enable large amounts of material to be housed and displayed with efficient use of space. They are designed to be used simultaneously by different visitors at each 'face' of the towers, while also encouraging a sense of exploration and discovery in each individual, who has to open drawers or pull out material to satisfy their curiosity.

The other micromuseum intervention is inside the space of the now empty black barrack hut of the former POW camp, the original use and material heritage contents of which have long been 'displaced'. Within the evocative atmosphere created by this sense of emptiness, the micromuseum installation not only replaces the lost heritage and recaptures the sense of place, but with its shining glass and steel-framed container-like structure, it provides a striking visual counterpoint to the bare wooden walls and floorboards. This 'container' is in fact a contemporary walk-in documentation and display space (see Fig 3.2).

FIG 3.2.   THE *ZEITHAIN* GROVE OF HONOUR MICROMUSEUM, 2011.

Before entering this space, visitors are drawn to walk around the exterior of this 'container', in the manner of a traditional display case. In doing so they see a series of objects and text panels, much as in a traditional museum exhibition. It is only on entering inside the space that visitors gain a sense of the multiple and personal individual narratives bound up within the history of this place. These narratives are layered within copies of photographs and documents which visitors are encouraged to discover for themselves inside themed folders placed beneath the object cases already seen from the outside. At the opposite end to the door into the space is a desk with a computer terminal where visitors can access further layers of the site's displaced history, such as the excavations after the camp's period of abandonment.

As a POW camp, Zeithain housed soldiers from Soviet Russia, Italy, Britain, Serbia, France and Poland from 1941 until 1945, in inhumane and unsanitary conditions which resulted in the deaths of between 25,000 and 30,000 Russian inmates and around 900 others, predominantly of Italian origin, whose remains were interred in four different burial areas, according to nationality (Stiftung Sächsische Gedenkstätten 2010b).

During the Soviet military administration of eastern Germany the decision was taken to erect a memorial to those who died at Zeithain. By 1949, the former Russian cemetery of Zeithain, on the site of the first mass graves from the POW camp, was turned into a memorial site where public commemoration of the site's past was focused. This part of the site can be visited today.

Although there were other cemetery areas for the Russian victims, these were within a closed Russian military zone and therefore not accessible to the public. There was no commemoration of the non-Russian victims, with their cemeteries being left untended and neglected, until gradually they were absorbed into the forest. The Soviet army used the area of the Italian cemetery for military exercises, destroying all obvious traces of it (Stiftung Sächsische Gedenkstätten 2010a). In the late 1970s, the development of the Zeithain memorial site was the subject of a collaborative Soviet–East German school project exploring the site's past. As a result of this project a decision was made to establish an interpretive memorial on site. However, during this process the particular past chosen for public presentation was highly selective. In line with the GDR's official narrative, as a state emerging from anti-fascist resistance, the role of anti-Nazi resistance fighters within the camp was highlighted out of all proportion, in contrast to the complete displacement of the histories of non-Soviet internees.

Following German reunification, Zeithain suffered from a lack of funding to support its activities, relying on volunteers to keep the site from closing completely. However, in 1997 a supporters' trust was established and by 2002 Zeithain had become the fifth memorial site within the Foundation for Saxony's Memorial Sites (*Stiftung Sächsische Gedenkstätten*), and the permanent exhibition was developed in 2003. With the withdrawal of Soviet troops from eastern Germany in the 1990s, all burial grounds where former inmates were buried can now be visited (Stiftung Sächsische Gedenkstätten 2010c).

The *Betonzeitschiene* – the Plattenbau Micromuseum

The second micromuseum deals with heritage displacement and loss: the loss of an industry (concrete panel manufacture) and architectural form (the Plattenbau) strongly associated with GDR society. The *Betonzeitschiene* (concrete timeline), also known as the Plattenbau Micromuseum, consisted of a physical intervention within a publicly accessible green space in the Johannstadt area of Dresden. It consisted of a spatial, architectural installation commemorating the history of the *Plattenbau* (prefabricated concrete panel apartment buildings) as an architectural form and local industry, and was sited within a public open space in the Dresden suburbs, on the site of (and incorporating aspects of) the former construction works producing the prefabricated *Platten* (panels). The factory was closed in 1990 and left to fall derelict, standing among numerous examples of the GDR *Plattenbau* mass-housing building style. A local interest group, the *Interessengemeinschaft Platte*, had wanted to preserve the factory building as a form of heritage, but O'Brien suggested the micromuseum intervention as both a more cost-effective and multi-faceted option to address the layered histories involved in the site. As such, the *Betonzeitschiene* represents the loss of this architectural form as being representative of the loss of social and cultural norms and identity markers for individuals who grew up during the GDR period.

For O'Brien the site was not just about the *Plattenbau* factory but about the use of the site further back into history and as a way of connecting to the losses that occurred over time within the same place; 'the museum is about the history of this piece of earth. This site tells us so many human things' (O'Brien quoted in Mead 2003, 33). Planned in 2003 (see Fig 3.3) and opened in 2004, his micromuseum intervention was made up of architectural elements, with concrete stepping stones forming a chronological route through a sculptural landscape (see Fig 3.4).

Described as having been 'conceived as an unusual fusion of museum and landscape architecture, it is on the edge of a site formerly occupied by a factory that made the concrete panels

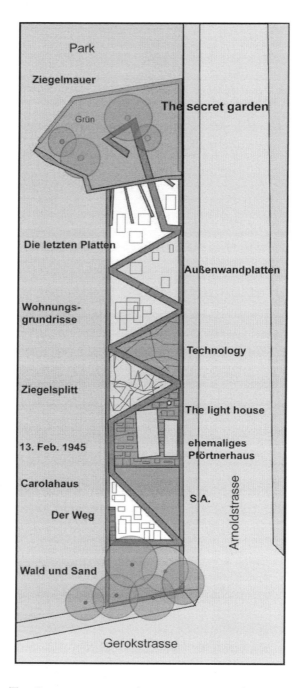

Fig 3.3.   Plan of The *Betonzeitschiene* (concrete timeline), also known as the Plattenbau Micromuseum.

FIG 3.4. THE *BETONZEITSCHIENE* (CONCRETE TIMELINE), ALSO KNOWN AS THE PLATTENBAU MICROMUSEUM, IN 2005. PHOTOGRAPH ALSO SHOWS THE 'ZIEGELSPLITT TRÜMMERSTEIN' SECTION OF THE MICROMUSEUM INTERVENTION.

and cladding for *Plattenbau* – the prefabricated buildings that were such a staple of the former East Germany' (Mead 2004, 8). With numerous remaining *Plattenbau* apartment blocks forming the backdrop to the green space, visitors entered the site of the micromuseum from the corner of Gerokstrasse and Arnoldstrasse, at the 'Forest and Sand' area. The whole micromuseum was a long strip of land (100m x 15m) running along the side of Arnoldstrasse, at the edge of the green space. The concrete slab 'stepping-stones', many with significant dates set within them, led the visitor along a zigzagging path through a series of wooden-framed triangular segments in the long strip of the micromuseum. The zigzagging path separated the installation into ten segments, each dedicated to a particular theme. In series, these segments were: *Wald und Sand* (Forest and Sand); *Carolahaus* (the name of a children's hospital that had been the first building on the site); *SA – Schule* (referring to the fact that during the Nazi period it became a training school for SA paratroopers); *Waage / Pförtnerhaus (ehemaliges Wiegehäuschen) / Fabriklampe* (weigh-house / gatehouse (former weigh-hut) / the light house); *Plattenreste aus Ziegelsplitt* (panel remnants made of brick fragments); *Technologie* (technology); *Wohnungsgrundrisse* (apartment layouts); *Außenwandplatten* (exterior panels); *Die letzten Platten* (the final panels); and finally, The Secret Garden, behind the ruins of the silo wall (return to Fig 3.3 to see the placement of the Secret Garden within the overall micromuseum intervention).

In relation to the micro-history, these triangular segments detailed stages in the fabrication process. However, the triangular segments also associated these aspects of the micro-history within the broader macro-history of the rebuilding of Germany following World War II. This was done by associating the segments with certain key dates of the macro-history, where these dates were cast into some of the slabs in the 'stepping-stone' path running between the segments. In addition, each of the segments had an appropriate title communicated on its wooden frame. One example of the alignment of the micro- and macro-histories within this site was the way in which the segment '*Ziegelsplitt Trümmerstein*' (brick fragments and rubble stone) was juxtaposed with the dates 1946 and 1948 cast within the slabs in the 'stepping-stone' path (see Fig 3.4 for this specific example). This was the time period (*Trümmerzeit*) when the rebuilding of German cities, after their destruction during the war, was the focus of individual and collective activity, where rubble from the bombing was used within the poured concrete panels of the new pre-fabricated buildings.

Originally designed to document, interpret and commemorate the lost histories of this site, the micromuseum survived *in situ* for four years (2004–2008), a period described by O'Brien as 'Phase I' of the *Betonzeitschiene* (O'Brien 2012a). Visitors to the site of the Plattenbau Micromuseum today will be faced with a derelict and vandalised collection of broken concrete fragments rather than the original interpreted encounter with the past, presented through a combined installation of architecture, nature and light, which drew international attention to this micromuseum a decade ago. However, O'Brien takes a longer-term and hopeful perspective, seeing the current situation as 'Phase II' of the *Betonzeitschiene* (O'Brien 2012b), leaving the door open to potential further 'phases' in its existence, such as reconstruction as part of the new housing project planned for the green space. This is interesting in terms of the repetitions of the sequential 'loss' and replacement of aspects of heritage (also seen in the history of the Frauenkirche), where the values of the heritage are (re)assessed in line with changes in society and in relation to what people wish to remember or forget, something which can be called 'selective social amnesia'.

INTEGRATION OF MACRO- AND MICRO-HERITAGE

Emerging from the multi-layered perspectives on history and heritage built in to the micromuseums described above is a sense that the micro and macro aspects of heritage cannot, and should not, be seen in isolation from one another. Simultaneously each is a product and a qualifier of the other, even when this is not necessarily acknowledged within the public or official discourses surrounding individual heritage sites. Neither micro- nor macro-heritage sites alone can communicate this interconnectedness of disaster and recovery; destruction and regeneration; trauma and reconciliation; loss and renewal; remembering and forgetting; place and time; past, present and future. Instead a mapping, networking and over-layering of all these may enable a richer saturation of both micro and macro evidence about the past, infused with knowledge and awareness of their shifting positions, meanings and values.

O'Brien's most recent architectural intervention, 'Slaughterhouse 5', is one example of an attempt to address this imperative to (re)integrate the micro and macro. This installation is a window within the cellar of the building where American author Kurt Vonnegut was held as a POW during the firebombing of Dresden, and which is now a conference centre. The glass for this window contains layers of text, maps and images within it in order to illustrate the complex interweavings of history, time and space, as O'Brien describes it:

The old map of the city of Dresden before the war and the new map of the city as we find it today are layered on top of each other which demonstrates powerfully the loss and the re-invention of space… I have also drawn a map depicting other important sites that I want to connect with each other, which creates a type of macromuseum guide to the city. This includes the micromuseum's [sic] I created for the famous German writer Erich Kästner and the industrial concrete panel factory (Plattenbaumuseum or Betonzeitschiene), which played such an important role in the rebuilding of the destroyed city. On the place where the famous Frauenkirche has been rebuilt I have placed the quote from Billy Pilgrim, the soldier narrator in Vonnegut's book who describes the city of Dresden after the destruction with the sentence, 'It was like the moon'. (O'Brien 2014)

This shows a significant step in meshing the macro and micro aspects of heritage and commemoration in a more integrated manner. It might perhaps be argued that this signals an emergent trend of drawing these aspects together into an 'official' heritage discourse shared by 'insiders' and 'outsiders', dwellers and tourists in Dresden and the surrounding area.

CONCLUSION

In eastern Germany each political period during the last century has had its dominant heritage discourse often to the detriment – or even 'displacement' – of what at the time were perceived as 'secondary' heritages, or heritages that needed to be forgotten. From a consideration of the issues introduced above and what has been drawn from the three examples, there are three significant points that have come through. The first of these is the imperative to acknowledge all aspects of heritage, particularly those that may be perceived to be difficult, due to their association with disaster, trauma and loss. Secondly, there is value in integrating macro and micro aspects of heritage in order to enrich the histories presented to a public consisting of both insiders and outsiders. Finally, the impact of a raised awareness of, and identification with, aspects of micro-heritage may already be influencing the 'official' discourses on heritage as the 'Slaughterhouse 5' example suggests. The strength of the micro-heritage examples explored lies in the successful integration of what has been described as 'the micro-architecture of memory' (Mead 2004, 28) with an additional layer of meaning created by their considered placement within associated places and spaces. The manner in which the 'visitor' becomes part of the story and the place comes through the way they each interact with this micro-heritage through an individual act of exploration. 'As you open a drawer or log on to the net, assembling the fragments of a life, a housing development or a city, you start to see both what is in front of you and beyond: the micro–macro relationships that gradually emerge' (Mead 2003, 37), which means that these previously 'displaced' relationships form a lasting impression.

BIBLIOGRAPHY AND REFERENCES

Allen, I, 2003 Editorial: Museum sector must heed lessons to ensure its continued success, *The Architect's Journal* 218 (3), 17 July, 20

Angermann, C, 2014 *Frauenkirche zu Dresden* [online], available from: http://www.frauenkirche.de/index.php [4 March 2014]

James, J, 2006 Undoing Trauma: Reconstructing the Church of Our Lady in Dresden, *ETHOS* 34 (2), 244–72

Lisiak, A, 2009 Disposable and Usable Pasts in Central European Cities, *Culture Unbound: Journal of Current Cultural Research* 1, 431–52

Mead, A, 2003 Memory Matters, *The Architect's Journal* 218 (3), 17 July, 27–37

— 2004 Dresden's outdoor museum fusion: Plattenbau Museum opens its doors in Dresden, *The Architect's Journal* 220 (12), 30 September, 8

O'Brien, R, 2012a *Betonzeitschiene* [online], available from: http://www.betonzeitschiene.de/index.php/phasei [4 March 2014]

— 2012b *Betonzeitschiene* [online], available from: http://www.betonzeitschiene.de/index.php/phaseii [4 March 2014]

— 2014 *Ruairí O'Brien artwork: Slaughterhouse 5 – Kurt Vonnegut – Story telling light sculpture from Ruairí O'Brien* [online], available from: http://www.rob-artworks.blogspot.de/ [4 March 2014]

Stiftung Frauenkirche Dresden, n.d. *Leben in der Frauenkirche, Geschichte, Mahnmal* [online], available from: http://www.frauenkirche-dresden.de/mahnmal.html [4 March 2014]

Stiftung Sächsische Gedenkstätten, 2010a *Gedenken nach 1945* [online], available from: http://www.stsg.de/cms/zeithain/geschichte/gedenken_nach_1945 [4 March 2014]

— 2010b *Gedenkstätte Ehrenhein Zeithain* [online], available from: http://www.stsg.de/cms/zeithain/startseite [4 March 2014]

— 2010c *Geschichte der Gedenkstätte* [online], available from: https://www.stsg.de/cms/zeithain/geschichte/geschichte_dergedenkstaette [4 March 2014]

UNESCO, 2009 *Dresden is deleted from UNESCO's World Heritage List* [online], available from: http://whc.unesco.org/en/news/522/ [4 March 2014]

4

Remembering the Queensland Floods: Community Collecting in the Wake of Natural Disaster

Jo Besley and Graeme Were

In January 2011, three-quarters of the Australian state of Queensland was declared a disaster zone, including Brisbane, its capital city, after a series of devastating floods caused widespread destruction. The floodwaters caused over two billion Australian dollars' worth of damage, wrecking properties and livelihoods, leaving almost 40 people dead or missing and countless thousands with a feeling of trauma or loss. This chapter engages with natural disaster collecting through an exploration of the two community collecting projects undertaken by the State Library of Queensland and the Queensland Museum in the aftermath of the floods. Focusing on the process of collecting, it will discuss the types of objects, images and stories selected, and how factors such as distress, destruction and damage (to people and things) raised issues for collecting institutions. This chapter contributes to an expanding museological discourse on the role of museums and collecting institutions as memoryscapes by offering a unique insight into the relationship between salvaged objects, traumatic memory and natural disaster events.

Disaster Collecting

Disaster collecting has only recently received due attention in museological literature. This may be because museums and collecting institutions have considered disaster collecting to remain outside the bounds of what is normally considered 'safe collecting'. This, however, is slowly changing as museums are increasingly recognising the importance of collecting salvaged artefacts to remember natural and man-made disasters (cf Williams 2007). In Australia, Dale-Hallett and Higgins (2010) describe how the Museum Victoria established the Victorian Bushfires Collection in the days after the 2009 Victorian bushfires in Australia. They explain how certain factors shaped the development of the collection, including the need to document the tragic events while materials were still available; the lack of collections from previous bushfires; how to respond to offers of donations from people directly affected by the fires; and to think through the role of the museum in community healing. Their work involved actively collecting stories, images and objects that captured the immediate impact of the bushfires, the community response, the aftermath, and the process of recovery and renewal, or what they called 'rescue collecting'. They also worked closely with a local primary school and collected artworks made by children directly experiencing the impact of the bushfires.

Gardner and Henry (2002) describe the issues and challenges of collecting artefacts in the aftermath of the events of 9/11, after which the National Museum of American History collected artefacts from the World Trade Center, the Pentagon and the field in Pennsylvania where Flight

93 crashed. They describe how two key questions emerged in debates about how to collect disasters: first, how are obligations met for future generations on collecting and preserving materials that tell stories that they will want and need to hear and tell? And, second, how are obligations met with people today in terms of telling the stories that they want and need to hear now, in the aftermath of such traumatic historic events? They describe how disaster collecting raised many challenging issues, including the fact that several of the participants had been eyewitnesses to the tragedy; some had lost friends or colleagues; some had found their institutions had participated in disaster response. The most emotive question raised was whether to collect at all, and if so, what artefacts to collect, and how and when this should be done (Gardner and Henry 2002, 38–9).

Shayt (2006) describes disaster collecting carried out by Smithsonian staff in the wake of Hurricane Katrina in 2005. He reveals how museum staff hastily put together a list of artefacts once the scale of the devastation in New Orleans had been established. Like the Museum Victoria's bushfire collection, artefacts were selected on the basis of documenting a material record of the natural disaster and which could support future museum exhibitions, public programmes, websites and publications. The selected artefacts documented the path of the flood, the people affected, levees (that had failed), the disaster response and recovery. Shayt claims that museum collections built from disasters, either natural or man-made, can be obviously unsettling; they are, he asserts, the core business of history museums.

If museums are to preserve a record of past events, then why is it that disaster collecting appears to be so under-represented in museum collections? The answer could be that it is a special sort of collecting: a form of 'extreme collecting' (Were 2012), as all objects carry with them a physical connection to the past, a biographical life, but the objects salvaged from disasters also carry with them an objective truth, a sort of scar that bears witness to traumatic events. Their association with difficult events makes the collecting of such artefacts especially risky and thus places them at the margins of conventional collecting practice.

In this chapter, we focus on the concerns raised in the wake of the Queensland floods, in which two collecting institutions involved the general public in processes of community collecting. We concentrate on the different processes of community collecting that occurred following Queensland's 'summer of disaster', instigated by the Queensland Museum and State Library of Queensland. Despite popular interest in floods as revealed in library catalogue searches, the two institutions faced a *tabula rasa*; an unprecedented scenario of urgent contemporary collecting where direction and policy had to be developed as the disaster and recovery unfolded. Each institution responded differently, and the various collecting and public programming activities that took place highlight the challenges of collecting and commemorating disaster in the digital age, of dealing with objects and stories related to survival and loss, and the ongoing implications of managing these 'extreme' collections.

COLLECTING THE QUEENSLAND FLOODS

Data relating to the use of the State Library of Queensland's catalogue shows that one particular topic consistently dominated clients' searches of the library's collection: floods. For years, 'floods', 'flooding', and '1974 flood' (in reference to the most dramatic 20th-century flood to inundate Queensland's capital city Brisbane) were the terms most consistently searched by visitors to the library. It is not surprising that floods should interest residents of Queensland, and

Brisbane in particular. In a place where the weather swings between periods of stifling drought and the shock of tropical cyclones and extreme flooding, people have a deep-rooted fascination with the power of water. Devastating floods punctuate Brisbane's post-settler history – yet this history is also distinguished by the tendency to forget, as each generation continues to disregard historical experience and intensively develop the city along the river and its numerous creeks and tributaries.

Brisbane is located within the catchment of a broad, sweeping river, with the central business district located on a sharp meander, with key civic buildings such as universities, courts and cultural institutions located just metres from the river's bank. The Aboriginal people of the region understood that this was a river subject to extremes and lived harmoniously with its shifting form and conditions, using fire stick farming methods to increase areas of grassland for hunting, giving some riverside regions the 'park-like appearance' that was commented upon by early European explorers (Gregory 1996, 2). Despite explorers' observations of evidence of both extreme flooding and drought, the penal colony was established in close proximity to the river in 1825. The 19th century's biggest flood swept through the area in 1841, just prior to the fledgling town opening for free settlement, to be followed by a succession of vast and destructive floods throughout the 19th century.

The 1974 flood, however, was the one to fully sear itself into the city's modern psyche. The flood peaked over the Australia Day weekend in late January, following an exceptionally wet spring and an unprecedented amount of rain brought by tropical cyclone Wanda. On the 27th, the Brisbane River broke its banks and caused massive damage. By the 1970s the city, as well as the neighbouring city of Ipswich, sprawled across the river catchment – more than 8000 houses and the central business districts of both cities were inundated and 14 people lost their lives (Bureau of Meteorology 2011). The flood and subsequent clean-up was the first to be depicted on television, and iconic images such as a rowboat of men saving beer from the local XXXX brewery captured the larrikin masculinity of the city's self-image. Significant flood mitigation and infrastructure works over the next few decades, including the building of a massive dam in the Brisbane Valley, prompted a new folklore that Brisbane was 'safe from flooding'. By the early 21st century, Brisbane was in the grip of a decade-long drought and floods, it seemed, were far from anyone's mind and comprehension.

There was perhaps a degree of collective amnesia in Brisbane and Queensland when large amounts of rain began to fall across the state from October 2010. Serious flooding began in early December and continued until the end of January; at least 70 towns were affected and three-quarters of the state was declared a disaster zone. In the second week of January 2011, an exceptional rain event occurred and the infamously misnamed 'inland tsunami' swept through Toowoomba and the Lockyer Valley, leading to severe flooding of Brisbane and Ipswich. Queensland residents experienced the floods in real time, whether first-hand as waters rose in their own neighbourhoods, or on television and across multiple digital platforms as crucial information and shocking footage proliferated from the media, eyewitnesses and even those trapped in the floodwaters. Amid the destruction, the state's major cultural institutions – the museum, library and art gallery – were also inundated. Collections were saved but buildings, infrastructure and the museum's prized World War I tank were damaged. At the same time, these institutions were immediately aware of their responsibilities to document, collect and interpret this extraordinary communal experience.

Natural Disaster Collecting and Preparedness

The most striking difference in the collecting practices of these two institutions is rooted in what could be described as their 'digital preparedness', which in turn was derived from each institution's core focus. While the Queensland Museum is strongly concentrated on traditional, taxonomic collecting around Queensland's natural and cultural history, the State Library of Queensland has been a leader in the use of digital technologies to increase access to its collections and develop a dialogue with its audience. The two institutions also have distinct collecting territories – while the museum collects three-dimensional objects, the library collects Queensland's documentary heritage and describes its role as 'co-creating Queensland's memory' (Philip and Takeifanga 2012, 37).

During the flooding, social media and digital platforms took on vital roles in communicating, linking and constantly updating the community. Early on, the State Library used the web to inform people how to protect, save and restore their own collections in flood conditions. As the library had an established digital relationship with its public, platforms such as Facebook and Twitter were extremely active and, during the flood itself, the library began to receive a huge amount of material. Upon re-opening, it launched a commemorative project where small-scale sandbags were available for visitors to inscribe and add to a communal installation. This recognition of the need for shared grieving and an emphasis on public programming and participation were to distinguish the State Library's continuing response to the disasters.

Just over a year after the disasters, the library presented two exhibitions in Brisbane, elements of which then toured the rest of the state. *Floodlines: A Living Memory* was a commemorative exhibition about the recent disasters with a strong emphasis on the use of multimedia and digital technologies, while *Floodlines: 19th Century Brisbane* explored Brisbane's longer flood history and drew on the collections of the library's John Oxley Collection in a more traditional exhibition format. In the six months leading up to the exhibitions, two participatory projects took place, creating several major elements of *Floodlines: A Living Memory*. One was 'Flood of Ideas', a web-based initiative to gather diverse and creative ideas from the community on how to better plan for and respond to future floods, with a selection of ideas ultimately exhibited (http://floodofideas.org.au/). The second was the *Flood and Cyclone Mosaic*, an interactive, constantly regenerating digital installation that continues to grow as members of the community contribute their images. These images become part of the artwork itself, as well as part of the library's collection. Other key elements of the exhibition included a soundscape, the *Political Mashup* montage of media coverage, the *Wall of Stories* compilation of digital stories and media footage, and *Augmented Reflections*, an app for iOS devices through which users could track the rise and fall of floodwaters in specific parts of the city, as well as the 'before and after' effects of Cyclone Yasi.

As this brief description indicates, the exhibition was a digitally dense and sensory experience, and by local standards was certainly groundbreaking in its use of technologies. High visitation demonstrates that it also went some way to meet the local community's need to reflect upon their recent experiences. It was intriguing that, apart from a few sheets of paper containing the handwritten notes of Premier Anna Bligh's iconic 'we are Queenslanders' speech, the exhibition contained no objects. The 19th century exhibition presented more objects but the vast majority of these were ephemera and photographs. Although there was a rich mixture of items, this exhibition was also largely an object-free experience. While this reflects the library's collecting domains, a conscious position was also taken by curators that as this was a disaster of the digital age, digital responses were the most fitting. In essence, the two projects exemplify the dual respon-sibilities the library has to its contemporary audiences: the need to commemorate and the need to interpret. With its total focus on interactivity and five central themes of Resilience, Reflect, Remember, Rebuild and Reconnect, *Floodlines: A Living Memory* was foremost an emotional exercise of memorialisation. *Floodlines: 19th Century Brisbane* instead sought to interpret, using historical sources and the perspective of time to address what Gardner and Henry describe as one of the oldest questions faced by historians: 'does history actually help us understand the present?' (2002, 51).

Salvaged Artefacts

In contrast, the Queensland Museum's focus was firmly on collecting objects. Unlike the library, which had a strong collection of material related to earlier flood events, it was found that there was just one item in the museum's collection related to Queensland's flood history – an archaeo-logical section that showed layers of compressed silt and mud from previous floods. This was not an object intended to tell flooding stories in particular. In revealing the multiple layers of river-side deposits, the section told many stories but its solitary presence in the collection highlights some of the dilemmas in collecting disaster. Which objects, and therefore which stories, allow present and future generations to understand how and why flooding occurs and the impacts and consequences it has on a community? How do objects convey the damage, destruction, loss and grief that flooding brings, not just for humans, but also for the natural environment and other species? In collecting about the 2011 floods, the museum was attempting to collect the present, with all the historical, social and scientific conundrums about judgement and signifi-cance that this brings. Engaging with the community to identify collections after such a collec-tively distressing event required sensitivity, yet timing was crucial. If the museum did not act quickly, much material would be lost, but were people ready and was there enough perspective to judge what was meaningful?

Perhaps one of the greatest difficulties with flood collecting relates to the type of damage a flood inflicts. The 'material culture' of flooding is sodden, muddy, disintegrated muck, and the first and strongest instinct is to clean it up. Arguably the biggest human story of the 2011 Queensland floods was the emergence of the 'Mud Army': masses of volunteers who, for weeks, cleaned up the cities and towns, often travelling great distances to do so. Working alongside the real army, the Mud Army took on legendary status as the grassroots heroes of the flood and, within a few days, had self-organised Mud Army t-shirts and caps (see Fig 4.1). Items such as these were, naturally enough, soon within the sights of museum curators but it was objects that bore the physical inscription of flooding that proved far more elusive, in part because the Mud Army were throwing them all away. Over time, stories emerged of structural bracing being stripped from houses, functioning appliances being dumped and just plain thieving; some of the gloss of mass volunteering faded. However, it was in this context of heroic cleaning that curators sought to locate meaningful objects.

Like the library, the museum initially used social media to elicit community donations. Both institutions were overwhelmed with offers of images and footage, but it did not prove to be a particularly fruitful avenue for identifying objects. More detailed coverage in the traditional media achieved some results. One way of understanding this is that people needed more expla-nation – examples, images – of what the museum was seeking before they could recognise that

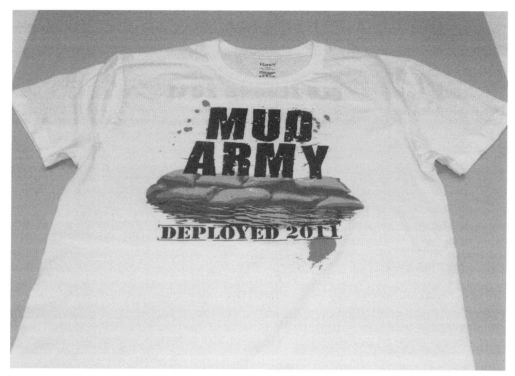

FIG 4.1.   MUD ARMY T-SHIRT, 2011. THE MUD ARMY WAS THE NAME GIVEN TO THOUSANDS OF
VOLUNTEERS WHO CLEANED UP DAMAGED AREAS FOLLOWING THE QUEENSLAND FLOODS.

their items may have significance. It is also likely that the community, like the museum itself,
was still dealing with the practical difficulties that flooding had brought. Keeping objects was
the last thing on most people's minds.

Ultimately, it was professional and personal networks that proved to be the most effective way
of identifying objects. Museum staff, both present and past, donated some of the most insightful
objects. Colleagues in government departments involved in flood and recovery operations, such
as the Emergency Services, were also quick to understand the task at hand. Curators had a 'wish
list' of the types of objects they would like to collect, developing themes to ensure coverage of all
aspects of the disaster: before, during and after. As it happened, such a level of choice was never
really possible, and the collecting process was a more opportunistic process than anticipated.
Three objects (or groups of objects), briefly described, highlight how post-flood collecting more
closely resembled salvage than strategy.

The first object is a mud-encrusted record player found in the wreckage of a riverside shed
used as a personal retreat by a middle-aged man – his den, with bar, music and river views.
Along with the mud-splattered lid, the turntable and base is coated with a layer of mud at least
an inch thick that has cracked like a tidal flat (Fig 4.2). The stylus is also coated in mud (Fig
4.3). This object's appearance speaks volumes of the power of the flood and the sheer quantity
of silt carried by the floodwaters that invaded people's homes. Its damage confronts those of us

Fig 4.2.   Flood-damaged turntable, Brisbane, 2011.

Fig 4.3.   Flood-damaged turntable (stylus detail), Brisbane, 2011.

who are aware of the delicacy and care with which a turntable is usually handled; what's more, this source of music and pleasure has been silenced. This is a visually arresting object, but it only speaks to generations familiar with vinyl records; to a child who knows only MP3 players, this is a totally mysterious object. A damaged piano was also offered to the museum. Swollen, muddied and bereft, the piano was probably more universally understood but its size and condition made it too difficult to accommodate. All such sludge-coated objects posed significant challenges in terms of contamination, storage and preservation, yet it is the mud itself that is the flood's legacy and that makes this object worth keeping.

The particular intensity of the onset of this flood led to an extraordinary rescue effort in the Lockyer Valley. As huge quantities of water swept through the valley, people resorted to sheltering on roofs. On 11 January, helicopter pilot Mark Kempton rescued 28 people while his own house in Brisbane was inundated by floodwaters. Through negotiation with Emergency Services Queensland, the museum was able to secure the rescue harness and strop that lifted people to safety, along with outstanding footage of the rescues shot from the helicopter. Kempton then donated his flying suit, helmet and gloves. This collection of items undoubtedly has historical significance and will endure as an important strand in the rescue narrative. Kempton subsequently recorded a lengthy video interview with the State Library where he described in detail his memories and feelings about the day, adding to the future value of these items.

It is often the fine-grained, less heroic stories that are the hardest to capture through objects. Yet their impact can be profound, as shown by an assortment of ephemeral, hand-made and intangible objects donated by two friends in Ipswich. These two long-term friends lived walking distance from each other, on either side of a creek. Both homes were flooded, the contents destroyed and buildings and gardens damaged. With the help of friends and volunteers, the women began to rebuild their lives. To express her gratitude, one attached a hand-painted banner to her fence, thanking all the helpers. A few months later, they organised a thank-you party, with hand-made invitations created from photographs 'spoiled' by the floodwaters to become intriguing new images. As gifts, they prepared bags of various things – many were unusual and humorous, including items donated to them during the flood, now 're-gifted'. They also wrote a song, thanking everyone individually for their particular contributions. The song was quirky, charming, bittersweet and beautifully sung. Some time later, the women recorded the song for the museum and donated the banner, an invitation and a gift bag to the collection. These objects articulate the personal dimensions and deep ambivalence of disaster, for with loss comes gratitude, and with destruction, creativity.

All of these objects, and more, were displayed in an exhibition on the first anniversary of the floods. Like the library's *Floodlines*, the *Bouncing Back from Disaster* exhibition also focused on resilience. Just a small section of the exhibition departed from this theme to examine the ecological and evolutionary impacts and advantages of natural disasters. Celebration and commemoration, it appears, is manageable territory for state cultural institutions. More time will need to pass before more probing questions can be asked. Will the assembled collections be able to sustain deeper interrogation? What new insights will they reveal with time? While the museum has not continued its collecting around disaster, the library has acquired significant new material, including a large body of oral history interviews.

COLLECTING FOR THE FUTURE

As the world experiences extreme weather events with alarming frequency – most recently the Tasmanian bushfires and the tsunami that struck Japan – it seems that museums and collecting institutions will increasingly need to implement collecting strategies to represent natural disasters for future generations. While Queenslanders may be well prepared, having responded to the great floods of 2011 and 1974, how well equipped are institutions for dealing with the grief and trauma as they become sites to recount suffering and loss? Again, in January and February 2013, Queensland experienced serious flooding. A number of regional towns and cities were devastated and the Mud Army set to work. A paradox is that if flood events take place in Queensland on a regular basis, then natural disaster collecting may itself become threatened as flooding becomes an integral part of everyday life and thus remains outside the scope of conventional museum collecting practice.

BIBLIOGRAPHY AND REFERENCES

Bureau of Meteorology, 2011 *Known Floods in the Brisbane and Bremer River Basins* [online], available from: http://www.bom.gov.au/qld/flood/fld_history/brisbane_history.shtml [27 January 2013]

Dale-Hallett, L, and Higgins, M, 2010 The Victorian Bushfires Collection: Symbols for healing, *Recollections* 5 (1), available from: http://recollections.nma.gov.au/issues/vol_5_no_1/notes_and_comments/the_victorian_bushfires_collection [26 November 2013]

Gardner, J B, and Henry, S M, 2002 September 11 and the Mourning After: Reflections on Collecting and Interpreting the History of Tragedy, *Public Historian* 24 (3), 37–54

Gregory, H, 1996 *The Brisbane River Story: Meanders through Time,* Australian Marine Conservation Society, Brisbane

Philip, L, and Takeifanga, N, 2012 Floodlines: two exhibitions interpret public memory of natural disaster in Queensland, *Museums Australia Magazine* 21 (2), 37–41

Shayt, D H, 2006 Artifacts of Disaster: Creating the Smithsonian's Katrina Collection, *Technology and Culture: the Journal of the Society for the History of Technology* 47 (2), 357–68

Were, G, 2012 Extreme collecting: dealing with difficult objects, in *Extreme Collecting: Challenging Practices for 21st Century Museums* (eds G Were and J C H King), Berghahn Books, Oxford, 1–15

Williams, P, 2007 *Memorial Museums: the Global Rush to Commemorate Atrocities*, Berg, Oxford

5

Displaced Heritage and Family Histories: Could a Foreign Family's Heritage in China Become an Ecomuseum 'Hub' for Cultural Tourism Management?

GERARD ESPLIN CORSANE

INTRODUCTION

This chapter introduces a personal (re)discovery of family heritage and history that has been 'displaced' and (re)placed in a number of ways in Wuhan over time. The key phase of the actual displacement of this family heritage took place around a period of human disaster, trauma and loss in the late 1930s and 1940s, brought about by war and conflict. Wuhan is the capital city of Hubei Province in the People's Republic of China. Located at the confluence of the Han and middle reaches of the Yangtze Rivers, the area has a long and rich heritage. Its significance over time is noted in many of the historical accounts of Chinese and global history. Its geographical placement has meant that it has been a natural meeting point and 'crossroads' in central China; a transportation hub and an economic, commercial, cultural and political centre. In 1949, Wuhan City was established by the merging of Hanyang, Hankou and Wuchan urban settlements. Positioned on the north-east bank, where the rivers meet, Hankou (or Hankow, as the Chinese name has previously been romanised)[1] remains a key port.

Two surviving buildings in Hankou, constructed in the first quarter of the 20th century, are linked to my family history and heritage. They have been connected to two businesses with which my paternal grandfather, Walter Hughes Corsane, was associated: the Hankow Ice Works and the Aerated Water Company. The first building, erected in 1918, was identified by Chinese researchers and the second, built in 1921, was identified by me (see below).[2] Both were located in the French Concession, one of the five foreign concessions in Hankow. For various reasons, these buildings

---

1 The Wade–Giles system for the romanised transcription of Chinese used 'Hankow'. When this system was replaced by the pinyin system, the spelling 'Hankou' became more common and official.
2 The placement of these two buildings into the family history and heritage has become very complex. The main reason for this is that the establishment of the 'Hankow Ice Works' pre-dates both buildings. MacMillan (1925) says that the 1921 building was rebuilt in that year, and this suggests that there may have been an earlier building at this site. The 1918 building is around a corner on the same street and has been identified by Chinese researchers as a building linked to the Corsane and Anderson businesses. It is also difficult to know if each building was only linked with one of the businesses. All of this has helped to further the sense of displaced heritage.

and their associated histories have undergone a certain amount of literal and figurative 'displacement' from the family heritage. The physical displacement of this family heritage took place following a period of human-made conflict, devastation, suffering and trauma that started in July 1937, with the invasion and subsequent occupation of China by Japan in the Second Sino-Japanese War. This human-made turmoil increased as this conflict became part of World War II and grew further due to internal civil strife and war within China itself, which continued until 1949.

This chapter outlines a personal 'journey' of recovering 'displaced' heritage. It covers some of the challenges associated with conducting family history research in today's world. The chapter includes initial findings in the (re)discovery of displaced family heritage related to the two buildings during a journey, which came more from chance and work-related opportunities through my positions at Newcastle University than by conscious design. It introduces some key historical details related to my paternal grandparents, Walter Hughes and Margaret Wallace Corsane (née Esplin). The chapter concludes with a suggestion that the 1918 building could potentially become the hub for an urban 'ecomuseum'.

Personally, the displacement of this family heritage was not simply physical. Descended from grandparents who were part of the 'Scottish diaspora' and the history of British colonialism, both my parents and myself were born outside the United Kingdom. With my parents emigrating to Southern Rhodesia, where I was born in 1962, there had been a distancing from the family history and heritage. I knew that my father, Charles Hughes Corsane, was born in Yokohama, Japan, in 1917 and that my paternal grandparents had connections to China. My mother, Jessie Lamond Knowles, was born in 1923 in Quetta (then in India), when my maternal grandfather was stationed there with the Black Watch Scottish Regiment.

These were the only facts of which I was fully aware. It was only during fairly infrequent visits back 'home' to Scotland and to Jersey in the Channel Islands that stories were recounted by relatives of the family's history in different parts of the world. During family reunions in Jersey, stories were shared by my paternal grandmother, Margaret Wallace Corsane, and by my uncle and aunt, Leonard Hughes Corsane and Pierrette Corsane. These included anecdotes about my grandfather's businesses and life in China, along with accounts of the trauma and suffering that my grandparents endured during their internment in the Japanese camps. Unfortunately, I never knew my grandfather, Walter, who passed away in 1950.

I regret not speaking to my father more about this heritage and history before he died in 1980. But, growing up in Africa, this history was so removed from the reality of life. However, since coming to the United Kingdom in 1999, I have started to reconnect with the family history and heritage. With my work experience in the heritage and museums sector and now as a university-based academic in heritage and museum studies, it seems strange to me now that I have only recently allowed myself more time to re-engage with the family's past. Yet, with experience in historical and heritage research, I may be better equipped to face the challenges normally associated with this type of research.

Family Historical Research and Heritage Displacement

When people are aware of their family histories, the closest heritage to them is what they have inherited from their own families. At a micro-level, our family histories and heritages can be our most important links to the past. They are often central to the processes of our identity construction, and they help to draw us into self-awareness of our family's long-term placements within history and how individuals from each generation were positioned in society and within specific 'communities'. They help to define our relationships to ontological notions of 'being', along with our affiliations with particular places and any associated constructed 'senses of place'.

Having worked in the South African heritage and museums sector in the 1990s, during the period of major transformation there, I became aware of how important people's family histories and heritages are to them. Generally, people approach cultural heritage organisations to register heritage resources, donate heirlooms and lodge family history information. Their expectations are that these are accepted and 'institutionalised' into archives and collections for posterity, entering into the 'official' record so that they do not become displaced over time.

At a local level, these micro-histories and heritages combine to form the fabric of community interpretations. Sometimes they become incorporated into more macro-level state or global historical representations. However, unless linked to individuals and/or events recognised as being historically significant, the majority of family histories and heritages traditionally tend to remain at a micro-level. They are often marginalised due to perceived inconsequentiality, thereby becoming 'displaced' and not even finding their way into the footnotes of the macro-histories.

Fortunately, with the expansion of social and local historical research, these micro-level histories and heritages have received increasing attention among historians, heritage practitioners, museum curators and archivists, and community-based history and heritage groups. In the information age, with advancements in computer and digital technologies, access to web-based online archives and heritage resources has increased, along with the ease of communication and knowledge exchange through email and social networking platforms. Consequently more data is being produced by both academic historians and the increasing number of people becoming 'amateur' historians, creating their own histories. This is particularly the case in developed societies, where people have more disposable income and leisure time to invest in studying family histories and heritage. For example, in the United Kingdom the number of people interested in tracing their genealogies and undertaking family history research is growing. This is confirmed by the popularity of television series such as *Who Do You Think You Are?*; the number of magazines and websites related to the topic;[3] specific software programs on sale; and niche market travel to different places to trace family histories.

However, with more people involved in family history research and more primary and secondary resources becoming accessible online, the challenges to this branch of studies have increased. The information explosion has resulted in competing sources of 'evidence', with contradictions in what is presented as 'fact'. There are questions about 'verification' and who confirms the information. This leads to the dangers of accepting what some may deem to be 'misinformation' – depending on their own viewpoints. In a postmodern world, people are more willing to select what they wish to believe and what they would like to forget. In relation to our family histories, we desire to present an affirmative picture with positive features, while glossing over anything negative and hiding family 'skeletons in the closet'. All of this can distort by further displacing aspects of a family's heritage, or by contributing to the construction of a sanitised and 'imagined' family history.

---

3 Magazine titles and associated websites include: *Discover Your History – Ancestors, Heritage, Memories* (www. discoveryourhistory.net); *Family Tree – Your Ancestors, Your History* (https://family-tree.co.uk/); *Your Family Tree* (www.yourfamilytreemag.co.uk); and *Who Do You Think You Are* (www.whodoyouthinkyouaremagazine. com). See also: www.ancestry.co.uk.

From the beginning of my journey in (re)placing my own family heritage and history in China, I have considered these challenges and reminded myself that my family's connections with China are likely to be perceived by some as being part of the negative history of a 'century of national humiliation' and the 'scramble for China by foreign devils' (see Bickers 2012).

HISTORICAL SOURCES AND STARTING A JOURNEY TO (RE)DISCOVER DISPLACED FAMILY HERITAGE IN CHINA

Following a series of serendipitous events in 2012, long-unexplored memories of having a family heritage linked to China were stirred. The first occurrence was the chance rediscovery of a family photograph album, with a page of photographs (see Fig 5.1) that became the initial source of evidence in '(re)locating' the family heritage in Hankow. These photographs not only depict this displaced family heritage, but the four 'landscape' format images also hint at upheaval and suffering. The top-left picture shows how the French Concession was barricaded against the Japanese, as it became one of the last defence strongholds. The other three landscape images show how the Hankow Ice Works provided water for the beleaguered population during the besieging and occupation of Hankow by the Japanese from October 1938 to January 1939. This is confirmed by the short captions written on the back of each picture.[4]

The building in the bottom-left photograph and in the two images on the right has now been identified as the 1921 building in Hankow (see Fig 5.4 and the details of how the building was finally identified). The building façade in the middle 'portrait' vertical photograph is more difficult to identify, as there is not enough detail in it to unreservedly link it to either the 1918 or 1921 building. As the establishment of the Hankow Ice Works pre-dates both buildings (MacMillan 1925, 185; see also Wang 1997), could the building depicted in this image be from an earlier location of the business?

The photograph album was accompanied by a memorandum, part of the office stationery of the 'Hankow Ice Works', providing the name for the first business linked to Walter. Searching through 1915 to 1990 issues of the *Supplement to the Edinburgh Gazette* and the *Supplement to the London Gazette*, one finds entries for 'Corsane, Anderson, & Co (Hankow Ice Works)'. In addition, the 1925 2nd edition of the book compiled by Allister MacMillan on the *Seaports of the Far East* contains an entry on 'Hankow Ice Works (Corsane, Anderson and Co, Proprietors), Rue du Maréchal Joffre, French Concession' and dates its establishment as 1904. After providing details on ice production, the entry goes on to note that: 'Messrs. Corsane, Anderson and Co, the founders and proprietors of the business, are also manufacturers of aerated waters by the most modern scientific processes. The plant of machinery for that purpose… was installed when the present new factory was built in 1921' (MacMillan 1925, 185).

Through online research I discovered an article featuring an interview with Anderson's son which makes references to Anderson establishing the ice works and the aerated water company (Evans 2008). Of particular interest are the similarities in the historical detail of Anderson and Walter being ship engineers who saw a business opportunity for manufacturing ice when onshore in Hankow. This is verified in the Corsane family shared memories (Corsane 2013, *pers comm*)

---

[4]   The importance of the area at this time in history is discussed in: MacKinnon, S R, 2008 *Wuhan, 1938: War, Refugees, and the Making of Modern China*, University of California Press, Berkeley and Los Angeles.

FIG 5.1. PAGE FROM FAMILY PHOTOGRAPH ALBUM, INCLUDING IMAGES SHOWING THE HANKOW ICE WORKS PROVIDING WATER FOR PEOPLE IN THE MONTHS SURROUNDING THE JAPANESE OCCUPATION.

and is also covered in Wang (1997), who, however, dates the establishment of the business as 1911 and locates it on Yuefei Street.

This memorandum also displays the two characters 和利 (which represent 'harmonious' and 'profit'), as the first two parts of the Chinese name for the business. These two characters are important in identifying the 1921 building in Figs 5.1 and 5.4. The identification was confirmed when the evidence in these images was 'triangulated' with the architectural details of buildings in two further historical photographs posted on a Chinese internet bulletin-board site (bbs.cnhan.com). Viewed together, these two additional photographs (circa latter half of the 1930s) show soldiers on parade outside two buildings. In these photographs, one building has distinctive arches and the other bears the characters 和利 on either side of the main entrance, with 'Hankow Ice Works' above the door. Using this evidence, the 1921 building (see Fig 5.4) was identified from the distinctive arches on the neighbouring building by me and a research support group on 3 September 2013, during a PUMAH ('Planning, Urban Management and Heritage') research trip.

Along with the photograph album and memorandum, there was a letter from my grand-father to my father. Written from 'Ash Camp, Shanghai, 9.2.46', the 1946 date of this letter (Corsane 1946) is extremely significant in relation to its content. The letter provides evidence that Walter was returning to the Hankow Ice Works that year. This detail helps to dispel some of the sense of 'displacement' of the family heritage, with contradictions in a couple of Chinese sources regarding the date of sale of the business/properties, discussed in more detail later in the chapter.

I took these three items to work one day in 2012 to create digital copies. On that day, I had a supervision meeting with Mohan Fang, a student from Wuhan, studying for an MA in Heritage Management. The material stimulated discussion, reiterating the point that this family heritage and history had become displaced from my personal consciousness. All that remained of this heritage for me were these few photographic and documentary sources, objects 'brought back' from China (now in the possession of various family members) and distant memories of the stories heard at family reunions in Jersey. One of the objects that could be significant is a 'Challenge Cup' trophy from the Hankow Golf Club, won by Walter in 1907, 1912 and outright in 1913. This shows that he was probably fairly established in Hankow from 1907.

At a personal level, most of the tangible and intangible cultural heritage had become displaced and 'lost'. However, with significant help from Mohan, especially when she was at home in Wuhan, a personal process of rediscovering and visiting these buildings began. Around these visits, there were email communications with my Aunt Pierrette in Jersey, who is the main living 'keeper' of the family's shared memories.

VISITING THE TWO IDENTIFIED BUILDINGS IN WUHAN

Since March 2012, I have enjoyed two opportunities to travel to Wuhan for business and research. The research was supported through a staff exchange network, led by the School of Architecture, Planning and Landscape at Newcastle University and funded by the Marie Curie 'International Research Staff Exchange Scheme'. By chance, just prior to 'rediscovering' this family history, I joined an exchange research project titled 'Planning, Urban Management and Heritage (PUMAH)'. Fortuitously, Wuhan University is a Chinese partner in this project. With documentary research and support in the field from Mohan and academics and students from Wuhan University, along with help from members of a local amateur historical association, the two buildings associated with my grandfather's enterprises were identified; I visited the 1918 building in 2012 and the 1921 building in 2013. In December 2012 I had the opportunity to carry out background research in Beijing with support from two students from Peking University, which is also a partner in the PUMAH project.

Coincidently, my first visit to the 1918 building (Fig 5.2) took place a few days before the Qingming Festival (Chinese 'tomb-sweeping day' holiday). This seemed appropriate as I could clean the two plaques placed on the building. The first plaque, placed by the Hubei Province Cultural Heritage Protection Unit and dated 27 March 2008, shows that the building has heritage designation from the governments of Hubei Province and Wuhan City.

The second plaque (Fig 5.3) names my grandfather and a person named 'Croucher' as part of a brief architectural description and historical interpretation of the building. A translation of this information raises several questions. Where was 'Corsane, Anderson, & Co (Hankow Ice Works)' located prior to 1918? Who was this person Croucher and why is Anderson not mentioned? In a list of British graves originally in the New International Cemetery in Hankow, there is an entry for T H Croucher buried in October 1936, with a note associating him with the Hankow Ice Works (Kress 1954). This is verified in family stories, where he is linked to the businesses. The plaque mentions the two businesses of the ice works and the bottled aerated soft drinks company, but does not mention the nearby 1921 building.

Finally, the plaque notes that the factory was sold in 1938 (see also Wang 1997). This information conflicts with the 1946 letter written by Walter, where he states that: 'I expect to get a passage to Hankow in about a week's time. The Ice Works has been badly looted by the Japanese… Should you be writing to me again the Ice Works will be the best place to send to'

FIG 5.2. PHOTOGRAPH OF THE 1918 BUILDING ASSOCIATED WITH THE BUSINESSES OF WALTER HUGHES CORSANE, AS SEEN IN 2012.

Fig 5.3. Heritage architecture plaque on the external wall of the 1918 building,
with references to Corsane and Croucher, 2012.

(Corsane 1946). This raises the question as to whether or not the 1918 and 1921 buildings were
sold together or separately and in which year(s).

During the first visit to the 1918 building with Mohan, the building 'caretaker' provided
photocopies of three Chinese articles (Ge 2009; Wang 1997; Xiao 2003). These provided useful
information, along with certain contradictions, with the date of the sale of the businesses being
the key example – maintaining the sense of heritage displacement. The caretaker also informed
us that the empty and unused building was owned by a businessman based in Hong Kong.

In November 2012, the (re)placement of the 1918 building within the heritage domain
continued when the Wuhan Land Resources and Planning Bureau (2012) issued a list of designated Industrial Heritage Sites in Wuhan. This includes 29 sites with three levels of protection,
with 15 designated as Level 1 sites. In this top level, three sites are designated with National
protection, three with Provincial protection and the remainder with City protection. The 1918
building is one of the three with Provincial protection. As such, the building cannot be demolished or altered without agreement from the government and the Hubei Province Cultural
Heritage Protection Unit.

The second building, built in 1921 as an 'extension' to the first, is around a bend in the street
from the 1918 building and not directly connected. This building was difficult to identify as the
adaptions to its façade meant that it could only be found based on the two historical photographs
from the Chinese website. In early September 2013, with substantial help from two students from
Wuhan University and four volunteers of Wuhan Humanities of cnhan.com, the building was
located (see Yu 2013). The building has been converted to house three smaller businesses, including
a hotel, and unlike the 1918 building, it is currently vital and alive with activity (see Fig 5.4).

FIG 5.4. 'EXTENSION' TO THE HANKOW ICE WORKS AND AERATED WATER BUSINESSES BUILT IN 1921, AS SEEN IN 2013.

(RE)DISCOVERING ASPECTS OF MY GRANDPARENTS' HISTORY

Alongside the (re)discovery of this family heritage has come the (re)discovery of family history linked to my paternal grandparents. This came mainly through a week of interviews with Aunt Pierrette Corsane in Jersey (Corsane 2013, *pers comm*). Many of the dates have also been taken from birth, marriage and death certificates and shipping passenger lists. In this chapter, and within the parameters of the theme of the book, there is only space for a very brief summary of the relevant parts of the history.

Walter Hughes Corsane was born on 24 November 1871 in St Andrews, Scotland. His mother Eliza Hughes was related to Sir Walter Watson Hughes, a wealthy adventurer and entrepreneur (Van Dissel 1972), and it was through this relationship that Walter acquired the capital to invest in the Hankow businesses. Having initially trained to go into the legal profession, Walter decided that he wanted to travel and started working in the shipping yards. Over time he worked his way up to being a ships' turbine engineer, and this is how he reached Hankow in China, where he co-established the two businesses. The years 1904 and 1911 have been given for the establishment of the Hankow Ice Works in conflicting accounts by MacMillan (1925) and Wang (1997). Other sources have even dated the founding of the business to the 1980s. However, the family narrative says that Walter bought the lease on land in the French Concession, as it had a good well for water, in 1900 and that building started around 1902. This would make the 1904 date more reliable.

In Dundee on 15 November 1916, Walter, as a widower, married Margaret Wallace Esplin (born 8 October 1896, Dundee), and in January 1917 they sailed to Yokohama. As Margaret found it difficult to settle in Hankow, they ended up being travellers, with my father being born in Yokohama in 1917 and my uncle Leonard in East Newport, Scotland, in 1920. They also spent a number of years in France when the boys were young and various periods in Jersey after the boys had been placed in the Victoria College boarding school on the island. With this constant travelling, Walter entered a period of semi-retirement, and T H Croucher (referred to earlier in the chapter; and see Fig 5.3) was probably brought in as a manager in Walter's absence. He was not a partner but had a free hand in Walter's absence as there were times when Walter was in Hankow alone, or he and Margaret were both away. One period when they were both away was from 1931, when Walter became seriously ill with an infection caused by advanced pyrea or periodontitis. On the advice of a local German doctor, they travelled to London so that Walter could see a specialist. It may have been sometime after Walter returned following this that the Aerated Water Company was sold.

In 1934 Walter and Margaret sailed to China for a longer period there together. They were in Hankow when it was besieged in 1938 by the Japanese and when the factory provided water for people between October 1938 and January 1939 at the start of its occupation. In 1942 life became very difficult for them and they were given refuge by some American missionaries. In March 1943 they were among those interned in the Civilian Assembly Centres, or concentration camps set up by the Japanese to detain foreign nationals (Leck 2006), starting in Lunghua camp on the outskirts of Shanghai. The Hankow Ice Works was taken over by the Japanese and turned into a military administration centre (Xiao 2003, 51).

Fairly soon after being released, Margaret returned from Shanghai to Jersey in March 1946. As already noted from the letter from Walter written in February 1946, he intended to return to Hankow, which he did. Unfortunately, in this year he was knocked down by an American lorry, had a stroke and suffered paralysis. Margaret went over to bring him back from China, and they sailed back from Woosung on the Empress of Australia, arriving in Liverpool in December 1946. Assisted by the firm Jardines, Margaret and Leonard (then working in China for British American Tobacco) began finalising the sale of what remained of the properties. By 1950, the Corsane heritage in Hankow was gone.

THE 1918 BUILDING AS A POTENTIAL HUB FOR AN URBAN 'ECOMUSEUM'

The displacement of the Corsane heritage in Hankow, as a result of human-made turmoil, is one small example of what the period of conflict and its aftermath brought on Hankow. However, over recent years, the value of heritage for potential urban regeneration in China has been recognised, and the governments of Hubei Province and Wuhan City have introduced measures to (re)place and protect designated heritage resources in the old foreign concessions. Added to this, there is increasing awareness among local community-based interest groups of the value of becoming involved in heritage and history projects. With all of this, there is the potential for developing an urban 'ecomuseum' plan to incorporate the rich industrial heritage and architectural history found in the old concessions. The 1918 Hankow Ice Works building might be worth considering for a possible gateway hub and interpretation centre for the area. Different from the traditional museum building, an ecomuseum approach is one that facilitates stakeholder-driven integrated heritage protection and management that can further promote the interconnected activities related to the safeguarding of historical and cultural environment, sustainable development, urban regeneration and heritage tourism. This may be the ideal time to consider an ecomuseum that will help to (re)place previously displaced family and community heritages that will draw people and investment back into the area.

CONCLUSION

Using a personal family history, this chapter shows how at many levels family heritage can become displaced, especially during times of human conflict, disaster, trauma and loss. With increased access to sources of evidence in the digital age, these heritages can be (re)located. However, with sometimes competing sources and contradictions in evidence, the processes of (re)placing lost heritage can be extremely complex. Yet, when it is replaced, this heritage can add to the historical interest of an area, draw tourists and be used in urban regeneration projects and effective cultural tourism management.

ACKNOWLEDGMENTS

I wish to acknowledge Aunt Pierrette Corsane for her invaluable help in providing information from the family histories. Thanks also go to my cousins Alan and Babette. I am extremely grateful for the continued support provided by Mohan Fang throughout. I was also given a lot of help by Professors Ming Zhang, Qingming Zhan, Shidan Cheng and Zhengdong Huang and the two postgraduate students Huan Kou and Fei Teng, all of Wuhan University. In Wuhan, I was also assisted with information and photographs by Hui She, Yangsheng Wang, Xuegong Wan and Hongzhi Hou, volunteers of Wuhan Humanities of cnhan.com (a local internet bulletin-board site about Wuhan Humanities). Professor Pengjun Zhao and the postgraduate students Bowen Ma and Ping Wen from Peking University provided much-needed help in Beijing. For help with translation, I would like to thank Dan Chen, Chunzi Zheng, Yage Huang, Yong Zhao and Yuan Gao. There are others who have helped at various points during this journey of (re)discovery, and I apologise if I have not mentioned everyone by name. Finally, I would like to acknowledge the Marie Skłodowska-Curie International Research Staff Exchange Scheme for funding the PUMAH project. The research leading to this output has received funding from the European Union Seventh Framework Programme ([FP7/2007-2013] [FP7/2007-2011] under grant agreement n° 295045.

BIBLIOGRAPHY AND REFERENCES

Bickers, R, 2012 *The Scramble for China: Foreign Devils in the Qing Empire*, 1832–1914, Penguin Books, London

Corsane, P, 2013 Personal communication (series of interviews on the Corsane family heritage and history in Hankow), 4–7 August, Jersey, Channel Islands

Corsane, W H, 1946 Letter from Walter Corsane to Charles Corsane, 9 February, Ash Camp, Shanghai

Evans, A, 2008 The life and times of a true man of the world, *South China Morning Post* [online], 9 March, available from: http://www.scmp.com/article/629178/life-and-times-true-man-world [4 March 2014]

Ge, L, 2009 The inhabitants of Wuhan and Hankow Ice Works, Chutian Daily, 14 August, 44. 葛亮："老武汉，和利冰"，《楚天都市报》第四十四版，2009年8月14日。[online], available from: http://ctdsb.cnhubei.com/html/ctdsb/20090814/ctdsb813010.html [4 March 2014]

Kress, 1954 *New International Cemetery, Hankow (Hankou)*, [Translated list from Kress by unidentified author], British Foreign Office file FO 369/5018, Public Records Office [online], available from: http://www.bristol.ac.uk/history/customs/ancestors/hankow.pdf [4 March 2014]

Leck, G, 2006 *Captives of Empire: The Japanese Internment of Allied Civilians in China, 1941–1945*, Shandy Press, Bangor PA

MacMillan, A (Compiler and Editor), 1925 (1907) *Seaports of the Far East: Historical and Descriptive, Commercial and Industrial, Facts, Figures and Resources*, 2 edn, W H and L Collingridge, London

Van Dissel, D, 1972 Hughes, Sir Walter Watson (1803–1887), *Australian Dictionary of Biography*, Volume 4, Melbourne University Press, Melbourne, available from: http://adb.anu.edu.au/biography/hughes-sir-walter-watson-3813 [4 March 2014]

Wang, Q, 1997 The ice making industry and soft drink industry in old Hankow, *Changjiang Daily*, 29 July. 王琼辉：“旧汉口的制冰与汽水业”，《长江日报》，1997年7月29日。

Wuhan Land Resources and Planning Bureau, 2012 *Industrial Heritage Conservation and Management Planning of Wuhan*, 16 November. 武汉国土规划局，《武汉市工业遗产保护与利用规划》2012年11月16日。[online], available from: http://www.wpl.gov.cn/pc-1507-41151.html [4 March 2014]

Xiao, Z, 2003 The Surrender Hall after the Japanese military occupation of Wuhan, *Wuhan Publicity*, September, 322 edn, 50–1. 肖志华：“受降堂与沦陷时的武汉”，《武汉宣传》2003年第9期，总第322期，第50-51页。

Yu, H, 2013 Hankow Ice Works was the first to make ice by machine in Hankow, *Changjiang Daily*, 7 September, 5. 余晖：“和利冰厂开启汉口机器制冰史”，《长江日报》第五版，2013年9月7日。[online], available from: http://cjrb.cjn.cn/html/2013-09/07/content_5215034.htm [4 March 2004]

6

Walls, Displacement and Heritage

Tim Padley

When the Tullie House Museum and Art Gallery Trust was given the opportunity to create a new gallery to reflect the Roman heritage of the area, they took the opportunity to explore the concept of the frontier. As one of the 'bookends' to Hadrian's Wall and with no visible site to interpret, the Trust was able to concentrate on the concept of frontiers, using both historical and contemporary references. This would enable visitors to explore the standing remains with an understanding of what the monument meant to the people living there at the time, both the army and the locals. Much of this concerns displacement of people from their homes, reasons for invasion, lack of a voice and relates to frontiers today. This chapter will show how the gallery illustrates each of these themes through display of artefacts from the museum's collections, quotations from Roman authors and the use of modern film and oral testimony.

Many of the inhabitants of the Roman Empire were removed from their homes through enlistment into the imperial army. The majority (some 72%) (Mattingly 2006, 166) of the Roman imperial army in Britain was made up of auxiliary troops, enlisted from the peoples who lived in the conquered territories. This served two purposes: it enabled the Empire to make best use of the characteristics and skills of the many different peoples under Roman control, while also serving to minimise rebellion by shipping these levies to far-off parts of the Empire. Thus while Dacians from modern Bulgaria/Romania were used to garrison Hadrian's Wall along with people from Belgium, Holland, France and Spain, the inhabitants of Britain were posted to Switzerland and North Africa. All of these people were conscripted for 25 years' service, at the end of which they were rewarded with full citizenship and allowed to marry. However, many were never able to return to their homeland. In the gallery this is represented by the story of the North African levies made by Emperor Septimius Severus at the beginning of the third century AD.

Severus was a native of Leptis Magna, near Tripoli in modern Libya. He became emperor after a civil war in which he defeated, among others, Clodius Abinus, the governor of Britain, who had used the troops stationed in the province in his army. When Severus became emperor, he uprooted many people from his home province to deploy them in Britain. Firstly, he did this to restore the number of troops in Britain after the civil war. Secondly, these troops were more likely to support the current emperor. Finally, he could embark on his project to conquer the area north of Hadrian's Wall and complete the conquest of Britannia.

This African legacy can be detected in the material culture the Romans left behind. In particular they brought with them two distinctive traditions that reflected their home province. The first of these were distinctive forms of pottery – round-bottomed casseroles, domed lids and convex-walled platters. Although pottery reflects cooking practices it does not necessarily reflect a movement of people as it can just be a matter of taste, such as having a wok today. The second piece of evidence, the use of ceramic tubes to provide the ribs of vaults, is more indicative of a movement of people. This technique is found in North Africa and Sicily and probably developed because there were few trees to provide the timber necessary to support building work in progress. This is not true in Britain, and the discovery of buildings that used these 'vaulting tubes' suggests that they were built by people who had brought their exotic building methods with them. The gallery illustrates the presence of these Africans by exhibiting examples of the pottery and vaulting tubes found locally. Also on display is the final piece of evidence that African people came to Britain; this can be seen in a tombstone from the fort at Birdoswald on Hadrian's Wall (Tomlin *et al* 2009, corpus number 3445). The inscription says: 'D(IS) M(ANIBUS)S(ACRUM) | G(AIUS) COSSUTIUS | SATURNINUS | [HI]P(PONE) REG(IO) M[IL(ES) LEG(IONIS) | VI VIC(TRIS)] P(IAE) F(IDELIS)[…]'. This can be translated as: 'Sacred to the departed spirits : Gaius Cossutius Saturninus, from Hippo Regius, soldier in the Sixth Victorious Legion Always Faithful…'. Not only do we have a name for the person, but also where he came from: Annaba in modern Algeria.

The fate of these soldiers can be tracked through the recovery of the above-mentioned cultural indicators from excavations. It appears that the recruits, who were Roman citizens and legionary soldiers, were assembled at York, where the African-style pottery was being produced in the settlement outside the fortress. They would have taken part in the campaigns north of Hadrian's Wall. These came to an abrupt halt with the death of Severus at York in AD 211 because his sons were more interested in securing the imperial throne and so left for Rome. The troops were then dispersed, including some along Hadrian's Wall (Swan *et al* 2009, 604). The use of vaulting tubes has also been found at Chester, the home of the Twentieth Legion, suggesting that some of the North Africans had been seconded to that legion as well as the Sixth, stationed at York (Mason 1990). Their future would be bound up with that of the Roman army in Britain, far away from their native province.

The Roman literature that survives from the first century AD does provide some information on what some Romans thought about the invasion of Britain. Quotations have been used to provide a Roman voice in the gallery. Tacitus, the 1st century historian (Ireland 1986, 23), in his description of Britain, stated that 'Britain bears gold, silver and other metals which are the prize of victory'. This sentiment is very similar to some suspicions about the reasons behind the 2003 invasion of Iraq. Tacitus also comments on the assimilation of the local inhabitants:

> … instead of loathing the Latin language they became eager to speak it effectively… our national dress came into favour… the population was gradually led into the demoralizing temptations of arcades, baths and sumptuous banquets. The unsuspecting Britons spoke of such novelties as 'civilization' when in fact they were only a feature of their enslavement. (Tacitus n.d.)

This quotation is written out and displayed with a collection of artefacts that show the way in which the Latin language and the Roman way of life is reflected in the archaeological remains. Baths and banquets are represented by the wooden shoes and oil flasks that were used in the baths and the fine tableware used for serving food and drink. This allows visitors to make up their own minds as to whether Romanisation was civilisation or slavery. This is backed up with a simple interactive display to help the visitor decide if he/she is more Roman than Iron Age or vice versa.

A Roman province was expected to pay its way. The cost of the Roman army and administration in Britain was enormous, some tens of millions of *denarii* per year. In order to represent this cost, a display was assembled that showed the resources extracted from the province (see Fig 6.1). These included mineral wealth, particularly silver, as well as farm produce and the two things noted

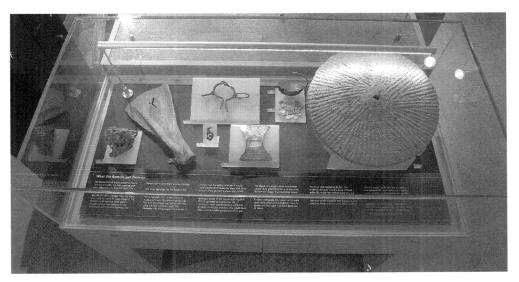

FIG 6.1.   RESOURCES EXPLOITED IN ROMAN BRITAIN (FROM LEFT TO RIGHT): LEAD ORE, AN OX
SCAPULA REPRESENTING FARMED ANIMALS, A SLAVE MANACLE, A SILVER INGOT AND SILVER COINS
WITH THE PURSE THEY WERE FOUND IN AND A QUERN REPRESENTING GRAIN.

by ancient authors as coming from Britain: slaves and hunting dogs (Strabo in Ireland 1986, 20).
To demonstrate the enormous amount of material needed, a simple weighing game was devised.
The aim was to balance a model tent on one side (representing the smallest unit of the army and
referring back to the full-scale tent elsewhere in the gallery) against the livestock, grain and coinage
required to support the soldiers for a year. By showing how many times this amount of leather,
grain and money was required to support the army, some idea of the enormous burden that the
province was under can be shown. This is backed up by the use of two Latin quotations concerning
the state of the Roman economy. At the beginning it was not seen as very profitable '... for even the
part that they (the Romans) do occupy is not very profitable to them'. But by the 3rd century the
position had changed: '... a land so productive of tax revenues' (Mattingley 2006, 492).

The local Iron Age people who were living in what is now Cumbria when the Romans arrived
were not literate, and so aspects of their relationship with the conquerors have to be viewed
through the items that they made. These can be divided into two areas – items made using
Roman materials and items that are Roman in concept and British in style and execution.

When the Romans arrived they brought glass vessels with them. Glass was a new material to
the local people and they often recycled it into distinctive jewellery. Deep blue glass, coloured
with cobalt, was popular and was made into bangles. In some cases the valuable blue glass was
used to cap less strongly coloured bottle glass, showing that the craftsmen understood this new
material and were able to maximise the effect they wanted.

Sculpture was another art form that developed as a blend of Roman ideas and local craftsmanship. The Romans brought their belief systems with them. Among these was the concept of the
Genius Loci, or 'Spirit of the Place'. During excavation of the fort in Carlisle, several statues of
this type of deity were found. They can be identified as *Genii Loci* by their attributes such as a

crown resembling city walls complete with gates (Coulston and Phillips 1988, numbers 472 and 473). One of them has classical proportions, but another does not. The latter has a large head that is out of proportion with the rest of the figure's body and large almond-shaped eyes that are typical of Iron Age sculpture. In the gallery these two figures are placed side-by-side, inviting visitors to make this comparison for themselves (see Fig 6.2).

Local craftsmanship and beliefs interpreting Roman ideas can be found in other items. For example, leaves made out of thin silver sheets were a common form of offering to temples. These often have either an inscription, an image of the deity or both in the centre. Two of these were found at Bewcastle, Cumbria, and are dedicated to the god Cocidius, a local god of hunting and war, who was equated with Mars. The figures on them are fairly crudely executed, as are the inscriptions. However, they are written in Latin, the language of the conquerors. Thus they can be seen to be local in style but are a Roman type of object fulfilling a Roman function (see Fig 6.3). These are displayed with other items showing local religious ideas interpreted in a Roman way, such as altars that are Roman in form with Latin inscriptions but are dedicated to local deities such as Ratis and Latis (Collingwood and Wright 1965, corpus numbers 1903, 2043).

Finally, there are items that were made as souvenirs for the occupying forces. Following the conquest of Britain, local people were forbidden from carrying swords by Roman law. The effect on the local craftsmen was to remove a source of patronage as there was no longer a market for military fittings among the tribal elite; they subsequently found a new outlet in producing jewellery and other items. Some of the jewellery they produced was based on pre-existing models and was in the form of elaborate 'safety pins'. These were decorated with patterns that had been used previously on pieces of military equipment. In particular they used enamel, a decorative technique that had

FIG 6.2.   TWO EXCAVATED SCULPTED HEADS DISPLAYED SIDE-BY-SIDE IN THE GALLERY.

FIG 6.3.   THE SILVER PLAQUE SHOWING THE IMAGE OF COCIDIUS AND THE LATIN INSCRIPTION.

played a large part in their pre-conquest work. A different form of brooch was the 'dragonesque', so called because early antiquaries thought that it looked like a dragon. It is now thought that the design is made up of elements found on other items. The main elements are two *cornucopiae* joined mouth to mouth at the centre and then heads formed from capped trumpets added at the ends (Hattatt 1982, 152–3). This is a British type of design and shows the flowing style that was favoured by local inhabitants. The style is also apparent on the decoration of the Ilam Pan. This is a small *trulla* (saucepan-shaped pan) that was found in Staffordshire. It has an inscription engraved onto it, but the main decoration consists of Iron Age swirling 'triple comma whorls' (Jackson 2012, 47). The patron was probably Roman and the inscription is in Latin, using the place-names given by the conquerors, but the way in which this souvenir of Hadrian's Wall was decorated is British in execution and manufacture (Jackson 2012, 46–51; Breeze 2012, 108).

One of the great challenges in developing the gallery was to make the Roman frontier relevant to a contemporary audience, particularly in terms of the displacement that the Roman frontier would have caused. One of the tools used to achieve this was the use of contemporary conflicts and frontiers. The aim here was to try to encourage visitors to make connections with events in the recent past and also with what they saw in newspapers and on their television screens each evening. However, it was important that 'universality' was seen, rather than singling out one particular conflict. In order to find the frontiers that could appear, they had to have a visible barrier. To emphasise the commonality of building barriers to defend borders and a way of life, film clips are shown on a series of monitors along a wall to show the geographical spread of these present-day 'conflict fences'.

Moving images were chosen to form a large part of the installation. This was partly to provide a contrast to static objects and partly to remind visitors that frontiers have an effect on people. The presentation of the information as moving images on television monitors was a deliberate decision in order to make a connection with seeing the world through the news every night. The second was the selection of iconic images to illustrate the theme. These were enlarged to give an impression of scale and to make them visible across the gallery. The images are presented as a collage with no captions, to try to draw attention to the universality of the theme of frontiers and the disruption they can cause, rather than focusing in on a particular case. The spread includes India/Pakistan, North and South Korea and Israel/Gaza, as well as Belfast and Morocco/Ceuta.

The design element, provided by Redman Design, was to create a background that resembled a security fence. This was made of cast concrete shaped like the sections used to create the Jerusalem Security Fence (see Fig 6.4). The monitors are set into this background and display moving images of various conflict fences. A crack in the wall was created to show large, still images.

The third element was oral testimony, to give a first-hand account of the effects of a conflict fence. These range from a description of growing up in divided Berlin to the lawlessness of the border area between Scotland and England, as well as the present-day Scots feelings about the border together with a first-hand account of displacement in Israel/Palestine.

Along with visual images, a selection of quotations was used to illustrate the idea of frontiers. These ranged from poets such as Robert Frost to graffiti from the fences in Belfast and along the Mexican border as well as a Czech proverb. The link to the Roman wall was provided by casting into the concrete the quotation from the Augustan History that 'Hadrian was the first to build a wall 80 miles long to divide the Romans and the Barbarians' (Scriptores Historiae Augustae Hadrian 11, 2, in Ireland 1986, 87) and adding a graffito saying 'we are still building them now...'.

The idea of being a conquered people is part of the computer-based game that is spread around the gallery. Based on a series of questions designed to assess how the visitor feels about

FIG 6.4. A VIEW OF THE 'LIVING FRONTIER' INSTALLATION SHOWING THE MONITORS AND THE CRACK FILLED WITH IMAGES.

FIG 6.5. TWO POSSIBLE 'IDENTITIES' IN ROMAN CARLISLE PRODUCED AS A RESULT OF A GALLERY COMPUTER GAME.

aspects of the Roman invasion they include asking whether the visitor would support the Romans in different ways. Although light-hearted, the intended outcome is to create an identity card showing what kind of person you might have been in Roman Britain. The idea of an identity card suggests, at a very low level, control within the Empire (see Fig 6.5).

In conclusion, the Roman Frontier Gallery at Tullie House has tried to demonstrate that the impact of the Romans was more than just a selection of artefacts and buildings. There are reflections on the displacement of people from their homelands by conscription and physical barriers. The gallery also shows how some local people embraced the culture of the invaders and blended it with their own. Events from the past are linked to similar, modern-day events to help people to 'fill in the gaps', and to make events that happened long ago relevant to frontiers today and to the ways in which people try to deal with the problems that they cause. Roman Britain is an important part of our heritage, and the gallery provides an interactive, thought-provoking environment for visitors to consider the role of frontiers and barriers in human history.

BIBLIOGRAPHY AND REFERENCES

Breeze, D J (ed), 2012 *The First Souvenirs: Enamelled Vessels from Hadrian's Wall*, Cumberland and Westmorland Antiquarian and Archaeological Society, Kendal

— 2012 Chapter 10 Conclusions, in *The First Souvenirs: Enamelled Vessels from Hadrian's Wall* (ed D J Breeze), 107–11

Collingwood, R G, and Wright, J P, 1965 *The Roman Inscriptions of Britain, Volume 1: Inscriptions on Stone*, Oxford University Press, Oxford

Coulston, J C, and Phillips, E J, 1988 *Corpus Signorum Imperii Romani, Corpus of Sculpture of the Roman World, Great Britain, Volume 1 Fascicule 6, Hadrian's Wall West of the North Tyne and Carlisle*, The British Academy, Oxford University Press, Oxford

Hattatt, R, 1982 *Ancient and Romano-British Brooches*, Dorset Publishing Company, Sherborne

Howard-Davis, C, 2009 *The Carlisle Millennium Project, Excavations in Carlisle 1998–2001, Volume 2: Finds*, Lancaster Imprints (15), Oxford Archaeology, Oxford

Ireland, S, 1996 (1986) *Roman Britain: A Sourcebook*, 2 edn, Routledge, London and New York

Jackson, R, 2012 Chapter 5 The Ilam Pan, in *The First Souvenirs: Enamelled Vessels from Hadrian's Wall* (ed D J Breeze), 41–60

Mason, D J P, 1990 The use of earthenware tubes in Roman vault construction, *Britannia* 21, 215–22

Mattingly, D, 2006 *An Imperial Possession: Britain in the Roman Empire*, Penguin, London

Swan, V G, McBride, R M, and Hartley, K F, 2009 The coarse pottery (including amphorae and mortaria), in *The Carlisle Millennium Project, Excavations in Carlisle 1998–2001, Volume 2: Finds* (C Howard-Davis), 566–660

Tacitus, n.d. *Agricola*, Book 1, chapter 21, available from: http://www.sacred-texts.com/cla/tac/ag01020.htm [5 February 2014]

Tomlin, R S O, Wright, R P, and Hassall, M W C, 2009 *The Roman Inscriptions of Britain, Volume III: Inscriptions on Stone Found or Notified Between 1 January 1955 and 31 December 2006*, Oxbow Books, Oxford

7

Remembering Traumatic Events: The 921 Earthquake Education Park, Taiwan

Chia-Li Chen

Introduction

Items of material culture are often preserved to sustain desired memories, or discarded or destroyed to suppress those that society wishes to forget (Foote 1988). Although few in number, there are museums dedicated to preserving, displaying and interpreting mankind's experience with natural disaster and tragedy. These include interpretive centres dedicated to interpreting earthquakes and volcanic eruptions, explaining conditions leading up to a disaster and its impact on the environment and society. Natural disaster museums help facilitate the reconstruction and interpretation of specific tragic events in order to help visitors re-experience and interpret the painful experience. Because of their organisational nature and created public space, natural disaster museums also provide an important venue for the public to reflect on the history, to consolidate national sentiment and to shape collective memories (Chen 2007).

Measuring 7.2 on the Richter scale, the Jiji Earthquake that struck Nautou County in Central Taiwan on 21 September 1999 led to 2415 deaths, injured 11,305 people and caused irreparable damage to 51,711 homes.[1] The scale of the disaster deeply affected the whole of Taiwanese society and formed a collective memory of loss and trauma. Cities near the epicentre (including Nantou, Taichung and Changhua counties) were most severely damaged, although other cities were also hugely affected (see Fig 7.1). In Taipei, far from the epicentre of the quake, buildings shook severely and five buildings collapsed, including the twelve-storey Tunghsing Building, with 87 people dying among the rubble of the building.[2]

In the wake of the so-called 921 Earthquake, the national Ministry of Education proposed to build an earthquake museum to preserve and display the remains of earthquake-ravaged Guangfu Junior High School in Wufeng, a township in southern Taichung County (Chou 2002). Today, the 921 Earthquake Education Park not only displays earthquake damage but also features educational exhibits designed to inform and console visitors. What in the social milieu makes it possible to establish a museum, and how do communities respond to such a decision? How does the 921 Earthquake Education Park employ and display exhibits to communicate earthquake

---

[1] For more information on the 921 earthquake, its epicentre and aftermath, see: http://en.wikipedia.org/wiki/921_earthquake [23 February 2014].
[2] The Tunghsing Building was later found to be of unsafe construction, with structural pillars and beams stuffed with plastic bottles instead of brick and concrete. A lawsuit was later brought against its architect and construction company.

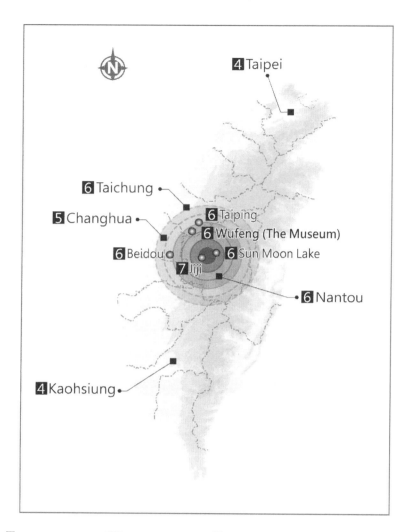

FIG 7.1.   THE SCALE OF THE 921 EARTHQUAKE IN TAIWAN.

information and interpret shared societal memories of pain and loss? How does the public, and those currently working as 921 Earthquake Education Park volunteers in particular, interpret museum-provided information and memories through exhibits? These are the research questions that will be explored in this chapter.

THE EARTHQUAKE MUSEUM AND DISASTER TOURISM

Following the earthquake, the government and academic circles were quick to approve the idea of creating a museum. After investigating several earthquake-ravaged sites, the Guangfu Junior High School was designated as the future location of the museum as it lies directly on the fault line and several school buildings and athletics tracks were severely damaged by the 921 earth-quake. However, the decision-making process failed to take into consideration the opinions of the local community, and the subsequent backlash reflected the outrage of those who had suffered greatly in the earthquake and were now being asked to host crowds of disaster tourists when facing potential relocation.[3] Questions were also raised over the appropriateness of plans to use TVBS Foundation relief funds[4] to support museum construction. Local historian Wu (2000) remarked critically:

> What exactly will this earthquake museum be? A place to memorialise and grieve lost friends and family; a place to remind future generations to remain vigilant; a place to witness a sorrowful page in Taiwan's history. However, if in creating this admonition to future generations we tread upon the lives of earthquake victims, we must stop and re-examine the earthquake museum's fundamental purpose. … The county government may not even be aware that the land adjacent to the ruined boundary walls of the Guangfu Junior High School is still awaiting stabilisation. We do not want to become a tourist attraction. [Author's translation]

Chou (2002) noted five principal reasons for community opposition to the creation of the museum. These included: anger over the occupation and removal of community living space; concern over the potential relocation of the entire community and school; a feeling that constructing a museum was not an appropriate disaster recovery priority; that constructing a museum in an earthquake zone would be dangerous and ironic; and finally, suspicion that the museum was an excuse to distribute political largesse. Community doubt and opposition began to dissipate only after the government had held several community meetings to explain the purpose of the museum, once a new dedicated access road for museum traffic had been incor-porated into the plans and following a government-led study trip for academics and community leaders to an earthquake museum in Japan.

UNDERSTANDING TRAGEDY THROUGH SCIENCE

Reflecting the hopes of its principal planner, the 921 Earthquake Education Park exhibits are framed primarily by science, humanities and historical facts. Chang (2003, 145–6) indicated: 'The memorial park first teaches visitors about the nature of earthquakes, then addresses their impact on human society before finally introducing the details of the 921 Earthquake as a way to foster empathy with earthquake victims and for visitors to learn the earthquake's important lessons.' This approach defines the 921 Earthquake Education Park exhibitions.

---

3 Soon after the earthquake, crowds of tourists visited the disaster area, which obstructed rescue efforts and provoked much criticism within Taiwanese society.

4 TVBS (Television Broadcasts Satellite) was founded in 1993, the first satellite television channel in Taiwan. The TVBS Foundation was established in May 1999 to raise funding for charity work in Taiwan. Through the broadcast of 921 earthquake news by TVBS, the foundation soon received many donations for the rescue and recovery work. It supported students of the Guangfu Junior High School in continuing their studies, and around 7% (NT$174,928,991) of the TVBS Foundation's 921 Earthquake Relief Fund total of NT$1, 202,582,242 was earmarked for the earthquake museum.

There are five exhibition halls in the 921 Earthquake Education Park. Chelongpu Fault Conservation Hall is the first stop on the museum tour. It describes earthquakes from an objective earth science perspective using display panels and interactive installations. In this hall, large reinforced plate-glass windows separate the scientific indoor displays from the cracked and fissured remains of the high school's outdoor athletics track; setting the two side-by-side graphically illustrates the destructive power of the 921 Earthquake (see Fig 7.2).

Next, the Earthquake Engineering Hall uses real-life examples to teach important engineering and construction principles for lessening an earthquake's impact. This hall comprises two sections, including a preserved earthquake-damaged classroom (part of the high school building) and a main exhibition area. The condition of the classroom illustrates actual earthquake damage, while displays feature images of pre-earthquake classroom life. The hall's main exhibition area was planned and designed by Taiwan's National Center for Research on Earthquake Engineering and teaches visitors how to prepare to stay safe during an earthquake and how to limit damage. The hall's interactive scientific displays are particularly popular with young visitors. On display are highly successful examples of how to select and use earthquake-resistant architecture and implement earthquake-preparedness measures in the home, helping to achieve the museum's stated objective of using exhibits to 'increase public awareness of the importance of safety building and further foster social responsibility' (921 Earthquake Museum of Taiwan 2009). Unfortunately, the 'social responsibility' facet today remains largely constrained to individuals' personal awareness of construction and earthquake principles, and has yet to be reflected on, examined critically and enacted in policies on constructing buildings on steep gradients, national land management and environmental protection policies.

The Image Hall, located in Guangfu Junior High's former Activity Center, was designed as a place to 'keep and display photographs and audio-visual records of the 921 Earthquake in order to preserve from humanistic and historical perspectives its emotive impact on society' (921 Earthquake Museum of Taiwan 2009). Its black-themed interior is designed to return visitors to the 921 disaster scene, with news video clips of collapsed buildings, damaged roads and emergency rescues (see Fig 7.3). Numerous video screens along the hallway all broadcast news programmes reporting on ongoing rescue work. Everything is designed to put visitors at the centre of this historic disaster. Stories of daring deeds to save lives and bring emergency supplies to those in need help to communicate the community spirit and humanity found throughout Taiwan.

Next stop is the Disaster Prevention Education Hall. In addition to introducing traditional legends about earthquakes from around the world and telling the story of disaster rescue dogs, this hall features a wall panel emblazoned with the slogan: 'What have we learned? To live in harmony with nature', written in various languages. While this slogan carries an important message, the museum leaves the visitor to contemplate its deeper meaning.

While the researcher was completing her research at the museum in the summer of 2009, typhoon Morakot caused a major disaster in southern Taiwan. One village was completely destroyed by a landslide, causing the deaths of 398 people. In response to this event, in 2011 the museum redesigned the Disaster Prevention Education Hall to include two additional sections. Although these did not feature as part of the author's research, they are mentioned here as they indicate ways in which the museum is responding to current traumatic events. *Flood Tribunal – Flood Prevention under Environmental Changes* deals with issues of climate change and rising sea levels. *Accident Prevention – Journey to the West* is a section aimed at young children that uses a famous legend about Tripitaka, a Buddhist monk chosen by Guanyin, the Goddess of Mercy, to travel to India (the

FIG 7.2. GUANGFU JUNIOR HIGH'S BUCKLED ATHLETICS TRACK ILLUSTRATES THE DESTRUCTIVE FORCE OF EARTHQUAKES.

FIG 7.3. THE IMAGE HALL IS DESIGNED TO RETURN VISITORS TO THE 921 DISASTER SCENE.

Journey to the West) to seek out a special Buddhist script, which would help prevent disasters and accidents in Taiwan. Tripitaka was accompanied by a monkey, a river monster and a pig, all four being presented with magic weapons and told how to behave in an emergency: be brave, calm and cooperative. These instructions, learnt by Tripitaka and his companions, were used to deal with the many hazards they faced on their journey. A contemporary version of this legend enables the same messages to be passed on to young visitors in this part of the exhibition.

Finally, the Reconstruction Hall chronicles the government's accomplishments in disaster rescue and public reconstruction works. The presentation in this area is comparatively low key and has become a platform for the promotion of governmental achievement.

LEARNING AND REMEMBERING: VOLUNTEERS' EXPERIENCES

Museum planners hoped that the five exhibit halls discussed above would help visitors experience awareness, understanding and empathy and, ultimately, raise public consciousness about earthquakes (Chang 2003). However, the question remains whether the public is indeed able to be made more aware, more understanding and empathetic. This question is particularly relevant to museum volunteers, whose role is both to learn and to guide. It is significant to understand how these volunteer guides gain an understanding of earthquakes through the exhibits and then recall and interpret their experiences. The author's research applied a qualitative interview method, interviewing a total of ten volunteers (five men, five women) from the 921 Earthquake Education Park. All participants were over 50 years of age and had at least one year's museum guide experience. Eight of the ten held a college or higher academic degree. Each participant is identified below by his or her gender abbreviation (M = male; F = female) and interview order number.

(a) Motivations to volunteer

Many volunteers stated that contributing to society and advancing learning were their key motivations for volunteering at the museum. However, for some the decision to work at this tragic site was not easily reached. F02 only enrolled as a volunteer after strong encouragement from her husband and friends:

> Both my Fonghuanggu Bird and Ecology Park co-workers and my husband kept encouraging me to come to the earthquake museum. I told them I was afraid of earthquakes. I'm sensitive to these things. … It's frightening for me to come here. But a co-worker asked me to come because they needed volunteers – especially on Wednesdays. So I said yes. It was about six months later that I finally came after I had agreed. (F02)

In contrast to those needing to overcome psychological obstacles, others thought it important to learn more about earthquakes because they lived in an earthquake-prone area. F06 reflected these sentiments when she said:

> My home district of Taiping was devastated in the 921 Earthquake, so I thought it important to learn more about earthquakes. I was also interested in volunteer work, so I signed up. (F06)

Still others came to volunteer through previous professional experiences. Because their jobs gave them first-hand appreciation of the severity of the 921 disaster, M04 and M09 were eager to share their experiences and reflections on the 921 Earthquake with a broader public. M04 explained:

> I was the head of a rescue team in the Sun-Moon Lake area. … Therefore I volunteered to be a guide to share my personal experiences of the rescue work. My major is civil engineering, so I can explain the causes and aftermath of earthquake in detail. (M04)

Similarly, M09 was the head the Shento Building Administration Office and had himself been dealing with the flood caused by the earthquake.

(b) Exhibit themes and education

Because volunteers occupy a dual role as both learner and guide, the research was designed to explore the cognitive learning and practical impacts of exhibits on volunteers. The researcher first asked participants about the causes of earthquakes. Eight replied that they were the result of plate tectonics and two replied they were due to the release of energy. The typical answer was as F02 explained:

> It is because of plate tectonics since our earth moves every day. What we need to learn is how to keep safe when an earthquake occurs. (F02)

A retired junior high school teacher majoring in geography, F03 gave a more diverse answer to the same question:

> The eruptions of volcanoes, the collisions of comets, the plate tectonics or even nuclear tests, all these might cause an earthquake. (F03)

When participants were asked what they thought was the main message of the earthquake museum, all responded that it was the dissemination of disaster prevention information. M09 commented on the importance of disaster prevention education, saying:

> We tell visitors to the earthquake park about preventive measures they can take to minimise damage during an earthquake, how to protect themselves, and what things they should prepare beforehand. (M09)

We found that learning about disaster prevention had affected participants' daily lives and habits. Five expressed their intent to assemble disaster preparedness packs; five said they would work on earthquake-proof items in their homes or avoid storing objects on high shelves; and four said they would organise emergency torches. When asked which of the museum's halls impressed them the most, eight cited the Earthquake Engineering Hall, based on its relevance to daily life and its installation of interactive exhibits to promote hands-on experience and observational learning. F03 commented on the advantages of using interactive displays:

> I like exhibit displays that are interactive. Hands-on learning and illustrations help visitors better comprehend earthquake-resistant facilities and technologies and how they help reduce the severity of earthquake damage… like that exhibit showing how a vase sitting on an anti-

slip mat is less likely to break than one sitting on a regular flat surface. Appropriate measures can lessen financial loss. These earthquake preparedness concepts are part of everyday life. The interactive process helps make explanations less dry and deepen the impact. (F03)

Three participants stressed the essential educational value of the preserved earthquake ruins, stating that the earthquake-warped remains of the school athletics track provided a direct and tangible display of the power of earthquakes. F07 said:

I particularly like the preserved athletic tracks because they exhibit clearly the tremendous change in area topography triggered by the earthquake. The two-metre difference in elevation is evidence. You can't experience it but it is right here to see. (F07)

Overall, the volunteers participating in this study had a relatively good understanding of the causes of earthquakes and unanimously believed that disaster prevention should be the main message transmitted to visitors.

(c) Recollection and interpretation

In examining how exhibits affected participants' personal recollections of the 921 Earthquake and how they interpreted and explained its images and history as presented in the Image Hall, all participants – with the exception of M10, who did not personally experience the earthquake event – were naturally drawn to memories of their own 921 experiences.

Of course… Of course I remember and think about it. Everything shook violently… up and down, side to side. We felt it, but we were in a traditional single-level sanho house and not a multi-storey building, so I wasn't too scared… not really nervous. The whole family ran out. Beidou was hit pretty hard too. (M09)

While the museum inevitably evoked earthquake memories, the passage of time and their iden-tification with earthquake-prevention education principles gave participants new perspectives on, and meaning to, their own 921 experiences.

I was really aware of it early on… but over time it has become somewhat dulled. There are some visitors who live nearby and visit the museum often. I don't think the museum is a painful place for them. This place can help them mourn and remember the past. I feel everything has been slowly turning in a better direction. True, it may be that those who were hit the hardest don't come here. They may not even have the courage to come to visit. (F01)

When asked which exhibit in the Image Hall was most the impressive, participants all gave different answers, as summarised in Table 7.1.

As shown in Table 7.1, the exhibits that most impressed participants were news reports and photographs related to people involved in the disaster rescue effort. F01 and M10 both chose the news report covering the rescue of two brothers from the collapsed Tunghsing Building.

Table 7.1: The Most Impressive Exhibit (and the reason given for this choice)

Participant	Most Impressive Exhibit	Reason
F01	Brothers Rescued (news clip)	Alert and inspiring
F02	Persimmons and Children (photo)	Lost people, children in trouble, unity
F03	Overseas Rescue Team (photo)	Caring without borders
M04	Jiufenershan Split Open (photo)	Shocking image of a mountain landslide
M05	A Reconstructed Building (photo)	Mutual aid in a time of crisis, deeply touching
F06	ID Card (enlarged image)	Historical artefact, testimony to blood and tears
M07	Collapsed Bridge and Building (photo)	Deep emotional attachment to the land
F08	A Child's Tears (photo)	Unambiguous testimony of natural disaster
M09	Post-Disaster Photo Series (photos)	Seeing images of this awful event encourages me to be prepared and vigilant
M10	Two Brothers rescued from Taipei's Tunghsing Building (news clip)	This highlights the relationship between keeping fit and surviving the disaster

I clearly remember many of the film clips, although some of the details are fuzzy. Like how it was many days until the brothers were finally found and that they were near a refrigerator… and things like that. Some are good for the general visitor regardless of whether or not they've experienced the disaster. Some may encourage heightened vigilance; some may provide inspiration. (F01)

A Child's Tears was another photo that captured the attention of some participants.

The one of a crying child is one that struck me deeply. I'm afraid to look at it. It makes me feel like… it affects me emotionally. It's something I'm not really comfortable dealing with. But it's a genuine record of the disaster. (F08)

Stories regarding foreign relief and rescue workers particularly touched F03 and F06:

Yes… Learning about the foreigners coming as rescue workers. Their compassion… that was what impressed me. (F03)

It was the one with the ID Card. I live in the 921 disaster area too, so looking at all those makes me feel that there are important real things preserved here in the Image Hall. … When I look at the dates printed on those ID cards, that's what most affects me. (F06)

A large-print version of a Taiwan national ID card is displayed in plastic in the exhibit hall (see Fig 7.4). The text narrates the story of Chao-liang Shen, a compassionate individual who took new ID photographs for disaster-area residents after the earthquake. This item also encourages visitors to think of a disastrous event cruelly destroying their own ID card and of the deaths of so many faceless ID card owners.

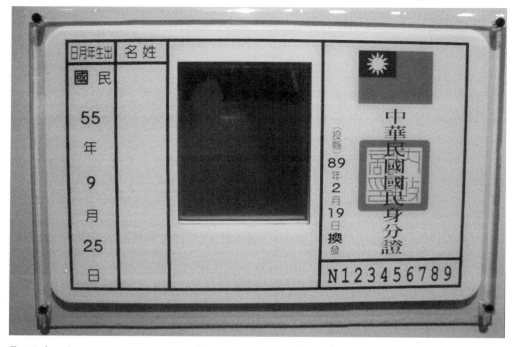

Fig 7.4. An enlarged copy of a Taiwan national ID card and story of the compassionate individual who took new ID photos for disaster-area residents.

Male participants expressed being particularly moved by images of reconstructed landscapes and buildings. M04 described his shock at the disfiguration of the landscape, while M07 stated his deep feelings for the land:

> Buildings collapsed, but why did the bridges fall? This question affected me deeply. … This photograph affected me deeply. … I experienced both the Baihe Earthquake of 18 January 1964 whilst a soldier in Chiayi and the 921 Earthquake. How did I feel? … I really have a hard time putting my feelings into words. Perhaps nowadays we have a greater emotional attachment to this land. … This is where I grew up. … I grew up in the disaster zone. My greatest wish is to see everyone here living in peace and safety with disaster-prevention awareness. … This is what I want the most. (M07)

Still others were particularly drawn to images of rebuilt buildings, landscapes and symbols of the collective effort rallied for reconstruction. M05 was especially inspired by the many architects who worked on reconstruction projects without remuneration:

> The way some buildings look today owe a great debt to those architects. Their involvement… their designs have touched our hearts. They put down everything else… many even worked without pay… just to help out. In the midst of a dangerous situation, everyone worked together. It was a moving experience, really. (M05)

In their analysis of disaster-reporting journalistic aesthetics, Tsai and Tsang (2003, 104) noted that, 'apart from reporting facts, disaster news reporting continuously evokes our emotional response'. Reactions induced by disaster reporting include feelings, empathy, identification and recollection. We found that the news reports on disaster relief efforts and miraculous rescues and photographs of children featured in the Image Hall induced greater emotive and empathetic responses from female participants than their male peers. Male participants had comparatively stronger fear and shock reactions to earthquake-induced changes to the landscape. Exhibited national ID cards served as reminders of the former lives and struggles of their owners, while reconstructed buildings symbolised post-disaster teamwork and camaraderie. Feelings of fear, mercy, encouragement and cohesion thus were intertwined throughout the exhibition narrative until finally coming together into a common expression of hope for the future, grounded in lessons learned from tragedy.

CONCLUSION

Tedeschi and Calboun (1995) advocate that the explanation and understanding of disasters is a way of reframing the perceived experience. They also promote achieving the manageability, comprehensibility and meaningfulness of tragic events through helping victims reassess their experiences from a more positive perspective, centred on personal well-being as a way of fostering personal self-respect and a proactive and positive outlook on life. Exhibits at the 921 Earthquake Education Park focus on providing science-based learning to the general public, helping volunteers understand earthquake origins and related preventive measures. Exhibit-triggered recollections also assist volunteers in transforming trauma into strength to support and assist others to comprehend disaster and manage their experiences.

Ferguson (2006) noted that, while some individuals regard events of five or more years before to be historical, others may need five or six decades of reflection before they are able to consider events as such. Volunteer interviews conducted for this study highlighted that memories of the earthquake tragedy remained vivid and complicated, reflecting the deep emotional impact of the event and its aftermath. Lo (2006, 6), in his discussion of the cathartic effect in Aristotle's *Poetics*, noted that apart from cleansing the spirit, the emotive responses of mercy and fear evoked by tragedy tend to promote altruism in those involved. This helps explains why those living in disaster-stricken areas are willingly returning to do volunteer work and allow the passage of time and the explanatory power of exhibits to turn personal tragedy and memories into motivation to help increase public understanding of the causes of earthquakes. As disasters continue to become an ever more common facet of modern life, museums should consider how to combine science-based knowledge and historical experiences to promote understanding among the general public on the origins and causes of various disaster events. Further studies should also explore issues of national land management and the impact of natural disasters on human culture and natural environment to encourage visitors' introspection and critical thinking on how mankind can live in harmony with nature.

ACKNOWLEDGMENTS

I am grateful to the volunteers of the 921 Earthquake Education Park, who generously provided valuable information in their interviews. Many thanks also to Ms Ching-Yen Liu and Director Wu De-Chi for their support. The conclusions reached and any subsequent shortcomings are mine alone. An earlier version was first published in Chinese in 2009, in the journal *Taiwan Wen Hsien* 60 (4), 439–69.

BIBLIOGRAPHY AND REFERENCES

921 Earthquake Museum of Taiwan, 2009 *921 Earthquake Museum of Taiwan* [homepage], available from: http://www.921emt.edu.tw/ [10 August 2009]

Chang, Yu-Teng, 2003 *Observations on the trends of museums*, Five Senses Art Management, Taipei

Chen, Chia-Li, 2007 *Wound on Exhibition – Notes on Memory and Trauma of Museums*, Artco Publisher, Taipei

Chou, Ching-Sheng, 2002 Case Study on 921 Earthquake Museum Planning and the Community, *Museology Quarterly* 16 (1), 15–25

Ferguson, L A, 2006 Pushing buttons: controversial topics in museums, *Open Museum Journal* 8, 1–30

Foote, K E, 1988 Object as memory: The material foundations of human semiosis, *Semiotica* 69 (3/4), 243–68

Lo, Nien-Sheng, 2006 Preface, in *Poetics by Aristotle* (trans Nien-Sheng Lo), Shanghai People's Publishing House, Shanghai, 1–16

Tedeschi, R, and Calboun, L G, 1995 *Trauma and Transformation: Growing in the Aftermath of Suffering*, Sage Publications, London

Tsai, Yen, and Tsang, Kuo-Jen, 2003 News Aesthetics: A Call for a Different Approach to News Research, *Mass Communication Research* 74, 95–119

Wu, Tung-Ming, 2000 The community of the earthquake museum weep [Author's translation], *Chinese Times* [online], 31 August, previously available from: http://www.wfcca.org.tw/modules/news/print.php?storyid=11 [1 August 2009]

Displaced Heritage:
Trauma, Confinement and Loss

8

Maze Breaks in Northern Ireland: Terrorism, Tourism and Storytelling in the Shadows of Modernity

JONATHAN SKINNER

Thirty-eight minutes sitting in a West Belfast cinema – entertained, informed and disturbed by Steve McQueen's (2008) brutal film *Hunger* about the life and death of Republican protester Bobby Sands – there is the representation of a cavity search and beating of blanket protesters in HMP Maze/Long Kesh. A man calls out 'Gerry!' The camera does not give a point of view, but is omniscient in its gaze and enduring in its lengthy profiling of faces and their powerlessness – whether prison officer, Provisional IRA volunteer or riot police. Though calling out to the character 'Gerry Campbell', there is a questionable presumption in the audience's mind that 'Gerry!' could refer to Gerry Kelly, incumbent Member of the Legislative Assembly for Northern Ireland, spokesperson for Policing and Justice for the Republican political party Sinn Fein, and former combatant who at the age of 19 was found guilty of conspiracy to cause explosion in 1973 and sentenced to two life sentences plus 20 years, much of which was spent in HMP Maze/Long Kesh. There, he was involved in the notorious blanket protests as he and fellow Republican prisoners refused to wear prison uniform, demanding political prisoner of war status, Special Category status that included rights to free association and not to wear convict clothing. This blanket protest began in 1976: prisoners refusing to wear prison uniform were confined to cells, stripped of personal possessions and given reduced rations. Colloquially, these conditions were known as to be 'on the blanket' and 'on the boards'.

Gerry Kelly was also later involved in one of the largest prison breaks: an escape from the Maze in 1983 with 37 other Republican prisoners. He shot one of the prison officers in the head during the escape via a hijacked prison meals lorry. He was rearrested three years later in 1986 in the Netherlands, caught with 100,000 rounds of ammunition. He served another three years before his extradition conditions led to his release in 1989 and his subsequent involvement in electoral politics and peace and reconciliation work in Northern Ireland that led to the Good Friday Agreement of 1998, an official ceasefire, the release of 428 remaining prisoners and effective demise of the HMP Maze/Long Kesh complex. The incarceration platform – the politics, the escapism, the big screen – is the perfect storm for the formation of violence archived by Allen Feldman (1991). Kelly is a living legend of a Foucauldian regime of Bakhtinian proportion. Kelly's story has now become one key part in the story and tour of the Maze prison, the Long Kesh. People want to see where he lived his political struggle as a violent Republican wanting a united Ireland and an Un-United Kingdom. In Dean MacCannell's (1999) Marxist-inflected assessment of *The Tourist* – a symptom itself of Modernity – there is an aura attached to such original objects and locations of action, to Kelly's 'boards' and 'Bobby's' death ground. The public, especially Republican sympathisers or former combatants and their families, want to view 'the Maze', 'the Kesh'; to tour the grounds, sample the cells and get a feel for the place, to re-remember or explore the parts previously off-limits. Their critics, largely a Unionist electorate and politicians, fear that these visitors might turn this site into a shrine to the Republican cause, and to the ten 'martyrs' who died by starving themselves to death for their political convictions.

There is a complex and conflicted or dissonant backstory to this site, a querying of whether it is indeed a tourist site or attraction at all. Storytelling, as the anthropologist Michael Jackson recounts, is never a simple reconciling of experience and point of view; storytelling acts like a catheter on a patient: it 'cathects' and 'recathects' the political subject, synthesising and subjugating (2005, 13). Telling the alleged Gerry Kelly story gives the blanket man and hunger striker a 'narrative imperative', order and coherence to his past. Telling a story also means not telling other stories. It is a selection as well as a silencing, just as Widlok (2008, 59) observes that walking a route or a trail opens a path and is simultaneously expansive and reductive, closing off a space around the path. Gerry's story is not the prison officer's story. It is his story, empowering to one but disempowering to another. Seductively, '[i]n telling a story we renew our faith that the world is within our grasp' (Jackson 2005, 17). The potential biographical linking of events in Kelly's life, from dissident protester outside and inside of incarceration, gives him a trajectory. There is *Machtgefuhl*, existential strength, a Nietzchean personal intensity and single-mindedness and conviction in this non-convict fulfilling his life project and transcending his circumstances (Rapport 1997; 2003), regardless of our support for him or not.

The prison officer shot by Kelly survived and lived to tell the tale – or, rather, not, seeing as we only see and hear from Kelly in film and other documentaries. Kelly is the proletariat socialist revolutionary taking his Irish Republican historical consciousness forward against the British regime. He is, from a sympathetic perspective, the armless underdog fighting for his cause with all of his might and all of his body. He resists the colonisation of his lifeworld, and *he becomes* by refusing (Crossley 2006). In the Maze, the state controls the slow life of his body but, in the hunger strike, Kelly controls the slow death of his body.

THE SHADOW TOUR

For De Certeau (1984, 115), '[e]very story is a travel story – a spatial practice', one that retains an 'everyday tactics'. Mapping out the theatre of Kelly's tour of incarceration – the questionable object of a questionable subject – we approach a contested space between violence and storytelling, one with the added tourism dimension of entertainment consumption. Such stories 'shadow us' to continue to invoke the work of Jackson (2005, 39). Both violence and storytelling occur in 'the contested space of intersubjectivity' (ibid). Where Kelly's resistance is dialectical in a Hegelian sense, his storytelling is dialogical: Arendt (1958) explains that a story told is a transposition from the inner self-centred world to an outer-shared world. In essence, a story's orality connects (Ong 1990). This connection is more soothing and reassuring than the intersubjective connection of a baton to a body. Can intersubjectivity really harbour in its shadows, we can ask?

Officially the Maze is known as Her Majesty's Prison Maze. Colloquially, depending upon your identity and politics, the prison is referred to as 'The Maze' or 'Long Kesh' (and as 'the H Blocks'). It was a prison built on the outskirts of Belfast to house paramilitary prisoners – Loyalist and Republican – between 1971 and 2000. It is an iconic set of buildings, prisons within prisons: two cell blocks within a compound with an interlinking and commanding corridor between them forming distinctive 'H' shapes, of which there are eight in total. It is an institution closely associated with the trials and tribulations of The Troubles at their height: the 1981 hunger strike when ten Republicans including the Member of Parliament Bobby Sands starved themselves to death; the blanket protest which in 1978 had over 300 protesters; the dirty protest when prisoners smeared their cell walls with the excrement of the Anglo-Irish conflict (Aretxaga 1995); murder, assassination, mass escape; and recently, the media portrayals of the big and small screen. Since 2000, the Maze/Long Kesh has been empty. Much of the site has been levelled, decontaminated and prepared for regeneration. Representative or emblematic parts of this 1.5km square area are now vacant listed buildings supported by politicians who were former combatants and who are now elected members of the national Legislative Assembly. Deputy First Minister and former Provisional Irish Republican Army leader, Martin McGuinness, has said that many people had 'looked at this place as a place where there was no future and there was just perpetual conflict. … Now we have come out of all of that, we have a lot to offer the world' (BBC News 2007).

The suggestion has been to include the Maze/Long Kesh as part of the legacy of The Troubles, the period of history between 1969 and 2001 that saw over 3500 people killed, with dissident action continuing to maim and kill. Political opponents to the preservation of the site object to this place of protest – a dirty reminder of the recent past – becoming a shrine. 'Raze the Maze' is the public and political position of the Ulster Unionist Party and the Traditional Unionist Voice, recently joined by UKIP (BelfastDaily.co.uk 2013). Whilst it is currently closed to the public (only accessible through Government-approved visits, heavily monitored and restricted with the permission of a Member of the Legislative Assembly), proposals have been put forward to turn it into a sports stadium or, most recently, a peace and conflict resolution centre. As a member of the Civic Society Reference Group to the Maze/Long Kesh Programme Delivery Unit (MLK PDU) for the Strategic Investment Board Limited (SIB),[1] I have been party to the plans, proposals and formerly successful bid from EU Peace III funding for £18.1 million for this latter proposal. As of March 2014, though, an escalation in cross-community tensions has led to a political deferment as to the future of the site and a redistribution of the funds elsewhere in Northern Ireland. None can 'gawk' at the seductive and enigmatic violent dramas (cf Taussig 2006) contained in the remnants of HMP Maze/Long Kesh, but the Maze/Long Kesh still *interpellates*. As such it is considered a 'dissonant' (Tunbridge and Ashworth 1996), 'difficult' (Macdonald 2009) heritage site where stories disinherit as much as they enshrine. The 'tragic' (Lippard 1999), 'dark' (Foley and Lennon 2000), 'thanatopic' (Seaton 1996) tourists, visitors, pilgrims engage with some of the buildings, sanitised to be a version of 'what they were', dressed with authentic props, surrounded by repainted walls.

Storytelling is a political act depending upon who tells the story, whose story it is and what words and terms are used. 'Graham'[2] is a tour guide of what we have heard so far to be HMP Maze/Long Kesh. He welcomes us – a group of local staff and students who want to see the place we have all heard about and 'where it all happened' – out of the car park into a set of modern temporary huts at the entrance to the complex, a vast flat open expanse, the grey rubble

[1] A professional advisory company within the public sector in Northern Ireland, fully owned by and accountable to the Office of the First Minister and Deputy First Minister (OFMDFM), working with the remit to accelerate the efficient delivery of major infrastructure programmes in Northern Ireland.

[2] Graham is a pseudonym.

of deliberate ruin. Demolition and clearance workers and trucks pass the windows. He will be our tour guide for the day. He sets the scene and lays down the terms of the visit. We can take notes but not record anything. We must sign ourselves in. We must not take any souvenirs. The way he sets himself up is as part mediator of the ruins, bringing them to life as an historian, and part pathfinder (Cohen 1985), a local with indigenous knowledge, though, and I will return to this, he stresses that he was not a combatant in The Troubles or prisoner of the Maze, but a person growing up in Belfast, a witness of The Troubles but not an eye-witness of the Maze. This distinguishes his tour rather like the distinction between professional guide and survivor-witness on Israeli youth tours to Poland (cf Feldman 2008, 67). Graham is also a professional guide with training and experience, and 'schtick' as Jackie Feldman (2007) refers to the banter and crowd-warming rhetorics of rapport at the start of a tour. This is our chance to see the story for ourselves.

> I'm Graham. Hello and Welcome. I'm a general know-all, being a man. I have an interest in Irish history. I'm knowledgeable. I grew up through The Troubles, so I know Irish history from the 1950s/60s: challenge me. Ask questions. I'm thick skulled. Now, the Maze, Northern Ireland. There's a problem here with names. You see, this place used to be an airfield before it was a prison. Long Kesh is its Irish name: it means the meadow or bog. Calling it the Maze or Long Kesh, or simply Kesh is a problem. It depends who you are, a Unionist or a Nationalist. The same with Ireland and Northern Ireland. This place, island, was partitioned in …

Graham continues his spiel. Later on, during a private personal tour of his Belfast, he shows me his tour guide notebook. He re-reads his notes just before each tour, and updates them after a tour or if he has learned, read or heard new details. Before taking us to HMP Maze/Long Kesh, he talks about its prehistory as a holding centre for internees, non-sentenced suspects who were held in cage compounds in the 1970s. The students take notes. He points to an aerial photograph of the site before much of it had been destroyed. Now only one of the eight H blocks remains as a representative sample. There are plans to move the last of the cages next to the last of the H blocks to make it a more compressed and efficient visitor attraction. It will still be as authentic as the real buildings, even if they have been cleaned, treated, painted over, sterilised, sanitised. We then leave our own modern heated hut for a minibus ride to the remains of the prison. Outside, we meet our driver who remains silent throughout the tour, opening and closing gates and doors for us, staying back from the tour group, remaining aloof from his tour guide colleague. It turns out that one of the students had recognised him from a football supporters' club and had struck up a conversation with him. He had been a prison officer at HMP Maze/Long Kesh. He had lived through the violence and now was the silent chauffeur-escort listening in to Graham's tour and the tourists' questions.

The tour of the prison is unscripted. Graham was told by the Northern Ireland Tourist Board and the Legislative Assembly to make of it what he wants so long as he does not take sides and remains neutral. He has some guidelines from civil service tours. It was that neutrality that we listened to, trying to figure out whether his personal stories and his professional histories identified his upbringing and his allegiances. The drive to the Maze points out the last of the internment cages, the massive walls of the Maze, a World War II air raid shelter, and then we are at the prison itself. There is a gate between the walls and the watch towers. If it can be given a 'look', it is worn-industrial. A sign tells us – or rather told us – to switch off engines in the air lock. We drive all the way through into the prisons of our mind, our media-saturated imagination.

H6: a scrubbed, sanitised prison without marking, graffiti, excrement. A snooker table which we can imagine snooker balls on; an Open University tutoring room where prisoners radicalised themselves with education, Irish language classes, the history of the suffragette movement – the original political hunger strikers. A solitary piece of Christmas tinsel is hanging from a bare ceiling. It is all so evocative and unpleasant. Students go into a cell; there is a bed to sit on, a pillow to lie on. One student wants to hear the door clang behind her, locking her in. It is like in the movies. Education crosses into entertainment. Graham points out the parallels between cruise cabins and HMP Maze/Long Kesh custodial cells. They are the same size but the ones here are free! We pass toilets, observation peep-holes, cells containing wooden boards for punishment. Each cell is brought to life with a commentary that mixes history books, television documentaries and tales from previous tours. These former tours come from former prisoners on peace and reconciliation visits, from former paramilitaries or even former prison officers, 19 of whom were killed on active service either in the prison or in their homes.

The Chapel: this is where communication took place between prisoners. It is a separate building outside of the inner gates and compounds. Like the cells, it has just been left. There are plastic flowers decorating the corners, a pulpit and faded lino carpet dating the building to a particular era. There is a consummation room, so we are told.

The Command Centre: this is the only two-storey building in the Maze complex. Inside, we see more dated prison furnishings: a bank of TV monitors, a hostage negotiation detail box, clipped-out wiring and prison officer graffiti. Here, Graham warns that the graffiti can be deceptive and come from subsequent visits rather than the last out from the working building.

The hospital: this, for many, is the high point of the tour. This is where the hunger strikers went to die. This is where their bodies decayed for a cause, shutting down slowly over the weeks: the loss of sight, the inability to move or communicate and finally the inability to breathe. Graham reminds us that the prison had its own rhythm, that we know only about specific events and key moments, but we should not gloss entirely over the everyday mundaneness of the place that we do not hear about. We see the visiting room for families. This leads Graham on to tell us about the mothers begging their sons to start eating again and choose to live for their cause rather than to die for it. We walk soberly to the room where Bobby Sands died. We file into the room and stand around its perimeter, heads bowed and respectful. The chatter of comments between students stops. The next room, Graham has portrayed more light-heartedly. It is a padded cell for someone who would self-harm. The pads are hard, and he jokes about the injury they could do, and the window being so high up that one former prisoner described it as akin to a religious cell with the light of God entering from on high. Finally, we visit one of the control rooms in the hospital where Graham points out more historical detail with a board detailing the types of prisoners by religious faith and sentence category. He tells us the story of the child prisoner who came in for stealing sweets and was held with the female prisoners until he was 16, who misbehaved and was subsequently sentenced for crimes within the prison, not leaving until his 40s. He once asked a prison specialist visiting the Maze if this story could still take place to this day. They had to agree, in theory.

Now, before we leave the hospital, I can note that, having been on a number of Graham's tours with my students, I have experienced changes to the script, to his storytelling. The first time we attended a tour, when we reached the Bobby Sands room – an empty cell room in the hospital like any room – we reached it cold. There was an element of shock to the visit: tingling, fearful of what had happened in this room. The next time, my attention was focused upon the student peeling some of the paint from the wall to take home a souvenir of the room, 'because I thought it would be kinda cool and I could give it to my friend, but they only said, "oh, its paint" and weren't bothered'. It was on the third visit that I noticed Graham cueing us in far more than on previous occasions. Each of the key narrative points in his tour was led up to in his rhetoric. This time he qualified his introduction to the room by saying that it was 'probably' the room that Sands died in, but that he might be told that he was wrong during a future visit of former prisoners or prison officers. As such, we were being chaperoned through the complex, but also fed tour guide lines to respond to in particular fashion. We knew what to expect with every twist and turn of the corridors. Graham has been developing and adding to his script on each tour occasion. Our pathfinder had found his path. Of significance, Graham added to the tour based upon previous tours: one anonymous former paramilitary leader talked about living in one of the cells, and communicating by flicking buttons across the corridors with fine thread attached. Graham added these rich details to the following tour. In other words, Graham's tour became a tour of tours. There was no set script as the tour guide interprets and the visitors interpellate. The script came from the tours themselves. This means that to some extent the tours changed in their delivery. It is also constituent dependent. A 'Unionist tour' downplays the Sands room for fear of applause. A 'Nationalist tour' adds details to this awful part of the narrative.

The tour had 'dialogic narration' akin to Edward Bruner's development of a Bakhtinian, polyphonic approach to the tour's tale (2004, 172). Certainly, in his explication of tours around Israel's Masada, Bruner makes the point that there is no one authoritarian account though a narrative core (ibid, 171) might coalesce, monolithic in the heteroglossia of the performance and empty meeting ground (MacCannell 1992) space or 'theatre of improvisation' (Bruner 2004, 18). This case study further extends the idea of the tour as 'a story in the making' (ibid, 24): this is not just as a story to tell after the tourist event, but this is a story emerging and evolving from the stories of others. It is not a set meta-story. Every tour is different depending upon the audience and the emotions, feelings and reactions of the tour guide.

Graham is king of HMP Maze/Long Kesh tours, and yet Graham was never a participant in the history of the place. His story of stories is a compilation lacking personal input. Some local former prisoners have set themselves up as tour guides in Belfast and they have criticised Graham for his position as Maze/Long Kesh tour guide. They feel that he does not have the authenticity, the experience, the personal touches to answer the various questions from tourists. However, Graham is the accredited professional tour guide, experienced and claiming to be impartial. They are former prisoners giving their biased versions of events. And yet, some of the students had become dissatisfied with Graham's answers to their questions with a recounting of stories from previous visitors he had toured. They wanted to hear from the driver. When Liam, a tour guide of West Belfast, former combatant and prisoner, gives a tour of the streets of Belfast, he lingers at the top of the Falls Road, gathers his audience around him and describes what it is like to be 'on the blanket'. Mouths fall open, the crowd draws ever closer in curiosity, and Liam goes through the physical movements of drawing a blanket around his cold body. We share this performance. Liam is not recounting a story but is restaging and re-remembering his story behind bars. Graham's version of accounts has distance to it. It is a third-person narrative for Liam's first-person narrative. As such, Graham's narrative has less narrative force to it. It comes across as less real. Graham simply was not there. This is the importance of 'being there' with its visceral and realistic qualities (Borneman and Hammoudi 2009). Graham's story is like a combination of 'surrogate ethnography' from others mixed with a textscape of readings and media viewings, but it is an absent experience all the same. Liam's narrative just is.

After viewing the hospital complex, we return to the minibus and are driven back the way we came, out of the Maze/Long Kesh. The way out follows the escape route taken by Gerry Kelly and 36 of his compatriots hiding in the prison food van. We leave near to a gate where the prison guards were attacked and overpowered. And we return to the huts at the entrance to the prison. There we sign our evaluation forms of the tour. Graham tells us to take more of an interest in local history and he talks about his love of walking the streets of Belfast, even offering to take some of us around for free, such is his passion for a good story.

MAZE MINEFIELD

We bond with the world. We inhabit it and are enclosed by its structures. Michael Jackson (2005, 17), in his *Existential Anthropology*, notes that personal identity and physical environment fuse; 'building and dwelling'. Writing generally about human nature, Jackson suggests that: '[v]iolence is the implementation of [a] logic of reversal' (2005, 41–2). It is a part of reciprocity in an exchange system that objectifies pain from one to an Other. Whether witch, gypsy, Jew or Catholic, 'scapegoating' objectifies and moves pain intersubjectively. There is frequently a proximity and intimacy in this: the violence is about nearness rather than distance (two families feuding in Northern Ireland; a prison officer and his family and the hunger striker telling the guard the names of his children and wife). The violence done to the Other is a false consciousness, Jackson adds, the culmination of a fantasy, an external symbolisation of inner discontent, often so strong that the fantasy motivation can outweigh any formative cause. In this chapter, the violence starts on the streets of Belfast between Protestant and Catholic but ends up as prisoner against prison officer, both seeing their close neighbours as metonyms for something else: the British state and the Provisional Irish Republican Army. Redemption, peace and reconciliation come, for Jackson, from breaking the cycle of violence, the 'habit of dialectic negation – transforming difference into alterity so that others embody what we cannot abide in ourselves' (2005, 47) – and, in this case, from the Good Friday Agreement of 1998 which decommissioned paramilitary arsenals; brought about a ceasefire from Loyalist and Republican terrorist groups; led to the controversial mass release of HMP Maze/Long Kesh prisoners and hence led to the closure of the prison itself.

CLOSING CURTAIN

At the end of *Hunger*, there is a round of applause and important conversations to be had. As we file out of the auditorium, there is the bleak waft of pizza leftovers from the back row. The Maze/Long Kesh was the epicentre of The Troubles, suggests local photographer Donovan Wylie (Beyfus 2010). Everyone is a part of this story, including the sacrilegious. And all visitors are involved in its telling, especially in disagreement but also in the attempted neutrality of the tour guide and even the anthropologist. The building is now one of the burdens of the past, *Vergangenheitsbelästigung* (Macdonald 2009, 9), needing renovating physically, and processing cognitively and affectively. At present it is not an inherited burden like the swastika-less 'dena-ziified' (*Entnazifizierung*) Nazi Party Rally grounds of Nuremberg. People still approximate the narratives of The Troubles. They still live them in a polarised province. As such, Kirk Simpson

(2009, 112) describes Northern Ireland as a transitional society where the memorialisation of conflict is often over-managed. The danger is that HMP Maze/Long Kesh becomes a political football, if not the football field it was once mooted to become – a muted token of remembrance. Simpson favours the diversity and complexity of a decaying ruination, a building ever changing; a crumbling and eroding mnemonic passing away with our memories. The British government will not allow this 'undesirable heritage' (Simpson 2009, 113) to be used or left or preserved, however.

17 April 2012: the Belfast Grand Opera House erupts to laughter throughout the local play, Martin Lynch's *Chronicles of Long Kesh* (2012). The Greek chorus explaining the history and events is a former prison officer caught up in the regime, trying to pay his mortgage, trying to keep himself and his family safe at work as well as at home. At the end of the play, he points out that, 'When you own a gun, it doesn't just sit in its pocket. It's in here [tapping his head].' There are no winners in violence, not even in its storytelling. 'I know things, I was there too', concludes the former prison officer. While the performance closes to a standing ovation, on the other side of the river Lagan, in the Titanic Quarter, the new £100 million Titanic Exhibition space opens to inject life into the empty docklands that were once the busiest and liveliest shipyards in the world, and once the preserve of a Protestant workforce. Time will tell how HMP Maze/Long Kesh will fare against the new Titanic building, and which indeed is the iceberg.

ACKNOWLEDGMENTS

My thanks go to the guides at HMP Maze/Long Kesh, sponsoring politicians, Briege Burns and Kirk Simpson, the SOAS MA Anthropology of Tourism class and audiences and discussants at several conferences and seminars. My apologies for any errors, presumptions and misrepresentations that are contained herein.

BIBLIOGRAPHY AND REFERENCES

Arendt, H, 1958 *The Human Condition*, University of Chicago Press, Chicago

Aretxaga, B, 1995 Dirty Protest, Symbolic Overdetermination and Gender in Northern Ireland Ethnic Violence, *Ethos* 232, 123–48

BBC News, 2007 *Maze must keep status: McGuinness* [online], 7 July, available from: http://news.bbc.co.uk/1/hi/northern_ireland/6279442.stm [17 March 2014]

BelfastDaily.co.uk, 2013 UKIP Join the Raze the Maze Campaign [online], *The Belfast Daily*, 9 May, available from: http://www.belfastdaily.co.uk/2013/05/09/ukip-join-raze-the-maze-campaign/ [1 March 2014]

Beyfus, D, 2010 Donovan Wylie: Deutsche Borse Photography Prize 2010 [online], *The Telegraph*, 18 January, available from: http://www.telegraph.co.uk/culture/photography/6988852/Donovan-Wylie-Deutsche-Borse-Photography-Prize-2010.html [3 March 2014]

Borneman, J, and Hammoudi, A (eds), 2009 Being *There: The Fieldwork Encounter and The Making of Truth*, University of California Press, Berkeley

Bruner, E, 2004 *Culture on Tour: Ethnographies of Travel*, University of Chicago Press, Chicago

De Certeau, M, 1984 *The Practice of Everyday Life*, University of California Press, Berkeley

Cohen, E, 1985 The Tourist Guide: The Origins, Structure and Dynamics of a Role, *Annals of Tourism Research* 12, 5–29

Crossley, N, 2006 *Reflexive Embodiment in Contemporary Society*, Open University Press, Milton Keynes

Feldman, A, 1991 *Formations of Violence: The Narrative of the Body and Political Terror in Northern Ireland*, University of Chicago Press, Chicago

Feldman, J, 2007 Constructing a shared Bible Land, Jewish Israeli guiding performances for Protestant pilgrims, *American Ethnologist* 342, 351–74

— 2008 *Above the Death Pits, Beneath the Flag: Youth Voyages to Poland and the Performance of Israeli National Identity*, Berghahn Books, Oxford

Foley, M, and Lennon, J, 2000 *Dark Tourism: The Attraction of Death and Disaste*r, International Thomson Business Press, New York

Jackson, M, 2005 *Existential Anthropology*, Berghahn Publications, New York and Oxford

Lippard, D, 1999 *On the Beaten Track: Tourism, Art, and Place*, New Press, New York

Lynch, M, 2012 *Chronicles of Long Kesh* (play)

MacCannell, D, 1992 *Empty Meeting Grounds: The Tourist Papers*, Routledge, London

— 1999 *The Tourist: A New Theory of the Leisure Class*, University of California Press, Berkeley

Macdonald, S, 2009 *Difficult Heritage: Negotiating the Nazi Past in Nuremberg and Beyond*, Routledge, London

McQueen, S, 2008 *Hunger*, Film4 Productions, Channel 4, Northern Ireland Screen, Broadcasting Commission of Ireland

Ong, W, 1990 *Orality and Literacy: The Technologizing of the Word*, Routledge, London

Rapport, N, 1997 *Transcendent Individual: Essays Towards a Literary and Liberal Anthropology*, Routledge, London

— 2003 *I am Dynamite: An alternative anthropology of power*, Routledge, London

Seaton, A V, 1996 Guided by the Dark: from thanatopsis to thanatourism, *Journal of Heritage Studies* 24, 234–44

Simpson, K, 2009 *Truth Recovery in Northern Ireland: Critical Interpreting the Past*, Manchester University Press, Manchester

Taussig, M, 2006 *Walter Benjamin's Grave*, University of Chicago Press, Chicago

Tunbridge, J, and Ashworth, G, 1996 *Dissonant Heritage: The Management of the Past as a Resource in Conflict*, John Wiley & Sons, Chichester

Widlok, T, 2008 The Dilemmas of Walking: A Comparative View, in *Ways of Walking: Ethnography and Practice on Foot* (eds T Ingold and J Vergunst), Ashgate, Aldershot and Burlington VT, 51–66

9

'We shall never forget, but cannot remain forever on the battlefield': Museums, Heritage and Peacebuilding in the Western Balkans

Diana Walters

Introduction

Since the break-up of the former Yugoslavia, the western Balkans has become synonymous with political chaos, war and ethnic tension. The area has been shaped and reshaped over centuries of displacement through occupation, migration and division. The resulting social, political, ethnic and religious complexities are both confusing and contested. The region has experienced an ongoing crisis; war, genocide, ethnic and cultural cleansing and mass migration have created a number of small countries, each now engaged in rapid nation-building. Economies and political structures are weak and corruption is endemic. Several substantial issues remain unresolved, notably the political status of both Kosovo and Bosnia Herzegovina, and tension around minorities and ethnicity in Serbia and Macedonia.

Destruction of cultural heritage was part of widespread ethnic and cultural cleansing as deliberate acts to obliterate traces of the enemy (Stone 2009, 320). Chapman (1994, 120) states: 'in a cultural war, the conquest of territories and the "ethnic cleansing" of settlements is insufficient. Nothing less than the destruction of past historical identities is needed.' He identified widespread destruction of cultural property including mosques, churches, cemeteries and museums. There were also calls for the removal of foreign or international words in order to purify languages (Goulding and Domic 2009, 97).

Identity, memorialisation and heritage have remained sharply in focus as part of the landscape of post-conflict reconstruction and cover a wide spectrum, ranging from appropriation for specific expressions of (contested) identity to efforts at reconciliation. Several international interventions in the region are supporting rebuilding and reconstruction of heritage, and there is an increasing focus on cultural tourism as an economic and social driver for recovery.[1]

This chapter considers examples of heritage-based peacebuilding in the western Balkans through ongoing projects organised under the auspices of the Swedish foundation Cultural Heritage without Borders (CHwB). These examples show how steps towards peacebuilding can be achieved when external facilitation allows for ownership, connections and trust between individuals and

---

[1] Notably the Regional Programme for Cultural and Natural Heritage in South East Europe (2003) and its successor, the Ljubljana Process: Funding the rehabilitation of heritage in South-East Europe (Council of Europe and European Commission); see: http://tfcs.rcc.int/index.php/projects/ljubljana-process [28 June 2013].

institutions to develop organically. At the time of writing the author is engaged in the projects and has drawn on her own observations and the views and opinions of colleagues and participants.

HERITAGE AND PEACEBUILDING

Internationally there are many examples of reconciliation heritage projects following crisis, often centred on intercultural dialogue and inclusion. Several international networks actively promote the role of museums for peacebuilding.[2] Other initiatives relate to the rights of peoples to their heritage and cultural expression by providing spaces for memory, dialogue and encounter after trauma, and are generally discussed as positive contributions to the enhancement of democracy and tolerance (see for example Crooke 2005; Utaka 2009). They are often cited as inspirational examples that show heritage as a transmitter of social justice and human rights. However, some commentators question the value of such developments to the local communities. Looking at examples in South Africa, Meskell and Scheermeyer (2008, 158) comment:

> While all tout national success and healing, there is an overwhelming sense that these edifices have little currency with ordinary South Africans, are attracting more foreign than domestic visitors, and are more about crafting a political pageantry than reflecting national heritage or building a multi-ethnic constituency.

Much of the criticism is based on the top-down nature of these initiatives and their limited engagement with people. For the Balkans, criticism is also levelled at the Ljubljana Process that started in 2003 (Council of Europe 2013). Research based in Serbia showed that the original aim of developing heritage for reconciliation has been weakened over time, with a shift towards creating bureaucracy and policy (Vos 2011). This debate is mirrored in peacebuilding generally, where a top-down approach is increasingly associated with 'liberal' or external interventions as opposed to bottom-up ones (Charbonneau and Parent 2012). Lehrer (2010, 272) argues that reconciliation 'is an organic process that unfolds in daily life, within and between aggrieved communities'.

Another criticism of external peacebuilding is that it can lead to a heightening of ethnic tension and insecurity. Heritage as part of nation-building may focus on strengthening specific constructions of identity to the detriment of minorities (Kostic 2007, 16). This at least in part explains why several major capital heritage projects are being developed across the whole region, driven by national or even regional statements of identity (see for example Ragaru 2012). Conversely, the opposite is true. In Bosnia Herzegovina an ongoing systemic political crisis has left some national cultural institutions stateless and without guaranteed sources of income. In October 2012 the National Museum of Bosnia Herzegovina decided to close its doors after protracted deadlock around management and governance and months of staff working without pay or basic resources. The fate of the museum and other national cultural institutions remains unclear.

Kostic's findings have resonance for the other western Balkan countries. His model of 'ambivalent peace' is characterised by a continuance of competing demands for ethnic rights and security

---

[2] See: International Network of Museums for Peace: http://inmp.net; International Coalition for the Sites of Conscience: http://www.sitesofconscience.org; International Human Rights Museums: http://www.fihrm.org [28 June 2013].

that are exacerbated by externally driven nation-building after conflict. He argues that this has consequences on both political and population levels; on the political level through 'ethnic politiking' and on the population through a 'strengthening of ethnonational cohesion and conservation of an existing ethnic conflict with all of its inherited incompatibility', thereby reducing the possibility of changing negative stereotypes and attitudes (Kostic 2007, 18).

Anyone visiting Balkan museums might be struck by the similarity of their exhibitions, regardless of location, typically comprising classical archaeology, ethnographic displays of folk art, crafts and textiles specific to each area and recent historical accounts of attacks by others. All are designed to strengthen a sense of identity based on ethnic construction. Scham and Yahya (2003, 405) observe that professionals 'who question or reject the paradigms of the past are vulnerable to accusations of providing aid and comfort to the enemy'. In a post-conflict phase this process seems inevitable.

Cultural Heritage without Borders

The Swedish Foundation Cultural Heritage without Borders (CHwB) is an example of an organisation initially created to support reconstruction of tangible cultural heritage damaged through conflict. Its mission is 'to promote cultural heritage as both a right in itself and a resource. CHwB works with civil society and institutions at all levels to strengthen peace building, sustainable socio-economic and democratic development and the realisation of human rights' (CHwB 2013). The organisation is based in Stockholm with three locally managed offices in Sarajevo, Pristina and Tirana.

CHwB was founded in 1995 as a response to the destruction of cultural property in the siege of Sarajevo in the spirit of the Hague Convention (UNESCO 1954). In recent years CHwB has expanded into urban and spatial planning, cultural policy, cultural tourism, museum development and interpretation – all based on long-term perspectives incorporating environmental and economic sustainability. In a period of almost 20 years, CHwB's work in the Balkans has moved from being crisis 'external' intervention, with a specific focus on protection and restoration, to broader societal and economic development. Operationally the shift has been from sending international experts into specific locations and projects to staffing locally managed offices with a broad portfolio of strategic heritage-based initiatives ranging from input to national cultural policy to small, community-based projects. Planned restructuring will enable the three Balkan offices to become independent NGOs, meaning that top-down is being replaced by bottom-up. As well as country-specific work there is a discrete regional programme constituting international collaboration between six different countries (Albania, Bosnia Herzegovina, Kosovo, Macedonia, Montenegro and Serbia). Three of these initiatives have peacebuilding as a core objective: the Balkan Museum Network, the Women's International Leadership Development programme and the Regional Restoration Camps. These projects have multiple funders and are designed to be open-ended, long term and strategic, making them different from many of CHwB's other project-based initiatives.

The Balkan Museums Network

The Balkans Museums Network (BMN) was initiated in 2011 following the '1+1: Life & Love' Simultaneous Exhibition project (Walters 2012). Eleven museums (including national institutions) had been collaborating since 2006 through a CHwB externally facilitated network.

Political and logistical barriers prevented the creation of a single travelling exhibition so instead they developed simultaneous expressions of their collaboration. For some museums this was the first contact they had had since before the war. The success of the 1+1 project seemed to indicate that a *bona fide* independent museum network based in and managed from the Balkans could be formed (Ljungman and Taboroff 2011, 18–22). However, once the project ended, obstacles reappeared mainly around structure, capacity, governance, cultural policy and political imperatives. Despite common needs and understanding of the benefits of cooperation, no institution or individual emerged to push the initiative forward. Conditions were not right and lack of internal and external resources prevented any development.

What followed was a short period where little active intervention came from the external agency, itself in a process of strategic review. BMN through CHwB continued to provide a shared platform in the form of a webpage that remained active and kept the notion of the network alive. Through this and through other social media, museums and individuals began to expand their own professional and personal links; from basic contact and sharing of information, to small joint projects and collaboration over research, publications and exhibitions. Personal friendships developed and deepened from the professional relationships of the previous five years. This continued reconnection on personal levels after conflict has been identified as an important first step for peacebuilding (Stover and Weinstein 2004, 324).

In spring 2013 a BMN conference was organised in Sarajevo. Funding was available to support colleagues wishing to attend, but this time the event was made available to all museums and cultural heritage organisations, not just the original 11. Individuals were invited to apply with a project idea relating to the conference themes of access and inclusion, cultural tourism and interpretation and communication. The result was 'Meet See Do', a three-day event attended by 60 people.

The event was significant for several reasons. The shift to a broader platform altered the dynamic, bringing fresh ideas and perspectives and a spirit of positive curiosity and enthusiasm, and revealing an impressive range of heritage-based initiatives in challenging circumstances. It was also the first time that many colleagues from Kosovo had felt able and safe enough to travel to Sarajevo, meaning that all countries were represented.

Meet See Do was also the first time that the issue of conflict had been openly discussed through the BMN. There was also a keynote speaker from Diversity Challenges,[3] an NGO based in Northern Ireland working with peacebuilding through dialogue and storytelling. The protracted nature of the Northern Irish conflict had resonance for many, and the vocabulary of 'the Troubles' was felt to be an apposite description of the Balkan crisis – an example of how the process of moving through conflict resolution is frequently shared with others (Crooke 2005). Dialogue about conflict also came through an *ad hoc* exhibition of personal objects from the participants (see Fig 9.1), creating object biographies where others could add their own commentary, much of which was insightful and empathetic. There was also a visit to an exhibition about the siege of Sarajevo at the History Museum of Bosnia Herzegovina. For several participants this was the first opportunity to see 'the other side of the story'. The exhibition had little or no interpretation (see Fig 9.2), allowing visitors to draw their own conclusions and following a model of similar displays in Northern Ireland (ibid, 133). One such visitor, an experienced archaeologist

---

[3] See: http://www.diversity-challenges.com/home-page/ [28 June 2013].

in her 40s from Serbia, commented that the experience was amazing, contentious and painful, and that it was extraordinary to see the same pictures but with different titles. Telling others she had visited the exhibition enabled her to observe a division between those who were interested and those who felt it would inevitably be biased. Reflecting afterwards, she identified feelings of guilt – not on a personal level, but more a sense of an injustice unresolved. As a child of Croatian and Serbian parents brought up in Tito's Yugoslavia, she did not feel a strong sense of national identity. 'Not at all', she commented, 'I identify more with my city, or my friends, or

FIG 9.1.   AN EXHIBITION OF PERSONAL OBJECTS FROM ACROSS THE WESTERN BALKANS SHOWED THE DIVERSITY AND THE CONNECTIONS BETWEEN THE DELEGATES.

my colleagues. I don't understand the idea of nation in that way. I understand people with similar backgrounds, or interests or age' (Anon 2013a, *pers comm*). Another participant who survived the siege felt she did not have the strength to see the exhibition with international colleagues for fear that the inevitable questions would require too much emotional strength. Later she felt that this reflected the lack of opportunity for dialogue and the attitude that 'it is easier not to talk about it as it is still very much present' (Anon 2013b, *pers comm*).

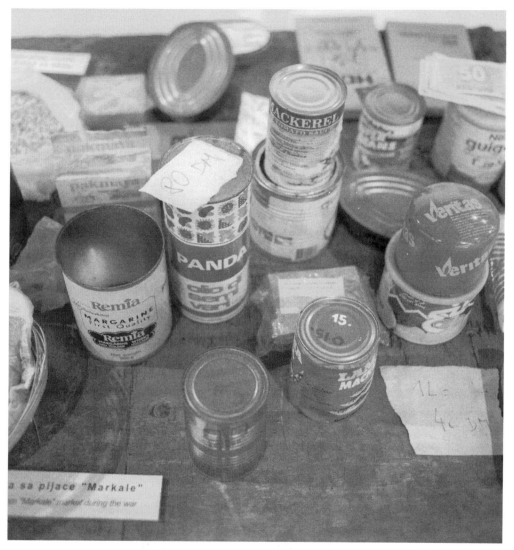

FIG 9.2.   EXAMPLES OF FOOD FROM THE SIEGE OF SARAJEVO. FOCUSING ON DAILY LIFE ENABLED ALL VISITORS TO CONNECT WITH THE EXHIBITION ON A PERSONAL, RATHER THAN A POLITICAL, LEVEL.

By the end of the conference, the need for international cooperation was being clearly vocalised, but there was a shift from previous years; individuals began to offer to drive the process forward themselves with action and commitment. The political and structural obstacles to such a development identified in 2011 remained, but the difference seemed to be that ownership and confidence was developing through professional and personal contact over a substantial period of time. External facilitation helped to create conditions where trust, dialogue and cooperation could begin, but the sustainability of this will only be achieved if the network is subsequently rooted within the organisations themselves.

Women's International Leadership Development (WILD)

WILD is a targeted capacity-building programme that has been adapted for selected leading women from across the heritage sector in six Balkan countries. It combines business psychology, psychotherapy, coaching and peacebuilding rooted in cultural heritage and diversity. WILD is creating a supportive environment for individual women to explore their personalities, their values and their approaches towards themselves, each other and their daily work. Conditions for trust and dialogue are created through a positive working atmosphere, total confidentiality and development of diversity as a tool for strengthening connections and working positively with difference.

As with the BMN, the WILD women have come together through an external agency and most of them had never met before the first meeting. A group contract was negotiated, including confidentiality, respect and active listening of others and shared values (trust, humour, responsibility, friendship, collaboration, respect and knowledge). Many themes have emerged. Several of the women felt their management style was shaped by challenging external conditions and that they wanted to focus more on staff development and partnerships. Institutions were regarded as moribund, hierarchical and inward looking, and the creative potential of cultural heritage was underdeveloped. All of the women spoke of the value of time for reflection and personal development, as well as having their more coaching approach to leadership endorsed and resourced.

The WILD group was diverse and comprised women from different countries, disciplines and age groups. The facilitators drew on management and change theory (notably Kubler-Ross 1969), countermanding the notion of difference as 'otherness' and a source of suspicion by an exploration of the contribution and strength of multiple voices and perspectives. The group began to bond, firstly through exploring possible collaborative work and subsequently through personal ties and friendship. After a generation of warfare, people previously from the same country were now meeting each other with different nationalities. 'I don't know anyone from Kosovo', one participant reflected, 'and I've never been there. Why would I go? But now, I would. We are colleagues doing the same job' (Anon 2013c, *pers comm*). Another experienced woman recounted how a younger colleague had asked several questions about politics between the countries in the former Yugoslavia. 'She has never met anyone like me before and could ask these questions. How else can we find out?' (Anon 2013d, *pers comm*).

The WILD initiative and its micro network is the core for multiplication of outcomes designed to cascade to other heritage professionals in the region. This is somewhat dependent on funding, but the value of these initiatives to the individual beneficiaries cannot be understated. Attitudes and fears shaped by decades of separation and political manipulation can only be changed through personal contact and collaborative working. The true legacy of WILD will be felt in the future development of institutions with empowered, strong and creative leaders unafraid of diversity and difference.

Regional Restoration Camps

The final example is based on restoration and conservation work in the southern Albania city of Gjirokastra, famous for its Ottoman-style architecture and for being the birthplace of former dictator Enver Hoxha.

For six years CHwB has organised twice-yearly Regional Restoration Camps (RRCs) in the World Heritage Site of the old town, providing approximately 45 students with practical and theoretical experience around ongoing conservation projects. The majority of participants come through affiliated Albanian universities, but individuals can apply and most Balkan countries are represented. Feedback shows that the main motivations are to meet and network internationally and to experience practical work on monuments.

In small groups led by specialist craftsmen, participants develop conservation skills in traditional techniques. From 2013 RRCs will also be held in Bosnia Herzegovina, Kosovo and Serbia. Two of the craftsmen, both from Kosovo, will participate in the Serbian camp. Both said that this was not something they had ever anticipated doing, and after reflection one commented 'but it's no different, just another job. I've worked on Serbian churches before and I'm proud of my work there. So much of the trouble is blown up in the media' (Anon 2013e, *pers comm*). The other agreed, though he stated that he would not go to Serbia if he had stayed in Kosovo during the war and not been a refugee. 'Experiencing war shapes people, of course it does. A nation that forgets its history cannot have a future' (Anon 2013f, *pers comm*). The importance of personal experience during conflict is a key factor in shaping response and can also lead to divisions within communities in post-conflict transition (Corkalo *et al* 2004, 149).

During the interviews the craftsmen reappraised their role and saw how it had developed over the years, realising that they were now actively peacebuilding through bringing people together and that this was a skill they had developed. Working together and problem-solving helped to build trust and connections, even where there was no common language (Anon 2013e, *pers comm*). Similarly, programme director Lejla Hadzic observed:

> Through the years of working with young professionals from all the countries of the region what we have realised is that no matter from which country you come from [sic], when you are in front of the 'problem' (façade to be plastered, or a window to be repaired) and you need to discuss with others in your group about the best approach... at this particular moment, when your professional ego speaks out you leave behind your nationality or the nationality of the other, you seek only for best practice, and as a group you seek for [sic] the best possible solution [see Fig 9.3]. Working practically with monuments does give you the opportunity to 'SEE' people through a lens different than national or ethnic [sic]. (Hadzic 2013, *pers comm*)

Reconstruction on a physical level is faster than on a social level and, for peacebuilding between peoples, personal relationships based on trust must be developed (Corkalo *et al* 2004, 159). The RRCs combine these two spheres using heritage-based restoration and conservation to foster dialogue and understanding.

FIG 9.3.   PARTICIPANTS WORK TO REPAIR A DOOR DURING THE 4TH REGIONAL RESTORATION CAMP IN GJIROKASTRA, ALBANIA.

CONCLUSIONS

These examples show how peacebuilding is a process that must be allowed to develop over time and that sustainability is best achieved when people connect themselves. Agencies can assist in creating the conditions for this, but they must also recognise that sustainable peacebuilding is more robust if it grows organically from the local or individual level.

Working with others from 'enemy' countries can be a huge personal risk. Agencies need to understand and plan for this when asking people to collaborate to satisfy their own agendas. International bodies such as the EU, ICOM or UNESCO can provide legitimacy and reduce a sense of isolation or abandonment, but their impact is rarely felt at local level. CHwB's work has connected people and institutions through projects and networks for the first time since the conflicts of the 1990s and thereby assisted in reducing fear and building trust. Through such approaches heritage can indeed contribute to peacebuilding.

Bibliography and References

Anon, 2013a Personal communication (interview with the author), Gothenburg, 7 May

Anon, 2013b Personal communication (email with the author), Gothenburg, 24 May

Anon, 2013c Personal communication (interview with the author), Bitola, 20 March

Anon, 2013d Personal communication (interview with the author), Bitola, 20 March

Anon, 2013e Personal communication (interview with the author), Gjirokastra, 1 May

Anon, 2013f Personal communication (interview with the author), Gjirokastra, 1 May

Chapman, J, 1994 Destruction of a common heritage: the archaeology of war in Croatia, Bosnia and Hercegovina, *Antiquity* 68 (258), 120–6

Charbonneau, B, and Parent, G, 2012 *Peacebuilding, Memory and Reconciliation: Bridging top-down and bottom-up approaches*, Routledge, New York

Corkalo, D, Ajdukovic, D, Weinstein, H, Stover, E, Djipa, D, and Biro, M, 2004 Neighbors again? Intercommunity relations after ethnic cleansing, in *My Neighbor, My Enemy: Justice and Community in the Aftermath of Mass Atrocity* (eds E Stover and H Weinstein), Cambridge University Press, Cambridge, 143–61

Council of Europe, 2013 *IRPP/SAAH: Ljubljana Process II*, available from: http://www.coe.int/t/dg4/cultureheritage/cooperation/SEE/IRPPSAAH/LjubljanaProcessII_en.asp [19 May 2013]

Crooke, E, 2005 Dealing with the past: Museums and Heritage in Northern Ireland and Cape Town, South Africa, *International Journal of Heritage Studies* 11 (2), 131–42

Cultural Heritage without Borders, 2013 *Strategic Plan 2013–2016*, CHwB, Stockholm

Goulding, C, and Domic, D, 2009 Heritage, identity and ideological manipulation: the case of Croatia, *Annals of Tourism Research* 36 (1), 85–102

Hadzic, L, 2013 Personal communication (email exchange with the author), May

Kostic, R, 2007 Ambivalent Peace: External Peacebuilding, Threatened Identity and Reconciliation in Bosnia and Herzegovina, PhD dissertation, Department of Peace and Conflict Research, Uppsala University, available from: http://books.google.co.uk/books?hl=en&lr=&id=3-qjT00IDIQC&oi=fnd&pg=PA1&dq=museum+objects+and+peacebuilding&ots=gN-x22vX1R&sig=sMrdX4iPKlTiMMflLxp7CrOBDUg#v=onepage&q&f=false [12 February 2013]

Kubler-Ross, E, 1969 *On Death and Dying*, Macmillan, New York

Lehrer, E, 2010 Can there be a conciliatory heritage? *International Journal of Heritage Studies* 16 (4–5), 269–88

Ljungman, C, and Taboroff, J, 2011 *Evaluation of Cultural Heritage without Borders in the Western Balkans*, available from: http://www.indevelop.se/publications/publication-evaluation-of-cultural-heritage-without-borders-in-the-western-balkans/ [8 March 2013]

Meskell, L, and Scheermeyer, C, 2008 Heritage as Therapy: Set pieces from the New South Africa, *Journal of Material Culture* 13 (2), 153–73

Ragaru, N, 2012 *The Political Uses and Social Lives of 'National Heroes': Controversies over Skanderbeg's Statue in Skopje*, available from: http://hal-sciencespo.archives-ouvertes.fr/docs/00/68/26/63/PDF/Political_uses_and_social_lives_of_national_heroes_Skanderbeg_NRAGARU.pdf [16 May 2013]

Scham, S, and Yahya, A, 2003 Heritage and Reconciliation, *Journal of Social Archaeology* 3, 399–416

Stone, P, 2009 Archaeology and Conflict: An impossible relationship?, *Conservation and Management of Archaeological Sites* 11 (3–4), 315–32

Stover, E, and Weinstein, H, 2004 *My Neighbor, My Enemy: Justice and Community in the Aftermath of Mass Atrocity*, Cambridge University Press, Cambridge

UNESCO, 1954 *Convention for the Protection of Cultural Property in the Event of Armed Conflict (Hague Convention)*, available from: http://portal.unesco.org/en/ev.php-URL_ID=13637&URL_DO=DO_TOPIC&URL_SECTION=201.html [5 February 2014]

Utaka, Y, 2009 The Hiroshima 'Peace Memorial': transforming legacy, memories and landscapes, in *Places of Pain and Shame: Dealing with 'Difficult Heritage'* (eds W Logan and K Reeves), Routledge, Abingdon, 34–49

Vos, C, 2011 Negotiating Serbia's Europeanness. On the Formation and Appropriation of European Heritage Policy in Serbia, *History and Anthropology* 22 (2), 221–42

Walters, D, 2012 The 1+1: Life & Love simultaneous exhibition. Cross-border collaboration in the Western Balkans, *Journal of Museum Education* 37 (3), 43–56

10

The Politics of Remembering Bhopal

Shalini Sharma

In 1984 the world witnessed one of its worst industrial disasters in Bhopal, India. Three decades later, survivors' poor health and a continuing lack of justice makes Bhopal a site of continued suffering. This chapter engages with the politics of remembering Bhopal. The state of Madhya Pradesh plans to establish a memorial museum to provide healing through closure and restore Bhopal's heritage. However the survivors wish to assert their moral right to their memory, and contest the notion of closure through commemoration: they plan to connect remembrance with vigilance by generating their own memorial. This chapter explores the politics inherent between the State's and the local community's actions towards remembering the disaster. It is based on the author's doctoral fieldwork (2010–11), her previous engagement with the survivor-led movement and her current role with the Remember Bhopal Trust, set up in 2012 to strengthen survivors' efforts towards creating a people's memorial museum distinct from that proposed by the State.

Introduction

The 1984 Bhopal gas disaster is known as one of the world's worst industrial disasters. At midnight on 2 December 1984, methyl isocyanate (MIC) gas leaked out from the Bhopal pesticide plant run by the American multinational Union Carbide Corporation (UCC), killing over 8000 people in the first three days alone; within the next two decades another 25,000 died, and more than 150,000 were left with chronic ailments (Amnesty International 2004). Groundwater contamination that began two years before the gas leak, due to UCC's routine toxic dumping,[1] has subsequently affected 18 communities (Moyna 2012; CSE 2009). The pollution of soil and water in Bhopal has led to a second generation of people being affected, caused by toxins entering the breast milk of gas-affected women (Agarwal and Nair 2002). Even as UCC's current owner, the Dow Chemical Corporation, rejects any liability, Bhopal remains a site of ongoing disasters and suffering.

In the last three decades, the State and the foreign multinationals UCC-Dow have acted together to attempt to dilute the memories of the gas leak and to subsume within it all memories of the water disaster that pre-dated it. Meanwhile, the 'mnemonic communities' of survivors (Zerubavel 2003) emerged as an organised global justice movement; 30 years later this movement remains dedicated to shaping the memory and perception of Bhopal by investigating and interpreting survivors' lived reality (Garbarino 1995). In India, where the State remains the custodian of public memory, the long and troubled history of Bhopal presents morally compelling

---

[1] UCC's internal telexes from 1982 (UCC01737; UCC01736) mention the leakage from its solar evaporation ponds used to dump routine toxic waste.

questions. The central question – which confronts the prevailing ideas of a sanitised 'official' history – is why is it important to preserve Bhopal's heritage of oppression and resistance, expressed by local actions, and how should this be done?

THE STATE: (DIS)REMEMBERING BHOPAL

The Indian State's[2] response to Bhopal has been that of 'remembering and disremembering' (Waliaula 2012). For the last 30 years, the State has resorted to systemic disremembering of Bhopal through the dilution and denial of facts, simply making symbolic rehabilitation using public relations gestures. There have also been many miscarriages of justice in and out of court-rooms, the negation of survivors' lived reality and violent suppression of survivors' peaceful movement whenever their demands confronted the State-corporate apparatus (Sharma 2013; see also Jones 1988; Dembo *et al* 1990; Scandrett *et al* 2009).

The State's remembrance of Bhopal has been ritualistic and symbolic, largely centred on the anniversary of the disaster. To remember and pray for the dead (but not the living survivors), the Indian Parliament and the Madhya Pradesh (MP) Legislative Assembly observe a two-minute silence on every anniversary, while the MP state government organises Shradhanjali divas or a commemoration day (Sharma 2013). The Indian Parliament has consistently ignored public demands to observe the anniversary as a national day of remembrance rather than simply declaring it an official holiday in Bhopal. Although the State repeats the need to remember and learn from Bhopal, it was only after 28 years of survivors' pressure that the disaster was included in the national curriculum (ibid).

Meanwhile, between 1985 and 2010, various members of the MP Legislative Assembly urged the national government – especially around the time of year when the disaster occurred – to build a memorial (Sharma 2011a). Principally there was no disagreement among Indian political parties on the relevance of a memorial, yet all plans were ignored until the 20th anniversary, when the MP government decided to build a US $25 million memorial using national funds at the now abandoned carbide factory site (ibid). This sudden change of heart provokes questions about the MP state's motivation to erect a Bhopal memorial.

WHY IS A STATE MEMORIAL IMPORTANT AND WHO WILL BENEFIT?

From the very beginning, the Indian State's idea of a Bhopal memorial was tied to its corporate interests. The State's vision was to promote tourism and, consequently, it regarded the Bhopal disaster site as a dark tourism facility (Foley and Lennon 2000) within its strategy. In the early 2000s, owing to survivors' protests, the Indian Ministry of Tourism had to halt its plans to transform the factory site into what it called a National Park, envisioning it to be a tourist complex with a memorial, crafts village, technology and amusement parks (see Bi 2000) and not as a National Heritage Site. However, from 2004 onwards the MP state government not only revived the Bhopal memorial project but also pursued it with unprecedented alacrity.

---

[2]   In this chapter, the 'State' refers to the State of India and the national government of India, whereas 'MP state' refers to the government of Madhya Pradesh state.

The context of this renewed interest in the Bhopal memorial was twofold. First there was a drive to boost domestic and foreign tourism to MP; it was crucial to resurrect public perceptions of its capital city. Second, conclusive evidence of ongoing contamination in and around the factory made it possible to sue Dow for its liabilities in 2004, making financial support for the project available. Dow put diplomatic pressure on the Indian government and threatened to withdraw investments in the country because of impending legal action (Sharma 2013; see also PMO-RTI 2007). Dow's main strategy, so far, has been to distance itself from UCC in the public domain. Unsurprisingly, it has been silent on the issue of the memorial. Also, while survivors and activists demanded that Dow pay for the clean-up of the contaminated site, the State did not demand that Dow pay for the memorial, possibly to keep it free of any corporate sponsorship. However, the memorial project – now with the backing of the MP state and the Indian government – was to go ahead, transforming the carbide plant into a safe memory zone and freeing the city from the haunting legacy of the disaster.

In 2005, the MP government organised a national competition to design the memorial on the UCC site; SpaceMatters, a New Delhi-based architecture firm with its proposal of a memorial complex that included an exhibition gallery and research facility was selected to design and build it. The MP government then expedited its fundraising drive for what it now called the memorial museum with revised grant proposals to the central government of India. It also went to the MP High Court seeking the court's permission to start building the memorial museum at the contaminated factory site.[3] To dispel public fears about Bhopal, the MP state planned to open the factory for public tours on the 25th anniversary; the intention was to indicate that the plant was safe and not a source of ongoing contamination (Gas Relief and Rehabilitation Minister Babulal Gaur, quoted in Laxmi 2009). When this plan failed owing to survivors' protests, the MP government paved the way for demolishing the plant by instructing the Indian Institute of Chemical Technology (IICT) to focus on 'dismantling' in its report (TOI 2013; IICT 2010). In August 2013, Jairam Ramesh, the then Environment and Forest Minister of India, who had earlier supported reopening the factory for public viewing and had dismissed claims of contamination as baseless (quoted in Singh 2009), now stressed that dismantling was essential owing to mercury contamination in the factory (quoted in PTI 2011a). This proposal ignored IICT's findings that it was impossible to detoxify it (IICT 2010; TOI 2013).

Clearly, while the Indian State considered a Bhopal memorial essential, it did not consider the carbide plant itself as a crucial element in any such memorial. The design for the memorial complex at the UCC factory site, devised by SpaceMatters to promote 'aspects of healing, remembrance and deterrence' (Joshi and Ballal 2011), now faced further scrutiny.

COMMEMORATION, CATHARSIS, CLOSURE

SpaceMatters had been thrown into the politics of Bhopal with limited choices; it could either distance itself from the project or mediate between the competing demands of the State and the survivors' movement. By handing over the problem to SpaceMatters, the State had found

---

[3] The MP government brought an application before the MP High Court, Jabalpur, in Writ Petition No 2802/2004 – related to toxic waste remediation at Carbide plant – which at the time of writing is pending before the High Court.

a way of advancing the memorial issue within the larger public domain without itself having to deal with survivor-activists directly. However, the survivor-led movement remained sceptical about the claims made for a State memorial, particularly those relating to healing and reconciliation.

Many in the movement, like survivor-activist Abdul Jabbar, saw State plans for conducting public tours at the carbide plant without ensuring an adequate clean-up as 'an eyewash' (quoted in Singh 2009), an attempt to distract the movement from its daily battle for 'real issues of inadequate compensation, shabby hospital care, pensions for gas-affected widows and rehabilitation of people with permanent disabilities' (quoted in Lakshmi 2009). Jabbar insisted that even if a memorial was to be built using public money, instead of financial support from Dow, it was essential that the government did not misuse it:

> Anyone who is familiar with the social and economic condition of Bhopal survivors will say that we first need water, food, medicine. Only someone with a full stomach talks about decorating his house. And government has a strategy because in civil constructions there is potential to mint money. (Jabbar 2011)

Since 2006, survivors have demanded a commission, empowered with financial and punitive powers, to supervise all matters related to the ongoing problems in Bhopal. Despite central government's agreement, progress continues to be stonewalled by reaching an agreement on funding. Many survivor-activists like Meera More (2011, *pers comm*) felt that the MP state preferred to create a memorial rather than empowering a commission; creating a structure would not allow survivors to lead a life free from the real and ever-present threat of poisons.

> [W]e were asking for a commission. But they [MP state officials] will respond only to a memorial. Because their interest lies in the memorial. Because inside the memorial they can do whatever they want with the toxic material – they can throw it away or dump it inside the factory. But this will not clean the poisons. ... Hence we want a commission while the politicians want a memorial. (More 2011, *pers comm*)

In 2009 SpaceMatters and survivors jointly – but with different motivations – urged UNESCO to nominate and preserve the carbide plant (see Fig 10.1) as an industrial World Heritage Site (Ballal *et al* 2011). SpaceMatters saw 'conservation as an act of social catharsis' (Moulsri Joshi, quoted in Lakshmi 2010), and incorporated the derelict factory structure in its proposal as the 'heart of the memorial museum complex' (Sinha 2011). Survivor leaders like Abdul Jabbar considered conservation of the plant as an act of vigilance and deterrence, comparing it to Nazi death camps and Hiroshima-Nagasaki, a place where 'remains of the structure were preserved so that people remain aware of the cost humanity paid for it' (quoted in PTI 2011b). Similarly, finding value in Bhopal's physical testimony as a 'case of heritage in terms of memory of a civilization' for its educational potential and 'what not to repeat' (Minja Yang, quoted in Rao 2009), UNESCO's representative in India urged the Indian government to study the case and make a nomination (ibid). As yet, no action has been taken by the government.

Survivors insist that remembering Bhopal is impossible without preserving its history, evidence and memories. Hajira Bi (2011, *pers comm*) observed that building a memorial without the carbide factory could erase Bhopal from public memory:

FIG 10.1.   UNION CARBIDE PLANT IN BHOPAL.

because if the factory is dismantled then the evidence (of Bhopal) will be destroyed. If memories of [it] are destroyed then people will gradually forget that there was a foreign corporation called Union Carbide that set up its factory in Bhopal.

Like Jabbar, she too considers Bhopal as an example for other communities to remain vigilant, using Carbide as shorthand for the disaster itself:

If at any other place a similar factory like [Union] Carbide is set up or if a similar disaster occurs then their (people's) resolve will be strengthened. First, never again [should] a similar set up be allowed that results into a similar disaster; if it ever happens where government encourages corporations like [Union] Carbide and allows them into the country then people will learn from Bhopal's experience. By remembering Bhopal they will learn that for our rights we should fight like Bhopal. [...] Preserving Bhopal memories along with [those of the Union] Carbide disaster is very important. Our memories of struggle as well as our memories of Carbide are lessons and inspirations for other cities. (ibid)

The SpaceMatters project hopes to build the memorial museum without waiting to resolve matters of environmental remediation or related liability while the movement wants Bhopal to see the project as an example of 'polluter pays'. Consequently, when instead of any decontamination specialists the central government of India invited architect firm SpaceMatters to present its preliminary case for decontamination, preservation and management of the factory plant (see Ballal *et al* 2011), survivor-activists were very displeased.

For survivor-activists like Rasheeda Bi (2011, *pers comm*), building a memorial without comprehensive toxic waste remediation and without involving survivors evokes memories of the 1989 out-of-court settlement where the State reached an agreement with UCC, so compromising corporate liabilities and justice. Many within the movement see this as yet another State attempt to exclude survivors from having a say in the matters that directly concern them. Such processes, the Bhopal survivor groups stress, 'does injustice both to the memory of those who have died and to the struggle of those who continue to survive' (ICJB 2006).

Survivors' fears about being hoodwinked by the State in the name of memorial and care are not unfounded. The Bhopal Memorial Hospital and Research Centre, built to provide care to gas victims and which houses a memorial, 'Homage and Hope', within its premises, is known to have conducted illegal drug trials on gas victims, resulting in at least 14 confirmed deaths and revenue of £140,000 from Western pharmaceutical firms (Lakhani 2011). Experiences of State betrayal and apathy are abundant: violence is legitimised in the name of suffering and healing (Das 1995). Hence, the movement is anxious that the Bhopal memorial does not become another commercial enterprise, producing suffering and injustices in the name of commemoration.

For the people's movement both the government's own shame and a critique of capitalism are integral parts of their heritage (Sarangi 2011, *pers comm*). SpaceMatters' efforts to 'expand the discourse and contextualise the tragedy within the shared heritage of Bhopal' (Joshi and Ballal 2011) at an international conference triggered important, but also difficult, questions. Many survivor-activists questioned the State's authority and motivation to build a memorial and its attempt to regard it as apolitical. It was felt that mediated academic exercises were of little practical value (Sharma 2011b).

In light of the issues raised by the protest movement, SpaceMatters acknowledged that the memorial and its healing potential should be linked with the process of commemoration (Sinha 2011) and hoped to provide a 'pragmatic, inclusive, urban vision' through a collaborative and participatory planning approach (ibid). However, unlike the State – which involves experts but excludes survivors – the justice movement proposes an alternative imagination for the memorial, one that privileges survivors as curators in order to ensure that the memorial is seen as a symbol of continuing struggle, not closure.

THE MOVEMENT MEMORIAL

The Bhopal survivor groups demanded full involvement in building a 'living memorial' (ICJB 2009), claiming that they have a 'moral right' to their memory (ICJB 2006). The first steps in creating a people's memorial came from survivors, when a statue (Fig 10.2) of a mother and child by Ruth Waterman (herself a child survivor of the Holocaust) was erected opposite the UCC factory in 1985. The Union Carbide factory wall, used for movement graffiti ever since the gas disaster, stands transformed into a memory-wall, exposing violence and injustices (Fig 10.3). The Yaad-e-Haadsa, or the 'Memories of the Disaster' museum, was set up by survivors on the 20th

FIG 10.2. STATUE OF MOTHER AND CHILD, ERECTED OPPOSITE THE UCC FACTORY, 1985.

FIG 10.3. THE UNION CARBIDE FACTORY WALL.

FIG 10.4.   ARTEFACTS AT THE YAAD-E-HAADSA MUSEUM.

anniversary with their artefacts such as portraits and personal belongings of victims in a room at the old Sambhavna Clinic building in Bhopal (Fig 10.4).[4] The science memorial exhibition set up by Dr Satpathi at Hamidia Hospital, using the aborted foetuses of gas-exposed women, was inspired by his vision that in addition to public education, the preservation of these foetuses was necessary in order to protect the DNA as essential proof for any scientific inquiry into the cause and effects of the disaster (Satpathi 2011, *pers comm*).[5] All of these survivor-curated projects were tools to spark other public interventions, commemoration, recognition, vigilance and deterrence, leading to personal and communal healing.

The movement had already built a memorial (statue of mother and child) and museum (Yaad-e-Haadsa) outside the UCC site. The movement's proposal of a memorial museum at the UCC

---

4   The old Sambhavna Clinic building, within walking distance of the UCC site, is a memory site itself: the first independent, community-based health clinic for survivors, set up by activists. From 2004 to 2011, after the clinic moved to a new location, this building housed the Yaad-e-Haadsa museum and the Chingari Trust (offering speech and physiotherapy to second-generation affected children). In 2011, when the latter relocated, this building came to house the Yaad-e-Haadsa museum, the Remember Bhopal Trust and the International Campaign for Justice in Bhopal. Thus, the interconnected notion of remembrance, healing and justice were always present in the movement's imagination.

5   The author first saw this exhibition in 2008 when it was positioned prominently outside Dr Satpathi's office at the Hamidia Hospital. In 2011, following many insistent requests to the administration, she was allowed access to the exhibition, which is now kept in store.

site offers a fluid, dynamic narrative shaped by the curatorial ideology of survivors at a trauma site that has been inaccessible to them so far. Unlike the MP state's vision of a memorial with a singular narrative – 'how the gas disaster happened and what we did thereafter' (Gaur 2011) – survivors like Rubi Parvez, mother of a 12-year-old boy with severe birth defects, insisted that the memorial museum should have multiple narratives:

> We would like to see a museum on all tragedies from that time to what we are facing now. [...] We want a museum on struggle by the affected from the beginning till now. We would want people from outside to visit the museum and we would want to see it too. We would want to see the story of suffering that we faced and check if our heart is strong enough to bear it now. Even today, when we talk, the scenes from that time unfold before us. It is a disaster that we can never forget. [...] This is what we want to see, to be strong, to never forget what we suffered and are still suffering. (Parvez 2011, *pers comm*)

Rasheeda Bi (quoted in Lakshmi 2011) wanted a true memorial to be 'one which lets us tell our stories of anger, bitterness, tears and courage'. Oral histories like those of Parvez and Bi, where survivors remember their memories of struggle, demonstrate that survivors see the movement's memories as central to the memorial narrative. Survivors consider themselves to be an important part of this memory structure and not merely an adjunct to the memorial complex.

It is planned that the movement's memorial – the wall, the statue, the hospital exhibition – currently dispersed in various places in the city will be incorporated in the memorial museum. The movement will provide important input created from survivors' oral histories and artefacts. Unmediated by experts, these raw memories from private and public struggle will be an integral part of the museum's narrative. This inside approach will enable inclusion of narratives that might escape a distant, expert imagination. For example, for Nafisa Bi, a piece of her broken bangle from the women survivors' scuffle with police at Tihar Jail stood testimony to women's street struggle; a perforated aluminium pan depicted their everyday struggle with poisonous water (N Bi 2011, *pers comm*). Similarly for Hajira Bi, memories of State betrayal and bureaucracy were indispensable to Bhopal's lessons for other communities facing similar contamination or State-corporate violations (H Bi 2011, *pers comm*). Many survivors, like Saeeda Parveen (2011, *pers comm*), who kept her husband's unfinished bottles of medicine, or Gangaram (2011, *pers comm*), who preserved the roller with which he painted the statue of mother and child, joined their personal histories with that of Bhopal. These stories and artefacts, collected by the Remember Bhopal Trust, will ultimately be incorporated into the museum at the UCC site if the State concedes to the survivors' demands for full involvement; otherwise they will be placed in a community museum.

The movement's input to the memorial museum will use plural narratives to link stories of past and present. Similarly it will enable survivors to be the narrators of their stories and experiences. Along with adults, the 'mnemonic imagination' (Keightley and Pickering 2012) of survivor children, many themselves second-generation gas and water affected, will draw on their personal experiences of suffering and struggle. For instance, for members of Children Against Dow Carbide such as Safreen Khan (2011, *pers comm*), their memories of children-led protests, their arrest for peaceful demonstration and subsequent police beatings, or their learning from visits to other sites of struggle were important to early group formation and education into the non-violent discourse:

So, if we tell that Bhopal suffered such a disaster we are saying that similar disasters can take place elsewhere too, not only because of gas leaks but also in other ways. For example, there are cement factories where through ash pollution people's health gets affected. We ask people to fight against those corporations which play with human lives. Through such stories we want to build awareness to ensure that what happened in Bhopal does not happen anywhere again. (S Khan 2011, *pers comm*)

In the words of Nawab Khan, ensuring that 'the story of Bhopal does not end with it' remains a priority task of the memorial museum for most survivors (N Khan 2011, *pers comm*); survivor-curated projects such as travelling bus exhibitions spark public interventions and create a landscape of collective memory for communities facing similar environmental and human rights injustices. In India, where mobile exhibitions were developed to act as surrogates for the National Museum (Phillips 2006), a survivor-curated mobile memorial takes the museum out of the building and to local communities. Similarly, by giving a voice back to the survivor communities the memorial museum has the potential to evolve into an 'outward-looking institution that participates in and helps to define the city's contemporary conflicts and contradictions' (Sandahl 2012, 478).

Conclusion

The Bhopal memorial museum is crucial not only to commemorate the painful events of Bhopal, but also for its position in the 21st century development discourse. But, if museums have largely been reluctant to depict that the industrial killings are a product of modernity and the modern bureaucratic state (Bartov 1996), how will the Bhopal memorial museum emphasise that an event like Bhopal should never occur again?

From a review of the politics of remembering Bhopal it is quite evident that survivors do not want the State to appropriate their pain and suffering because of its active role in creating and sustaining their trauma for the last three decades. Having been betrayed repeatedly in the past, the survivors do not wish the same fate for their memories. By contesting the State's proposal of healing through commemoration, the movement reasserts its voice and agency in rethinking healing as a matter of social justice and human rights. Bhopal should not be presented as the 'biggest example of human rights violations in the world' (Verma 2009) but as a means of countering the 'production of rightlessness' (Baxi 2006).

A survivor-curated project challenges and offers opportunities for India's postcolonial memory project. It confronts the prevailing hierarchy of the museums and their celebratory narratives of modern India. It offers possibilities to rethink the social and communicative role of museum spaces by opening them to new audiences, particularly to those that may be facing similar alienation.

Bibliography and References

Agarwal, R, and Nair, A, 2002 *Surviving Bhopal 2002: Toxic Present – Toxic Future. A Report on Human and Environmental Chemical Contamination around the Bhopal disaster site*, Shrishti, New Delhi, For the Fact Finding Mission on Bhopal, available from: http://bhopal.net/oldsite/documentlibrary/surviving bhopal2002.doc [3 March 2014]

Amnesty International, 2004 *Clouds of Injustice: Bhopal Disaster 20 Years On*, Amnesty International, Oxford and London

Ballal, A, af Geijerstam, J, and Joshi, M, 2011 Contemporary, painful pasts as heritage – looking ahead in Bhopal (Conference Report), *TICCIH Bulletin* 53, 3rd Quarter, available from: http://ticcih.org/wp-content/uploads/2013/04/1315373416_b53.pdf [30 October 2011]

Bartov, O, 1996 *Murder in Our Midst: The Holocaust, Industrial Killing and Representation*, Oxford University Press, Delhi

Baxi, U, 2006 Protection of Human Rights and Production of Human Rightlessness in India, in *Human Rights in Asia: A Comparative Legal Study of Twelve Asian Jurisdictions, France and the USA* (eds R Peerenboom, C J Peterson, and A H Y Chen), Routledge, London and New York

Bi, H, 2011 Personal communication (interview between Hajira Bi, Rama Lakshmi and the author), Bhopal, 8 January

Bi, N, 2011 Personal communication (interview between Nafeesa Bi and Rama Lakshmi), Bhopal, 16 September

Bi, R, 2000 Joint letter by survivors' organisations to Mr Suresh Prabhu (then Minister of Chemicals and Fertilizers), 1 February, typed copies of the Ministry of Tourism Advertisement and survivors' letter available from: http://www.bhopal.net/oldsite/amuse.html [5 March 2014]

— 2011 Personal communication (interview between Rasheeda Bi and the author), Bhopal, March

CSE, 2009 Contamination of soil and water inside and outside the Union Carbide India Limited, Bhopal, Report by S Johnson, R Sahu, N Jadan, and C Duca, Centre for Science and Environment, New Delhi, available from: http://www.cseindia.org/userfiles/Bhopal%20Report%20Final-3.pdf [3 March 2014]

Das, V, 1995 *Critical Events: An Anthropological Perspective on Contemporary India*, Oxford University Press, Delhi

Dembo, D, Wykle, L, and Morehouse, W, 1990 *Abuse of Power: Social Performance of Large Corporations. The Case of Union Carbide*, Apex Press, UK

Foley, M, and Lennon, J, 2000 *Dark Tourism: The Attraction of Death and Disaster*, Continuum, London

Gangaram, 2011 Personal communication (interview between Gangaram, Rama Lakshmi and the author), Bhopal, 8 May

Garbarino, J, 1995 Growing up in a Socially Toxic Environment: Life for Children and Families in the 1990s, in *The Individual, the Family and Social Good: Personal Fulfilment in Times of Change* (ed G B Merton), vol 42, Nebraska symposium on motivation, University of Nebraska, 1–21

Gaur, B L, 2011 Personal communication (interview between Babulal Gaur and the author), Bhopal, 16 May

ICJB, 2006 *Remember Bhopal*, pamphlet produced and disseminated by the International Campaign for Justice in Bhopal during survivors' footmarch to Delhi, available from: http://old.studentsforbhopal.org/Assets/13RememberBhopal.pdf [1 August 2013]

— 2009 *Opening up Bhopal's Death Factory is a Dangerous Publicity Stunt: Survivors* [online], available from: http://studentsforbhopal.org/node/255 [3 March 2014]

IICT, 2010 *Detoxification, Decommissioning and Dismantling of Union Carbide Plant (Union Carbide India Limited, Bhopal)*, Indian Institute of Chemical Technology (Council of Scientific and Industrial Research), Hyderabad, India, for the Directorate of Gas Relief and Rehabilitation, Government of Madhya Pradesh, Bhopal, India, available from: http://chemicals.nic.in/Reports%20of%20NGRI,%20NEERI%20&%20IICT/IICT%20-Part1.pdf [1 October 2013]

Jabbar, A, 2011 Personal communication (interview between Abdul Jabbar, Rama Lakshmi and the author), Bhopal, 15 May

Jones, T, 1988 *Corporate Killing: Bhopals Will Happen*, Free Association Books, London

Joshi, M, and Ballal, A, 2011 Bhopal Gas Tragedy: Dissonant History, Difficult Heritage, *Context: Built, Living and Natural, Special Issue on Museums* 8 (2), 7–15

Keightley, E, and Pickering, M, 2012 *The Mnemonic Imagination: Remembering as Creative Practice*, Palgrave Macmillan, Basingstoke

Khan, N, 2011 Personal communication (interview between Nawab Khan and the author), Bhopal, 29 December

Khan, S, 2011 Personal communication (interview between Safreen Khan and the author), Bhopal, 5 May

Lakhani, N, 2011 From tragedy to travesty: Drugs tested on survivors of Bhopal, *The Independent*, 15 November, available from: http://www.independent.co.uk/news/world/asia/from-tragedy-to-travesty-drugs-tested-on-survivors-of-bhopal-6262412.html [2 August 2013]

Lakshmi, R, 2009 India's plan to open chemical disaster site brings protest, *Washington Post*, 27 November, available from: http://articles.washingtonpost.com/2009-11-27/world/36788753_1_bhopal-victims-satinath-sarangi-worst-industrial-disasters [11 August 2013]

— 2010 India moves to clean up site of deadly 1984 Union Carbide gas leak, *Washington Post*, 9 July, available from: http://www.washingtonpost.com/wp-dyn/content/article/2010/07/09/AR2010070903209.html [3 March 2014]

— 2011 The Morality of Memory, *Exhibitionist*, Fall: Museums, Memorials, and Sites of Conscience, 66–70

Laxmi, S, 2009 Plan for Bhopal tours causes outrage, *The Telegraph*, 11 November, available from: http://www.telegraph.co.uk/expat/expatnews/6543345/Plan-for-Bhopal-tours-causes-outrage.html [8 August 2013]

More, M, 2011 Personal communication (interview between Meera More, Rama Lakshmi and the author), Bhopal

Moyna, 2012 Groundwater contamination around Union Carbide factory is now officially confirmed, *Down to Earth*, 26 September, available from: http://www.downtoearth.org.in/content/groundwater-contamination-around-union-carbide-factory-now-officially-confirmed [5 August 2013]

Parveen, S, 2011 Personal communication (interview between Saeeda Parveen, Rama Lakshmi and the author), Bhopal

Parvez, R, 2011 Personal communication (interview between Rubi Parvez, Rama Lakshmi and the author), Bhopal

Phillips, K K, 2006 *A Museum for the Nation: Publics and Politics at The National Museum of India*, published PhD thesis, University of Minnesota, UMI Number: 3243370

PMO-RTI, 2007 Scanned copies of internal correspondence between Dow, Indian Government and Indian industrialists, obtained by Bhopal activists through Indian Right to Information Act 2005, previously available from: www.legacy.bhopal.net/pmo.html/ [15 July 2013]

PTI, 2011a Centre mulls over demolition of Union Carbide factory in Bhopal, *DNAIndia*, 25 May, available from: http://www.dnaindia.com/india/report-centre-mulls-over-demolition-of-union-carbide-factory-in-bhopal-1547395 [30 August 2013]

— 2011b Bhopal victims oppose Union Carbide plant demolition, *The Hindu*, 27 May, available from: http://www.thehindu.com/news/national/bhopal-victims-oppose-union-carbide-plant-demolition/article2053760.ece [30 August 2013]

Rao, R, 2009 Survivors urge UNESCO to make Union Carbide factory a heritage site, *The Indian Express*, 27 February, available from: http://archive.indianexpress.com/news/survivors-urge-unesco-to-make-union-carbide-factory-a-heritage-site/428787 [5 August 2013]

Sandahl, J, 2012 Disagreement Makes us Strong?, *Curator* 55 (4), 467–78

Sarangi, S, 2011 Personal communication (conversation with the author), Bhopal, 13 June [same issues raised by S Sarangi at *Bhopal 2011: Requiem and Revitalization International Workshop and Symposium*, organised by SpaceMatters in Bhopal, 3 February 2011, Bhopal]

Satpathi, A, 2011 Personal communication (interview with the author), Bhopal, 8 January

Scandrett, E, Mukherjee, S, Sen, T, and Shah, D, 2009 *Bhopal Survivors Speak: Emergent Voices from a People's Movement*, Bhopal Survivors' Movement Study, Word Power Books, Edinburgh

Sharma, S, 2011a Notes from archival research conducted at Indian Parliament and Legislative Assembly (for PhD thesis), School of Oriental and African Studies, University of London

— 2011b Notes from participant observation at *Bhopal 2011: Requiem and Revitalization International Workshop and Symposium*, organised by SpaceMatters in Bhopal, 23 January–4 February 2011, Bhopal, India, see: http://www.bhopal2011.in/ [5 March 2014]

— 2013 Media and New Social Movements: The Case of the Justice for Bhopal Movement, unpublished PhD thesis, School of Oriental and African Studies, University of London

Singh, M P, 2009 Bhopal Gas Disaster: Disaster Tourism, *Frontline* 26 (26), 19 December 2009–1 January 2010, available from: http://www.frontline.in/static/html/fl2626/stories/20100101262603800.htm [11 August 2013]

Sinha, S, 2011 Bhopal Gas Tragedy Memorial, in *Bhopal 2011: Landscapes of Memory* (eds A Ballal and J af Geijerstam), SpaceMatters with Norwegian University of Science and Technology (NTNU), New Delhi, India, 113–18

TOI, 2013 UCIL site contamination a threat, *Times of India*, 2 August, available from: http://articles.timesofindia.indiatimes.com/2013-08-02/pollution/41005263_1_cse-ucil-bhopal-gas-tragedy [5 August 2013]

UCC01736, 1982 *UCC Telex on leakage from Solar Ponds*, 10 April, excerpt available from: www.bhopal.net/oldsite/excerpts.html [15 May 2013]

UCC01737, 1982 Excerpts of *UCC Telex on leakage from Solar Ponds*, 25 March, available from: www.bhopal.net/oldsite/excerpts.html [15 May 2013]

Verma, J, 2009 Bhopal Disaster, Biggest Example of Human Rights Violation, *Outlook India*, 5 December, available from: http://news.outlookindia.com/items.aspx?artid=670577 [15 May 2013]

Waliaula, K W, 2012 Remembering and Disremembering in Africa, *Curator: The Museum Journal* 55 (2), 113–27

Zerubavel, E, 2003 *Time Maps: Collective Memory and the Social Shape of the Past*, University of Chicago Press, Chicago

11

Animating the Other Side: Animated Documentary as a Communication Tool for Exploring Displacement and Reunification in Germany

Ellie Land

Introduction

In 2007, I completed my animated documentary, *Die Andere Seite (The Other Side)*, which features oral history recordings of former West and East Berliners describing their memories of life in Germany before and after the fall of the Berlin Wall. The film emphasises personal perspectives of displacement and reunification from both sides of the divide. Participants tell of their curiosity about what they thought lay on 'the other side' and then describe their feelings and reactions on visiting for the first time, once the Wall was opened. One of the film's main aims was to create a document of life in a divided Germany, in order to preserve for future generations personal experiences of post-World War II Germany, using interviews from participants who were born after the erection of the Wall.

This chapter will look at how animation is used to communicate personal experiences of division and reunification in Germany. It will focus on *Die Andere Seite*'s use of an animation device known as 'metamorphosis' to visually represent memories associated with reunification after division, and how communities were affected by the changes that resulted. The chapter also explores the use of space as a visual means to communicate change and transition.

Project Background

The making of this film has personal significance for me. As a child, I lived in West Berlin from 1986 to 1987 after my mother decided to relocate the family from the UK to be with my stepfather, who at the time was serving with the British army stationed at Spandau. We lived in a typical Berlin apartment outside of the army barracks and I went to a German-speaking *Grundschule* (primary school). On days out, we would often visit the tourist sites of West Berlin, including Checkpoint Charlie, which back then, before the Wall came down, was quite different to the shining, ordered place of remembrance constructed there today. My lasting impression of Checkpoint Charlie and that section of the Wall was that it seemed to be an area in the heart of the city that had been deserted. It was tensely quiet; the personnel moved more slowly and deliberately while on the other side of the street West Berliners would be busy about their daily business. I remember the crumbling concrete pavement and weeds growing out of the cracks. I remember the profusion of graffiti sprawling the length of the Wall, as far as the eye could see. I was mesmerised by this graffiti. It was exciting and dangerous, and something I had been told was forbidden. On reflection, perhaps this graffiti symbolised the sense of danger and mystery surrounding my own understanding and feelings about the Wall.

I also remember strongly my intense curiosity about what and who were on the East side of the Wall. I remember thinking that everything on the other side consisted of various shades of brown and cream and that no one had a television! I am not sure where those ideas came from because in reality they simply were not true. They were more likely my child-like simplifications of the facts. For example, my idea that no one in the East had a television was probably my interpretation of state-controlled access to media. I brought these ideas with me when I returned to the UK in 1987, and they remained until 1997 when I returned to Berlin on a school trip. In 1997 I was able to witness the other side for myself. I was immediately struck by how my ideas of what I thought was on the other side did not match up to what I was experiencing in reality. Of course, my experience was somewhat diluted by the decade of progress since reunification and could not be compared to that of someone who had experienced 'crossing over' in 1989. Nevertheless, the experience got me thinking about how those living in divided Berlin might similarly have had ideas about life on 'the other side' that were subsequently challenged or revised when the Wall came down. This was the beginning of my journey toward creating a film that documented the experiences of reunification by contrasting Berliners' beliefs and ideas about what lay on the other side with the reality of what they found. Making the film also enabled me to explore my own experiences, which felt somehow unresolved and incomplete, due to my moving back to the UK. I eventually began making *Die Andere Seite* in 2007, 20 years after I had left Germany.

The film-making process took approximately one year. I began by making notes of what I remembered about my own experiences of living with the Berlin Wall. I trawled family albums to find photos to aid in this process and provide visual cues and references. I read books based on true stories of living in divided Germany, such as Anna Funders' *Stasiland: Stories from Behind the Berlin Wall* (2003). Books like Funders', while not ostensibly or solely about the Wall, nonetheless helped form a picture of life during that time.

I then made a trip to Germany, to gather primary research, beginning with visual artefacts. I travelled almost the entire length of the Wall, starting just north of Mauerpark in Prenzlauer Berg and ending in Heinrich-Heine-Straße, finding along the way bits of wall still standing and numerous monuments and parks erected where other parts had fallen. I visited Berlin's numerous museums in order to gain a thorough understanding of historical events. I used a sketchbook to record my visits and experiences, which became a valuable source of visual references for the film.

Following this preliminary research, I was ready to interview people about their experiences. At first, I used personal networks to find those who were willing to share their stories with me. Later, I collaborated with a film festival, Interfilm Berlin, who used their networks to source further participants. In the end I interviewed ten German people, both in the UK and in Germany. Material from four participants was used in the final edit of the film. However, the process of interviewing all the individuals involved enhanced my knowledge of the lived experience of life before and after the Wall, and enriched the visual and thematic tapestry of the film.

Traditional documentary techniques inspired by film-maker and theorist John Grierson were used to plan the interviews. As Brian Winston (1995, 92) comments on the theories of Grierson: '"Treatment" or dramatization (also sometimes reinforced as "interpretation") reflects the documentarist's desire and willingness to use actuality material to create a dramatic narrative'. In *Die Andere Seite* the actuality material is the recorded interviews of the participants' experiences, and the selection of participants depends on the dramatic qualities of how they tell their stories and how their stories could be edited together. First, I met with each participant to discuss their experience in an informal environment. This enabled me to observe whether the participant was comfortable being interviewed about their personal experiences and also if their personality was demonstrated through drama in their interview, as this enables the audience to identify with and 'warm' to the person being interviewed. The second decision-making factor was balance between stories representing experiences from each side of the Wall. Lastly, it was important to explore a variety of experiences, which meant that if people had similar experiences, I had to choose which to include. This decision usually came down to considerations of dramatic qualities.

These criteria allowed the selection process to be open in that I was led by the experiences of the participants in establishing the main themes to be explored in the film. This meant that participants had a degree of control in the film-making process because their stories guided the structure of the film.

This method of allowing participants to lead the shaping of the narrative is similar to life story interviewing and can lead to the film-maker being overloaded with material, the management of which then becomes problematic. However, in this case, the method formed part of the creative process in that the subject area was led by the participants. I would then use this as a stimulus to shape further questions for the other participants and ultimately to structure the film.

After recording the interviews, I reviewed the material and used the stories in the creation of an animatic, a device that places storyboard images into a moving time line, where they are synced with sound from the actual interviews. The animatic is crucial in allowing the film-maker

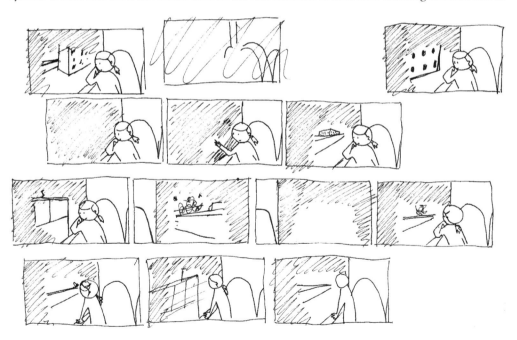

FIG 11.1. STORYBOARD IMAGES PREPARED FOR THE ANIMATIC STAGE IN MAKING *DIE ANDERE SEITE*.

to understand the material they are working with in a time-based environment, and is the expression of raw creative thinking. It is where the film begins to take its final shape and where a visual language is established, as is the interrelationship between image and sound. It is where the ultimate effect on its audience can be engaged with, explored and refined. Practically, this involves arranging and rearranging the storyboard images to the sound in order to understand how visual ideas work with the audio from the interviews. For this film, I began editing an animatic before I had collected all of the interviews. This allowed me to identify any gaps in the representation of the material, for example if there was an over-representation of stories from the West compared with those from the East, and vice versa, as it was crucial for the film that each side was represented fairly.

I opted to use an animation style known as 'straight ahead animation'. Animation is often created using the 'key frame' method, which means the beginning and end of a movement are defined as 'key frame one' and 'key frame two', with 'in between' frames inserted to ensure smooth movement between key frames. In hand-drawn animation this allows the image to retain a consistent size and shape. Straight ahead animation takes almost the opposite approach. Its starts with a key frame, but then the animator keeps on animating until they reach the last frame, without the second key frame to ensure consistency of line. The result is a more fluid, organic-looking product where characters and objects do not retain a consistent shape and size. Straight ahead animation is less time-consuming than key frame animation and enabled me to quickly and freely respond to the audio I had gathered. In addition, the fluidity of the resulting image was perfectly suited to the visual expression of the intangible, fleeting nature of memories, ideas and impressions.

The drawings used in the film were created with a dip pen and ink on small A5 pieces of paper. These were then filmed under a rostrum camera, the resulting images replacing the story-board images used in the animatic.

DISCUSSION OF RESULTS

This section will highlight some of the animated properties and filmic strategies used in *Die Andere Seite* to communicate personal experiences of displacement and reunification arising from the erection and subsequent demolition of the Berlin Wall.

Metamorphosis is a device unique to animation and is intrinsically bound up with the fabric of animation itself. Metamorphosis happens when one image fluidly transforms into another image, and this can happen with any form of animation: drawn, stop motion or CGI.

Metamorphosis is used throughout *Die Andere Seite*. When 'Kat', our second speaker in the film, says 'I tried to remember the first time I saw the Wall, I really didn't know it as a child because we didn't talk about it, I learnt it at school really. I think my parents didn't show us, to not make us afraid of it', the images begin as a point of view shot of someone walking alongside the Wall. The Wall then metamorphoses into white space from which the point of view shot changes to a normal viewing position from where we see a bleeding figure of a small child meta-morphosing out of the white space, and hear the sound of her blood landing on the floor. In most animated films metamorphosis occurs between two images, one object or character shape-shifting into another. In *Die Andere Seite*, metamorphosis occurs when the image breaks down completely into white space, out of which another image appears.

Metamorphosis serves two functions in this film. Firstly, it serves as a visual representation of memory. The images moving fluidly in and out of white space, as in Kat's story, suggest the recollection of distant memories in a way that the audience recognises and identifies with. Further, the use of metamorphosis in reconstructing memory in this way strengthens the audience's emotional relationship with the film, providing a deeper understanding and a human connection with the narratives presented.

Secondly, metamorphosis is the signifier of change. The key change in *Die Andere Seite* occurs when the Wall opens and the barrier is no more, at which point the background is rendered in colour for the first time. This change is also illustrated by less movement in the frame, reinforced by a static camera. The occurrence of metamorphosis in *Die Andere Seite*, of one thing becoming another, reflects both the process and the result of reunification. As Paul Wells (1998, 69) describes in his book *Understanding Animation*, 'metamorphosis can resist logical developments and determine unpredictable linearities (both temporal and spatial) that constitute different kinds of narrative construction'. Not only is metamorphosis used in this film to evoke memories of change, it is itself also a different kind of narrative construction, one that serves to signify change and unpredictability. Animating change in this way serves to embody the actual experiences of the interviewees during the time of reunification.

The location, setting or space of an animated film is traditionally used to visually represent the thematic backdrop of the story. Traditional animation design involves creating characters who perform actions within a certain location, predetermined by a script. In *Die Andere Seite*, the visual palette, location and characters are designed to foreground the voiceover and personalities of the people interviewed as the core elements of the film. Unlike traditional animation, the film does not encourage the audience to form a relationship with the drawn characters on screen, but rather to engage with the 'characters' of the interviewees represented by the voiceover. The same can be said for the design of the location, which is deliberately sparse and undecorative, such as when showing rows of typical Berlin apartment blocks. While visually representative of parts of the real Berlin, this technique also controls the amount of visual stimuli received by the audience in order to direct their focus on the audio. Like radio, which engages audiences' imaginations without images, the deliberately measured visual language of *Die Andere Seite* allows the audience to use their imagination to explore and visualise the content of the voiceover for themselves. As Aylish Wood (2006, 6) states, 'space begins to gain meaning of its own when it no longer solely serves a supporting role by giving meaning to the actions of its characters'. The reconfiguring of

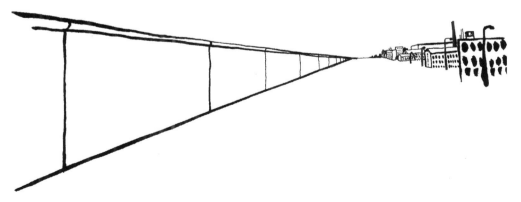

Fig 11.2.   Space design for *Die Andere Seite*.

the role of space in *Die Andere Seite* serves to visually communicate with the audience in a way that enhances the impact of the voiceover. I use the term space here, because it denotes something that is often not well defined, and may be contested or notional, as opposed to location, which is more fixed in its meaning. Using the term space also enables a discussion of the visual properties of the filmic space, which are not limited to descriptions of visual representations of the 'real world'.

In *Die Andere Seite*, the border between East and West Germany has been temporarily re-erected, albeit in a form arising from the animator's mind; a drawn, colour-void document, rather than an attempt at objective representation, such as a photograph or realist painting. The film illustrates stories from those that lived on either the East or West side, by visually placing them on the side from which they originally came. The film 'traps' these characters on their respective sides of the Wall; they are unable to cross the border, which is usually represented as a concrete wall or wire fence. For example, the film shows a little boy from the former West looking through a fence at the 'death strip' and a girl from the former East riding a train, her only view an endless vista of the Wall. Crucially, the audience is permitted to cross the border. Whenever the narrative switches to the other side, the camera leaves the side it has been viewing and travels through the Wall. In permitting the audience to travel across the border in this way, the film not only illustrates an action that in reality was virtually impossible, but also allows the audience to engage in a state of authoritative viewing, in that they are permitted to go where the characters in the film are not.

In *Die Andere Seite*, the border has once again trapped our characters, their stories held by the side they originated from. Yet the border has allowed the audience to cross. Then, in the final third of the film, the border opens and the audience witness a transformation of space. The solid, impenetrable border breaks and opens like a two-sided gate to reveal a vividly colourful West Berlin. This transformation of filmic space visually represents the real-life transformation experienced by the interviewees. 'Dieter' from the former East recalls: '… and there were so many different things to eat on the street and I thought, why should I eat on the street? Normally I eat at home! This was something I had never seen before.' Aylish Wood (2006, 8) states: 'these unexpected transitions allow a viewer to also experience moments of uncertainty, of being in between in an encounter with transforming space'. Witnessing this transition not only provides the audience with a representation of actual events but also leaves the viewer with a lasting visual impression of another effect of the reunification: the feeling of uncertainty.

Die Andere Seite employs space in a way that differs from traditional animation in that it allows an act of authoritative viewing by enabling the audience to cross the border in a thematically rich visual re-enactment of the opening of the border, while at the same time evoking the sense of uncertainty present in the experiences of the interviewees and their fellow Germans.

OUTCOMES

Die Andere Seite was finished and released in 2007. It was selected to screen in over 20 film festivals worldwide, including Interfilm Berlin, London Short Film Festival and Animafest Zagreb. It was also selected to screen in a programme curated at the Institute of Contemporary Art, London. In Autumn 2009, *Die Andere Seite* was screened as part of celebrations of the 20th anniversary of the fall of the Wall. Programmes were curated by the British Council and by film festivals in Germany and Austria. *Die Andere Seite* was included on a DVD collection of short films about Berlin, called 'Kurz in Berlin', which includes commentary by the directors.

CONCLUSION

Die Andere Seite has used animation methods and documentary storytelling techniques to portray a difficult time in Germany's recent history. By foregrounding the real stories of people who experienced a divided and then reunited Germany, the film is an important document for the

FIG 11.3.   STANDING NEXT TO THE BERLIN WALL AT CHECKPOINT CHARLIE, ELLIE AND LLOYD LAND, 1986.

feelings of displacement felt by Germans as a result of changes to their countries' economic, political and cultural structures.

The use of metamorphosis and of non-traditional concepts of space in *Die Andere Seite* has engaged the audience in a new form of viewing and understanding. Metamorphosis visually denotes constant change and states of flux while also providing a time-based structural element that thematically signifies the process and result of change. By engaging the audience in an authoritative viewing experience, *Die Andere Seite* elicits feelings of disorientation and change, connecting them to the experiences of the participants in the film. The reconstruction of the Wall as a thematic device creates a visual boundary for the audience to cross, echoing the experiences of entrapment and subsequent liberation at the heart of the film.

BIBLIOGRAPHY AND REFERENCES

Funders, A, 2003 *Stasiland*, Granta, London

Land, E, 2007 *Die Andere Seite*, available from: https://vimeo.com/8240123 [10 July 2013]

Wells, P, 1998 *Understanding Animation*, Routledge, Abingdon

Winston, B, 1995 *Claiming the Real*, British Film Institute, London

Wood, A, 2006 Re-Animating Space, *Animation* 1 (2), 133–52

12

Restoring Gorongosa: Some Personal Reflections

Rob Morley and Ian Convery

Introduction

Mozambique has a rich, diverse natural heritage. Nowhere is this better represented than Gorongosa National Park (GNP), for decades a symbol of Mozambique to the outside world (see Fig 12.1). Situated at almost the geographical centre of Mozambique, GNP was once one of southern Africa's most richly endowed protected areas. At independence Mozambique had only three terrestrial national parks; GNP was the largest, most well-known and easiest to access. In its heyday GNP was the playground of the rich and famous; visitors included American astronauts and Hollywood stars. The Apollo 16 astronaut Charles Duke, visiting the park in 1971, is reported to have remarked that visiting GNP was as 'thrilling as landing on the moon' (Hanes 2007, 87). However, the park became synonymous with a long and bloody civil war, and its destruction mirrored that of the country as a whole.

GNP is now returning to its former glory after decades of war and under-investment, thanks in part to an innovative public–private partnership between the Government of Mozambique and the Carr Foundation. Much has been written about the success of this project, so in this chapter we want to tell the story of an earlier, less celebrated (but no less important) initiative to restore GNP. In doing so we offer personal reflections on our time working in GNP as part of the 1996–1999 IUCN/ETC[1] African Development Bank-funded project to restore the park (hereafter referred to as the ETC project). The project was started only a few years after the 1992 Rome Peace Accord, at a time when working in Mozambique was both challenging and potentially dangerous.

In a 2009 National Geographic film, Carlos Lopes Pereira (then Head Ranger at GNP) states that 'if Mozambique loses GNP, it loses its soul' (Nat Geo WILD 2010). We would agree with this sentiment; it is a breathtakingly beautiful landscape. Our account is based on memory and perception; these are personal experiences, and while we make reference to literature, in the tradition of storytelling and following Roland Barthes' suggestion that narrative texts are not one thing but a weaving together of different strands, we can only hope to tell one of many truths about restoring GNP and the cultural importance of lions (*Panthera leo*) to both the park and the surrounding community. In doing so we also offer some thoughts on the management of natural heritage, both during and after armed conflict.

---

[1] International Union for Conservation of Nature, ETC is a not-for-profit organisation whose stated aim is to 'empower people, organizations and institutions and contribute to their sustainable development processes' (ETC 2014).

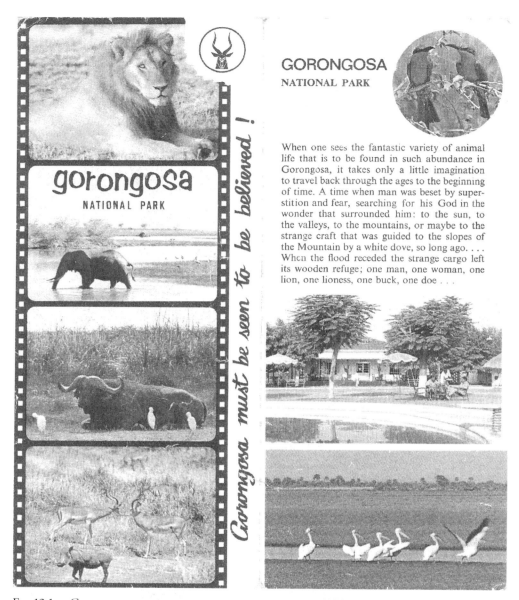

Fig 12.1. Gorongosa tourism leaflet from the early 1970s.

Conflict in Mozambique: a Brief Background

In July 1960 the colonial government of Mozambique declared Gorongosa a National Park. It had been a 1000km² reserve since 1921, further expanded to 3200km² in 1935 and expanded again to 5300km² when National Park status was achieved in 1960. The Independence War (1964–1974) had almost no impact on GNP and its wildlife; most fighting occurred north of the Zambezi River. All was to change, however, with the 1977–1992 civil war, fought between the government forces (FRELIMO) and a rebel group, REMANO.[2] The conflict was largely centred in central Mozambique and particularly around GNP, where RENAMO had its headquarters and main supply bases. As a result of this conflict and its immediate aftermath, park infrastructure, together with much of the large animal population, was decimated. After the war, from 1993 to 1996, commercial hunters added to the losses endured during the armed conflict by hunting the remaining wildlife to supply meat to the provincial capital, Beira, and to other local towns and villages.

As Cahen (1998) notes, 'Mozambique has in many respects suffered from an unfavourable geographical position, bordering on Rhodesia, which attacked guerrilla bases in Mozambique and was destroying Mozambique's infrastructure as early as 1976, and South Africa, which backed RENAMO from 1980 onwards'. Underlying cases of the conflict have been the subject of much controversy and debate, which has tended to polarise around two opposing ideological positions (Baden 1997, 6). The first is that the war in Mozambique was an externally sponsored project of destabilisation against the FRELIMO government. This started as a Rhodesian response against FRELIMO providing sanctuary and training bases for Robert Mugabe's ZANU guerillas, and then continued and expanded under the South African apartheid regime's political strategy of destabilisation of 'Front-line States' and conservative Western (mainly USA) concern about a Marxist-Leninist government providing a role model for other African states. 'There would have been no war and no violent conflict had it not been for South Africa's policy of destabilising its neighbours' (Hanlon 2000, 4). In this view, RENAMO are seen as a puppet force, set up and sustained by external support, with no real political programme or domestic power base (Baden 1997, 6).

An alternative view is that the causes of the war were mainly internal, a product of FRELIMO's failed socialist experiment, particularly their alienation of rural communities and traditional leaders and of forced resettlement, political re-education and imprisonment of dissidents. Most commentators now broadly accept that both internal and external factors were involved (Baden 1997, 7; Abrahamsson and Nilsson 1995).

What is undisputed is that the effects of the civil war (1977–1992) were devastating. One million people died and over five million (one-third of the population) were forced to flee their homes. Damage to Mozambique's infrastructure was estimated to exceed US $20 billion. Most rural shops, schools and health posts were destroyed or forced to close (Hanlon 1995). The war attracted considerable international attention as one of the cruellest contemporary wars. In the absence of a clear frontline, control over civilian populations became a key issue in fighting. Both sides, but particularly RENAMO, pillaged local communities and the war was characterised by attacks on the civilian population and their social and cultural foundations (Nordstrom 1991). Attacks were made on road and rail traffic, with people publicly burned alive in buses to instil fear of travel; attacks were made on schools and clinics (important sources of FRELIMO popularity), to make teachers and nurses afraid to work in rural areas and to make ordinary people afraid to use these facilities. RENAMO burned villages and laid landmines to prevent the

---

2   FRELIMO and RENAMO are acronyms that have become proper nouns. The former stands for *Frente de Libertação de Moçambique*, or Mozambican Liberation Front; the latter is *Resistencia Nacional Moçambicana*, or Mozambican National Resistance.

FIG 12.2.   GORONGOSA PARK ENTRANCE GATE, 1997.

population accessing fields, roads and other resources, creating a legacy Mozambique continues to endure despite two decades of de-mining activity (Nordstrom 1991; Hanlon 2000; Human Rights Watch 1992). As Sideris (2003) indicates, in this sort of warfare, society unravels as social arrangements and relationships that provide people with inner security, a sense of stability and human dignity are broken down. While FRELIMO also committed acts of atrocity, RENAMO operated on a larger scale and were much more systematic in their targeting of civilians (Human Rights Watch 1992).

Unsurprisingly, the war also had a significant impact on natural resources. According to de Soysa and Gleditsch (1999), most post-Cold War armed conflicts have been concentrated in regions heavily dependent on agriculture such as South Asia, parts of Latin America and Central Africa. Indeed, natural resources often constitute the 'prize' for controlling the state, and have played a conspicuous role in the history of armed conflicts (Le Billon 2001; Brück 2000). From 1984 onwards, the Mozambican war became focused on RENAMO's attempts to either capture or destroy the most productive lands or at a minimum to deprive any regular access to agriculturally important areas. Put into context, between independence in 1975 and 1983, the proportion of land under cultivation in Mozambique rose to 10%; by 1989 the war had reduced this figure to only 4%. Forests in particular are often the site of guerrilla warfare, denying local people access to firewood, building materials and food sources (Le Billon 2001; Luckman *et al* 2001). Valuable timber is also often logged to fund fighting. As a local villager remarked during fieldwork: 'During the war many trees were destroyed by timber poachers. The government was too busy with the war and didn't do anything about it' (Field notes 1998).

More than 80% of the country's cattle were lost as a result of the war, and insecurity of travel prevented farmers from selling their produce at market and cut them off from vital inputs

(Luckman *et al* 2001). The almost total loss of domestic stock meant that wildlife resources – from large mammals down to small birds and even insects – became an obvious replacement source of animal protein.

Dudley *et al* (2002) state that historically, the no man's land created by conflict often protects wildlife by limiting human incursions and human population densities within disputed territories. They state, however, that modern wars typically have a detrimental effect on wildlife in less developed countries, with the resulting breakdown in law and order, availability of arms and ammunition and lack of alternatives to bush meat, resulting in large-scale depletion of medium to large mammals.

GNP was particularly badly affected by the conflict. For RENAMO, GNP represented National Government, both in colonial times and as a symbol of the post-colonial government in distant Maputo. In December 1981, the park's headquarters at Chitengo was first attacked by RENAMO. In 1983, due to an increasing number of attacks, the park was abandoned and declared closed to visitors for safety reasons. From 1983 to 1992, the park was the stage of some of the heaviest fighting between FRELIMO and RENAMO, which alternately occupied the park. Pre-war, Chitengo was a sophisticated and extensive camp, with staff and tourist housing, training and research centres, workshops, post office, restaurants and a school. RENAMO spent significant resources destroying the camp, with a distinct focus on buildings that most represented government, ie the school, post office and clinic.

It was during this period that most of the wildlife slaughter took place. Both sides in the conflict used automatic weapons to shoot thousands of head of game. Extensive herds of elephants and hippo were decimated, with reports of RENAMO exchanging ivory for weapons with the South African military. Prior to the war there were around 7000 elephant and 5000 hippo in GNP. An aerial survey in 1997 (the Dutton Survey)[3] estimated both elephant and hippo to number around 20 individuals per species. With the loss of domestic livestock and severely limited access to other foodstuffs, wildlife became a major food resource, which was rapidly and wastefully depleted. As a local chief (known locally as *Régulos*) noted during our fieldwork: 'During the war, everything stopped, it [hunting lodges] was all destroyed, the animals were killed, no you don't see them so often in the forest' (Field notes 1997).

The second wave of large-scale wildlife loss in GNP occurred during the transitional (and at times lawless) years between the signing of the 1992 Peace Accord and the election of a new government in 1994. During this time it was impossible to enforce official rules on natural resource management, and commercial poachers from Beira and surrounding districts were able to access the park to hunt animals for quick profits (Watson *et al* 1999, 12). This period also coincided with a series of droughts, with historically low flow levels recorded in the bordering Pungwe River. This allowed trucks and off-road vehicles easy access, and animals were readily found as drought forced them to concentrate at the few remaining water sources. Poachers (many of whom were ex-combatants waiting to be disarmed and demobilised) used both modern equipment (eg semi-automatic rifles and high power spotlights) and traditional techniques (eg snare

---

3   Paul Dutton is an ecologist with a long-standing connection to GNP prior to the civil war. Though the survey had serious limitations (as is often the case when the pilot has to observe and the rear-seat observer has limited training), it was significant that Dutton was involved as his involvement provided a link back to GNP's golden era.

lines, traps and spears) to hunt. The absence of any game rangers or conservation personnel meant they were able to massively reduce remaining populations of medium to large mammals in a relatively short space of time. The situation was exacerbated by the fact that soldiers on both sides were able to take advantage of the situation.

GETTING THE PARK WORKING AGAIN

Long and Brecke (2003) state that in the aftermath of violent conflict, many facets of life are reconfigured: relations of power, techniques of government, modes of organisation, livelihoods, identities and the relationship between people and places. GNP had also been reconfigured, becoming a protected area without protection, infrastructure or significant wildlife. Walking into Chitengo in 1996 was like walking into a war zone (see Figs 12.2, 12.3 and 12.4). In 1995 the EU funded the IUCN to complete a partial restoration of game rangers' accommodation and staff offices; while this work was a very important first step in getting the park functioning, much of the site felt derelict and evidence of conflict was everywhere (see Figs 12.3 and 12.4). There were bullet marks on most buildings, with graffiti covering many walls, scribbled by combatants of both sides at various stages of the conflict. The authors had both previously worked in conflict areas, including Zimbabwe, Ethiopia and Nagorno-Karabakh, and so were used to seeing evidence of war, but the scale of destruction within GNP was shocking.

The ETC team members were mainly young, white and European, and we became known as 'Zucula's babies', after the Project Manager Paulo Zucula, a former Agriculture Minister in Mozambique, later to become the Transport Minister. There was also inspirational leadership from Roberto Zolho, Warden of Gorongosa National Park. Roberto, a leader in natural resource management in Mozambique, had been in the park as a wildlife student before the civil war started and was resident in Chitengo when the camp was first attacked by RENAMO. Along with several classmates he escaped and walked to the nearest town (Nhamatanda). Roberto was back in GNP working as a government wildlife conservationist when he again came under attack, and he again walked for several days through the bush to safety. Trained in Mozambique and Tanzania, Roberto believed he was part of the 'Gorongosa ecosystem', a feeling shared by his staff and by all who worked with him during this time.

Many of the surviving park rangers who had worked in GNP prior to the war returned to the park during the ETC project. This was incredibly important as they provided continuity, experience and commitment to the project (particularly as the rangers often went without salary for weeks at a time yet continued to work). Moreover, some of the park rangers had fought on opposite sides during the war, yet somehow were able to work together. Our experience was that they integrated well together, despite any formal trauma counselling activities. Ranger training (led by ETC member Brit Reichelt) allowed new and existing game rangers to feel valued and to integrate as conservation professionals. We would sometimes ask how it was possible for them to work as colleagues so soon after the peace agreement, to which they would frequently shrug and say 'things happened in the war and the war is finished'. As we reopened the road network, old rangers and old foes discussed war stories and spoke of attacks, ambushes and skirmishes. There was no obvious sense of malice, just a general sadness and acknowledgment that the war had destroyed so much – communities, park and wildlife. We say this not to underestimate the personal trauma and pain many of these rangers doubtless felt, but we witnessed a deeply humbling sense of commitment from park staff to get the park working again.

FIG 12.3.   PUBLICITY MATERIAL (PRODUCED BY THE SAFRIQUE SAFARI COMPANY) SHOWING THE CHITENGO RESTAURANT AND POOL IN THE 1970S.

FIG 12.4.   GORONGOSA SAFARI TOUR BUS, 1997, WITH DESTROYED TOURIST VILLA BEHIND.

One ranger (Perera) would often speak of his role in the war, how he was responsible for infil-trating enemy units, gaining their trust and eventually leading in government forces to capture or kill them. He was one of the best-regarded people by all the staff, and knew every square metre of Gorongosa. Both during the war and in the immediate post-war period, as an experienced hunter Perera led poaching raids in Gorongosa. Recruited as a ranger by the IUCN programme when they returned to the park in 1995, he became a key member for anti-poaching and intelligence gathering and guided the team that reopened the road network, as he knew where all the road markers were.

The ETC project was part of a broader opportunity to reintroduce community and govern-ment control in a post-conflict landscape. It was inclusive, which was at the time novel for resource management projects in Mozambique. It recognised the importance of active participa-tion by local communities; indeed, participation was a necessity as there was no money or desire to build a retaining fence. The only practical option was to negotiate a human fence. Doing so would require a participatory project that was economically, as well as ecologically, sustainable. An example of this approach was a project developed with the Nhambita community, which borders the park to its south.

Nhambita was established as a relocation area for people who were removed from the park in 1948. In keeping with the 'fences and fines' approach to protected areas that dominated at the time, human settlements were considered by the Portuguese government to conflict with wildlife management. Though many people remained in Nhambita *Régulado*[4] during the war, there had still been internal population movement to safe places within the forest and significant suffering occurred within the community. There, resentment remained (among the community elders) that when GNP was established in 1948, local people were deprived of hunting rights and a depend-able source of animal protein. Together with Debra Howell (an ETC staffer and social scientist), we spent much time trying to understand resource management in Nhambita and in particular how the community interacted with GNP (Convery *et al* 2000). Understandably there was some suspicion regarding our work: 'why do you ask these questions, are you from FRELIMO?' There was also an enormous amount of goodwill and recognition that getting the park working again was an important part of the post-war reconstruction process for the community.

Lions: Physical and Spiritual

Bertelsen (2003) argues that the war changed many traditional beliefs, either through violence or politics. He cites the example of the wild goats that grazed Cabeça do Velho, a mountain outside Chimoio (a small town located further up the Beira corridor). Locally, it was believed that the goats were related to ancestral spirits, and were not eaten by local people apart from on ritual occasions. 'In communal memory the goats were slaughtered and killed by Frelimo soldiers during the war; for many people this epitomised Frelimo's perceived antagonism towards tradition' (ibid, 273).

Similarly, the fate of the Gorongosa lion is bound up in tradition, warfare and politics. Elsewhere we have written about the significance of lion spirits and local traditions in Central Mozambique (Convery 2006). Bertelsen (2003, 274) also highlights the importance of lion spirits

---

4   A village or grouping of dwellings.

and quotes a *chirenge* (rainmaker) who told him 'the *chitengwa* (traditional lion of Gorongosa) lives there in Gorongosa. Oh yes, it lives there! And it has a lot of power.'

It was therefore significant that a lion was retained as the emblem for GNP, as the heritage and culture of both the park and the surrounding communities is deeply connected to lions and lion spirits. In Nhambita, lion spirits play a particularly important role in relation to governance and resource use. This became apparent on numerous occasions during fieldwork in the community. For example, on the way back from some fieldwork accompanying two women from Nhambita to the health post, one of the women became increasingly anxious to return before sunset so that they would not have to walk through the forest at night:

> It is better to get back before sunset, that way we will not offend the ancestors. ... If we have to walk through the forest at night it is important to clap hands all the way through the forest until we reach home ... they might undo all the work from the health post and we would have to return. (Field notes 1997)

When asked how walking through the forest might offend the spirits, she stated that it was the lion spirit she was particularly concerned about, but there might also be other 'evil spirits or unhappy ancestors, and it was not good to be there at night, particularly without any "offerings"'.

The *Régulo* in particular appeared to follow a rigid series of conventions while in the forest. When collecting firewood he would clap his hands all the way, to 'ward off bad spirits' and allow the 'good spirits to help him find firewood'. As we walked through the forest he discussed various aspects of forest law:

> When someone commits a crime in the forest, the spirits (in the form of a lion) will seek vengeance. If someone mates in the forest, the lion may come and take grass from his roof during the night while he sleeps ... the lion will not bite that person, only scare them.
>
> (Field notes 1997)

In such scenarios, the *Régulo* should perform a ceremony involving hand-clapping in order that the lion does not visit that person again. After the ceremony, 'the lion will roar to show that things are back to normal' (Field notes 15 August 1997). The *Régulo* described the spirit lion as follows:

> The lion is a spirit lion who lives nearby [Gorongosa National Park]. The lion has 'colours around its neck, it is lucky' and will only come out if there are problems; there are different kinds of lion, bad and good spirits, but the good lion protects ... long ago, when the [good spirit] lion used to attack and kill buffalo, it would push the carcass close to families who needed it. (Field notes 1997)

The *Régulo*'s ancestral name was Chicari, a named shared by the first *Régulo* of Nhambita, meaning 'lion' in the Sena language. Our fieldwork highlighted the importance of the lion 'spirit guardian' in Nhambita, as a source of protection, justice and healing: 'When somebody commits an offence, the lion may make that person "sick", make their spirit sick. This can be a great problem to cure – it needs the work of both the curandeiro [local healer] and me' [*Régulo*] (Convery 2006).

Fig 12.5.   Chief Chitengo in Gorongosa, 1998.

Similarly, the Chief and *Régulo* of Chitengo village was, according to local tradition, a lion made human, who would return to a lion spirit on death (see Fig 12.5). One of his predecessors (also Chief Chitengo), who ruled during the early 1900s, was believed to be a magician who transformed himself into a white lion when he died. His story is told on the GNP website (GNP 2014):

> long ago, they knew the secret of how men become lions when they die. But the new generation does not know. The tradition of transforming into a lion had its own secrets, and the 'nyakwawas', or chiefs of Mbire, knew it. A 'nyakwawa' is according to lineage. When someone dies, it is his son's turn, and then his son's.

During our time in Gorongosa, Chief Chitengo would tell us that he possessed both the physical and moral strength of a lion. He saw lions as part of the community and said that a lion would never hurt him as they were related. On visits with the Chief to the Urema flood plain area of GNP, formally prime lion area, he explained that the lions would watch out for his people, to make sure unmarried couples did not have sex outside (he said any sexual activity needed to take place under a roof). In this way lion spirit beliefs function as a mechanism for maintaining village social and moral structures (Convery 2006), with the traditional lion roaming the woods as a manifestation of the ancestor spirits and a moral guardian (Bertelsen 2003).

Lions clearly have huge significance for local communities, both as a threat to life and livestock, but also as a part of spiritual traditions associated with animalism and ancestor rituals. In

GNP, lions became an umbrella species, the conservation of which enhanced the overall richness of the protected area. During the 1950s and 1960s, lions were abundant in GNP (with possibly the highest concentrations in Africa) owing to prey density on the Urema floodplains. Given their abundance, charisma and totemic importance for humans (both locals and visitors), the lion became the emblem for GNP.

The mythology of Gorongosa lions, at least as far as the park is concerned, dates back to the 1940s, when GNP was still only a reserve. A new headquarters was built on the floodplain close to the Mussicadzi River, but shortly after completion the site had to be abandoned owing to heavy flooding. What happened next helped to create a major tourist attraction. A pride of lions moved into the empty building and over time this became a major tourist draw. The houses had spiral staircases to the roofs and whole prides would use this excellent location to look for prey across the floodplain; the old house subsequently became known as *Casa dos Leões*: the lion house.

In this way, both the spiritual lion and the physical lions of Gorongosa represent a convergence between local cosmology, conservation, national territory and, as Bertelsen (2003, 274) puts it, political rhetorics of tradition. The decision to rebrand GNP with a lion acknowledged the traditional values of both the park and the surrounding communities.

LIONS AND OTHER ANIMALS

While the spiritual lion survived the conflict, the physical lions of Gorongosa were largely missing during the ETC project. The Dutton Survey estimated numbers to be very low if not extirpated, and re-establishing numbers of wild animal species within GNP and adjacent areas was an important consideration during the early stages of the ETC project. Increasing wildlife numbers either through translocation or breeding programmes provides a major tourism boost. Survey and monitoring programmes were established, and the local IUCN office, which had been instrumental in the early 1993–1994 restoration process, worked alongside ETC in this survey work. As expected, the Dutton Survey found very few large ungulates, though small numbers of buffalo, wildebeest and hartebeest were observed during the survey. Waterbuck, impala, oribi and warthog had however shown considerable increases since a previous aerial survey, in 1993. Perhaps most significantly, elephant and hippopotamus, both 'ecosystem engineers' who can change, maintain or modify their habitat and influence the availability of resources to other organisms (Jones *et al* 1997; Soulé *et al* 2003), were at their lowest historically known numbers.

The low numbers of animals was a concern. From an economic perspective, low animal numbers, combined with GNP's remote location, meant a return to previous levels of tourism was impossible. Ecologically the loss of large bulk grazers such as hippo and buffalo, and of mixed feeders such as elephant, had an impact on vegetation, leading to the expansion of woodlands (especially fever-tree Acacia (*Acacia xanthophloea*)), which increased dramatically on the floodplain edge. Floodplain drainage was affected by the loss of hippos, as these animals had in the past kept channels open and formed new channels.

The reintroduction of animals was suggested, but at the time it was felt that until poaching was under very tight control this would be, according to Zolho, 'stocking someone else's larder'. Given serious constraints on financial and human capital, relocation was not possible; we simply did not have a clear enough understanding of what wildlife remained. However, once protection measures started to take effect, we observed a rapid increase in species such as sable, impala, waterbuck and warthog, and there were occasional (but increasing) observations of elephant, hippo and lion. This was borne out by research in the later phases of the Gorongosa restoration under the Carr Foundation.

CONCLUSIONS

The ETC project was highly ambitious. The project helped to reduce the continued illegal over-exploitation of natural resources and identified that some of the worst offenders were not local communities, but rather senior local government figures. Basic tourist accommodation and camping facilities were re-established and visitors started to return to the park; although numbers were very small during the first couple of years (visitors had to be totally self-supporting, including food and fuel), most became very effective ambassadors for the park. The project also reintroduced the 'rule of law', but law in a new format: more inclusive, less confrontational. As a result there was considerable buy-in from local communities, and the rebranded GNP lion emblem reflected a shared heritage, both physical and spiritual.

There were also considerable failings, particularly associated with the adequate and timely provision of financial resources and equipment, as well as the misdirection of funds and material to the capital, Maputo, rather than to the field. Nevertheless, local, provincial and national government institutions started to work together on natural resource issues, and Mozambican managers and technical staff were able to rebuild, or in most cases create, frameworks that had never previously existed. Perhaps the most important outcome was that the community saw that people at governmental, societal and NGO level were interested in partnering with them, with an understanding from all sides that 'their' natural heritage could help them to rebuild Mozambique, and reignite pride in themselves as custodians of Mozambique's natural heritage, and its 'soul'.

BIBLIOGRAPHY AND REFERENCES

Abrahamsson, H, and Nilsson, A, 1995 *Mozambique: The Troubled Transition*, Zed Books, London

Baden, S, 1997 *Post-conflict Mozambique: Women's special situation, Population Issues and Gender Perspectives*, Bridge Report No 44, Institute of Development Studies, Brighton

Bertelsen, B E, 2003 The Traditional Lion is Dead: The Ambivalent Presence of Tradition and the Relation between Politics and Violence in Mozambique, *Lusotopie* 10, 263–81

Brück, T, 2000 *Macroeconomic Effects of the War in Mozambique*, Working Paper Number 11 (QEH Working Paper Series, QEHWPS11), University of Oxford International Development Centre

Cahen, M, 1998 *Nationalism and Ethnicities: Lessons from Mozambique*, Contemporary Portuguese Politics and History Research Centre, Lisbon, Portugal

Convery, I, 2006 Lifescapes and Governance: The Régulo System in Central Mozambique, *Journal of African Political Economy* 109, 449–66

Convery, I T, Howell, D, Schwarz, A, Zolho, R, and Zucula, P, 2000 *Combating Environmental Degradation Around a National Park: A Human Fence in Mozambique*, University of Northumbria, Division of Geography and Environmental Management Departmental Occasional Papers, New Series No 33 de Soysa, I, and Gleditsh, P, 1999 *To cultivate peace: agriculture in a world of conflict*, International Peace Research Institute, Oslo

Dudley, J P, Ginsberg, J R, Plumtre, A J, Hart, J A, and Campos, L C, 2002 Effects of War and Civil Strife on Wildlife and Wildlife Habitats, *Conservation Biology* 16, 319–29

ETC, 2014 *About ETC* [online], available from: http://www.etc-international.org/about-2/about-etc/ [14 May 2014]

Gorongosa National Park (GNP), 2014 *Gorongosa's Lions* [online], available from: http://www.gorongosa. org/explore-park/wildlife/lions-gorongosa [2 April 2014]

Hanes, S, 2007 Greg Carr's Big Game, *Smithsonian*, May, 84–92

Hanlon, J, 1995 *Supporting peasants in their fight to defend their land*, Christian Aid, Maputo, Mozambique

— 2000 *Mozambique: Will Growing Economic Divisions Provoke Violence in Mozambique?*, Swiss Peace Foundation, Institute for Conflict Resolution and The Federal Department of Foreign Affairs

Human Rights Watch, 1992 *Conspicuous Destruction: War, Famine and the Reform Process in Mozambique*, Human Rights Watch, New York

Jones, C G, Lawton, J H, and Shachak, M, 1997 Positive and Negative Effects of Organisms as Physical Ecosystem Engineers, *Ecology* 78, 1946–57

Le Billon, P, 2001 The political ecology of war: natural resources and armed conflicts, *Political Geography* 20, 561–84

Long, W, and Brecke, P, 2003 *War and Reconciliation: Reason and Emotion in Conflict Resolution*, MIT Press, Cambridge MA

Luckman, R, Ahmed, I, Muggah, R, and White, S, 2001 *Conflict and poverty in Sub-Saharan Africa: an assessment of the issues and evidence*, IDS Working Paper 128

Nat Geo WILD, 2010 *Africa's Lost Eden*, premiered on National Geographic Wild channel (USA), 12 April

Nordstrom, C, 1991 Women and war: Observations from the field, *Minerva: Quarterly Report on Women and the Military* 9, 1–15

Sideris, T, 2003 War, gender and culture: Mozambican women refugees, *Social Science and Medicine* 56 (4), 713–24

Soulé, M E, Estes, J A, Berger, J, and Del Rio, C M, 2003 Ecological Effectiveness: Conservation Goals for Interactive Species, *Conservation Biology* 17, 1238–50

Watson, E, Black, R, and Harrison, E, 1999 *Reconstruction of natural resources management institutions in post-conflict situations*, Working Paper No 1, School of African and Asian Studies, University of Sussex and Centro de Experimentação Florestal, Sussundenga, Mozambique

13

The Last Night of a Small Town: Child Narratives and the *Titanic*

John Welshman

Introduction

The story of the *Titanic* – its construction, its sinking on 14 April 1912 and the discovery of the wreck in 1985 – is one whose worldwide appeal was amply demonstrated by the flood of new books and television documentaries on the 100th anniversary in 2012. This chapter engages with the theme of displaced heritage by reflecting on the process of writing my own book on the subject and by exploring what we might term 're-narrated heritage' (Welshman 2012). One of my original premises was the belief that in the recent emphasis on myth, and on the causes of the disaster, the focus on individual passengers and crew members, and their stories, had been lost (Howells 1999). What seemed to me most interesting were the personal narratives of both passengers and crew, and the light that the human detail of their stories sheds on the worlds they came from. The book was conceived as a trade book, a popular narrative that depicts the disaster through the personal stories of 12 people. I chose my 12 as a cross section of those on board my 'small town' (Lord 2005, 153). In that respect the 2012 book's approach is similar to that of my previous book on the experiences of wartime evacuees (Welshman 2010). The book covers the stories of passengers and crew and reconstructs individual experiences, as well as surveying broader social changes. I felt the small town metaphor was helpful in conveying the sense that all aspects of society were on the ship: rich and poor, male and female, old and young, generous and selfish. The book seeks to rebalance the passengers and crew depicted in Walter Lord's book of 1955; employs a 'life history' approach to uncover the lives of these people before and after the disaster; and aims not just to offer a minute-by-minute retelling of events, but also to explore key themes – the construction of the ship; migration; nationality and place; the wider histories of technology and radio; age and gender; and social class.

The book thus weaves the stories of 12 passengers and crew into the narrative. These narratives can be grouped into three types: those published shortly after the disaster, in the interwar period, and much more recently. Moreover, they present a range of challenges to the historian who attempts to use them. The book and this chapter feature the stories of three people who were children or teenagers at the time of the disaster: Frank Goldsmith, Eva Hart and Edith Brown. Few of the many books about the *Titanic* have dealt with the 109 children on board (Geller 1998). The approach in this chapter is that of a social historian reflecting on the methodological challenges involved in weaving three autobiographical or biographical accounts of childhood into a popular narrative. It is not primarily about textual analysis but it does seek to trace the contours of what has been recently called 'creative non-fiction', and to explore the relationship between history, memory, biography and fiction (Bainbridge 1997). The chapter thus reflects on the construction of this *Titanic* 'archive' of survivor accounts or eye-witness testimonies, the more recent of which constitute re-narrated heritage. It relates these to the book's theme of displaced heritage, arguing that these three representations of the *Titanic* disaster reveal not the displacement or loss of people and cultural heritage, but their reconstitution and transmission through processes of individual and collective memory.

FRANK GOLDSMITH

Frank Goldsmith (aged 9) travelled Third Class on the *Titanic* with his parents Edith and Frank, and their friends Thomas Theobald and Alfred Rush. The family were from Strood in Kent, England, and were migrating to live in Detroit. Frank and his mother survived but his father, Thomas Theobald, and Alfred Rush, drowned. Frank died in January 1982, and his account was published after his death by the Titanic Historical Society. Frank describes his book as his 'autobiography and account', and he said he wrote it for his sons, grandchildren, and great-granddaughter (Goldsmith 1991, 31). Frank argues that his recollections are 'as sharp as if that event only happened yesterday' (ibid, 33). Frank's later life is related by his wife Victoria in a Preface (ibid, 19–30), which adds a further dimension to the book as a historical source.

The book opens with one of Frank's earliest memories, a Christmas Day get-together held after his fifth birthday. He describes his excitement on hearing that the family were booked on the *Titanic*; the journey that the family made from their home to Southampton; and the early days of the voyage. He recalls waking on the night of Sunday 14 April to find that the ship had collided with the iceberg, and his excitement to find that they were going to be allowed to get into the lifeboats. When the family arrived on the Boat Deck, they found that only women and children were being allowed through to the lifeboats, and Frank's father hugged him, saying 'So long, Frankie. I'll see you later' (ibid, 46). Thomas Theobald takes off his wedding ring, and gives it to Frank's mother, saying, 'If I don't see you in New York, will you see that my wife gets this?' (ibid, 50). Frank and his mother are in the Engelhardt D lifeboat, and his mother holds his head so that he cannot watch the *Titanic* sinking. On the *Carpathia* rescue ship, Frank is befriended by Samuel Collins, a fireman and fellow survivor, and on arrival in New York he and his mother are looked after by the Salvation Army. Frank and his mother are helped by the Titanic Relief Committee. On the one hand, Frank's account effectively ends with the arrival of the *Carpathia* in New York. Much of the rest of the book consists of photographs and transcribed documents, particularly letters from the Salvation Army and other *Titanic* survivors (ibid, 64–128). On the other hand, there are incidents that Frank does not include in his account, including his being taken by Eva Cowan, the 12-year-old daughter of Salvation Army workers, to the Siegel-Cooper Department Store on Sixth Avenue in New York, riding on the escalator and being bought an ice cream soda. We learn this from the letters and newspaper articles included in the rest of the book (ibid, 73).

As in the two other accounts of younger survivors, which follow below in this chapter, Frank's story has been filled out by later research undertaken by himself. Chapter Two, for instance, has a lengthy section on the construction of the *Titanic*, Captain Edward Smith, the number of lifejackets and lifeboats that she carried, and the passengers and crew. In checking the time that he and his mother climbed a steel ladder located to the rear of the ship's fourth funnel, Frank notes that 'timewise, according to official investigation records, it was 1:40am' (ibid, 51). Similarly when the two got to the lifeboat, Frank writes that 'unknown to me at the time, of course, D was the last lifeboat to be lowered' (ibid, 51). When Frank and his mother are in the lifeboat, they spot the lights of what looked like another ship, and Frank notes 'there has been much conjecture as to the identity of this vessel… at that level, it had to be some ship other than the *SS Californian*, which was probably twice as far away, and lying at rest because of being surrounded by dangerous ice floes' (ibid, 55–6). He notes that 'according to the official records', the first lifeboat had been picked up by the *Carpathia* at around 4am, and there was none of the early panic 'which we read about' (ibid, 50). It was 'later' that Frank learned from his mother of Thomas Theobald giving her the wedding ring for safekeeping. It was a visit in 1973 to another survivor, Edwina Troutt, that confirmed that Frank's mother had used her skills as a dressmaker to make emergency clothes onboard the *Carpathia*.

It should also be noted that Frank's account is not linear, in that he occasionally jumps ahead to the sinking of the *Titanic* and rescue by the *Carpathia*, and similarly goes back to earlier child-hood experiences. He does not try to create suspense. For example, at the moment of getting into the lifeboats, Frank goes back to narrate how the Goldsmiths came to be going to Detroit, how his aunt had been living there and urged the family to join her. As in many other accounts, Frank claims that people predicted the disaster (ibid, 44). Nevertheless, Frank does manage to convey the excitement that he felt as a nine-year-old: trying to climb onto a baggage crane; watching the stokers and firemen in the boiler rooms; his respect for their travelling companion Alfred Rush (aged 16) who opted to stay with the men and perished; the lowering of the lifeboats; and how he was befriended on the *Carpathia* by Samuel Collins, the fireman and fellow survivor from the *Titanic* (ibid, 39, 50, 52).

EVA HART

Eva Hart (aged 7) travelled Second Class on the *Titanic* with her parents Esther and Benjamin. The family were from Seven Kings, east of London, and were migrating to start a new life in Winnipeg, Canada, where Eva's father planned to work in the construction industry. Eva and her mother survived, but her father Benjamin was drowned. Eva's book was first published in 1994, based largely on transcripts of tape-recorded interviews between Eva and her friend Ronald Denney. Eva claimed that as she grew up and during her adolescence, she tried to forget about the *Titanic* disaster. But she also admits that over the years she has frequently told her story, at meetings, on radio and television and in the newspapers (Denney and Hart 1994, 13). Compared to Frank Goldsmith's account, Eva's is more professionally produced, with a List of Illustrations, a Preface, 12 Chapters, an Epilogue, an Appendix, a Bibliography and an Index.

In the book, Eva describes her early life in the London suburb of Seven Kings; her father's decision to emigrate; and the booking of their passage on the *Titanic*. She describes the early days of the voyage; her parents' response to the collision; and their journey up to the Boat Deck (ibid, 40). From Lifeboat 14, Eva and her mother watch the *Titanic* sink, and they are eventually rescued by the *Carpathia*. Much of the book is concerned with Eva's claim that her mother had a premonition that the *Titanic* was going to sink. Eva and her mother only stayed in New York for a short time and returned to England in April 1912. They received help from the Titanic Disaster Fund. The rest of the book covers Eva's later life and her growing involvement with the new-found interest in the *Titanic* story following the release of the 1958 film 'A Night to Remember', including newspaper interviews and radio and television broadcasts.

It is clear that, like Frank Goldsmith's account, the book is informed by much further research concerning the ship, which breaks up Eva's story (ibid, 186–7). Eva admits of the departure from Southampton that she was totally unaware that they were in the company of some of the wealthiest men in the world. And, like Frank Goldsmith, she fills in details of the construction of the ship, its equipment and crew, and other details. She notes of the cabins that 'photographs taken at the time show the first class cabins with brass bedsteads, marble washstands, armchairs and sofas and with fans in the ceilings' (ibid, 23). At another point she notes that the passengers did not know that the *Titanic* had been receiving radio reports of icebergs on Sunday 14 April 1912, and elsewhere 'what none of us knew at this stage was that the iceberg had torn a narrow gash some three hundred feet long below the water-line on the starboard side of the ship' (ibid, 33, 35). The book has occasional footnotes giving other factual details (ibid, 25, 37, 39, 50–1).

Eva's account starts with the sinking of the ship, before going back to retrace her early life in Seven Kings. She therefore gives away the story of the disaster early on (ibid, 11). She interrupts her account of her friend Nina Harper to note that while Nina's father was drowned, Nina and her aunt survived, and Eva and Nina re-established contact 65 years later. Eva notes of the dog she befriended on board that it was only in 1955, when she read Walter Lord's book *A Night to Remember*, that she realised it was a French Bulldog (ibid, 26). Moreover Eva is keen to enter into some of the key debates about the disaster – over the position of the *Californian*; the question of the last tune to be played by the band; the actual sinking of the ship and whether it broke in two; and over the treatment of the survivors by the various disaster appeals (ibid, 38–9, 41–2, 45; Ballard 1987). This is a case of Eva projecting her later interests as an adult onto her younger self, and is an example of re-narrated heritage; at the time of the Inquiries, Eva would still have been only seven.

One of the main themes in Eva's account is her claim that her mother had a premonition of disaster (Denney and Hart 1994, 17, 19). Nevertheless a letter that Esther Hart wrote, on Sunday 14 April, does not mention the premonition (ibid, 30–1). This raises the question of what Eva did actually experience some 82 years before the publication of her account. Certainly her account of the sinking is remarkably similar to that of Second Class survivor Lawrence Beesley (Beesley 1960, 35, 47; Denney and Hart 1994, 36, 41). Eva does not try particularly hard to portray herself as a child. The book is an absurdly complete account from someone who was seven years old in 1912 (ibid, 49). But, like Frank Goldsmith, Eva does manage, even though her account was published 82 years after the disaster (when she was 89), to recover herself as a girl of seven. She writes, for instance, about the dog she befriended; playing with her friend Nina Harper; her favourite doll; and the terror of becoming separated from her mother (ibid, 26, 31, 45).

EDITH BROWN

Edith Brown (aged 15) was the eldest of our three young survivors represented in this chapter. She was born in South Africa and was travelling with her parents Elizabeth and Thomas; the family were emigrating to Seattle, where Thomas planned to open a hotel. They were travelling Second Class. Edith and her mother survived, but her father Thomas drowned. Edith Haisman, as she then was, died in January 1997. Unlike the accounts of Frank and Eva, this is a biography by her son David Haisman rather than an autobiography, and is narrated in the third person rather than the first. David writes that 'throughout her life, the author's mother had told her family vivid stories of the fateful night of 14 April 1912' (Haisman 2009, 144). Much of the book is informed by the experiences of David Haisman in the Merchant Navy. The claim that the *Titanic* did not have enough professional seamen is repeated at least three times. Haisman writes of his mother's life story that 'as a family, we always believed what our mother told us over the years. She had no reason to tell us anything other than the truth as she remembered it' (ibid, xvii).

The book opens with the collision with the iceberg, but also offers different perspectives, from crew members and others. Chapter Two then rewinds to describe the early lives of Elizabeth and Thomas Brown in South Africa, their journey to England and stay in London, and their journey to Southampton. The next five chapters describe the first five days on the *Titanic*, 10–14 April 1912, and the collision with the iceberg. Chapter Eight then resumes the narrative begun in Chapter One, with Edith and Elizabeth being parted from Thomas and getting into the lifeboat. As their lifeboat descended, Thomas said to them 'I'll see you in New York' (ibid, 65). Haisman describes the rescue of Edith and Elizabeth by the *Carpathia*; their arrival in New York; their journey to Seattle; interviews they gave to newspapers; and their eventual voyage back to South Africa. Edith married Frederick Haisman in Johannesburg in June 1917. After he died in November 1977, Edith attended various *Titanic* conventions in the 1980s, even going on a cruise to the wreck site in 1985.

As in the other two accounts, this account opens with the collision with the iceberg, but compared to them it is the most sophisticated, shifting as it does across alternative viewpoints. Although David Haisman does not cite sources or provide a bibliography, it is clear that it is informed by much research; for instance, he claims he had met Lookout Frederick Fleet in the course of his career in Southampton. There are detailed descriptions of the departure of the ship from Southampton; the work of the Captain; and the near-collision with the liner the *New York*. In the course of the early days of the voyage, the Browns meet, or are aware of, many of the famous passengers on board. This means that there is much that David Haisman has to reconstruct. While the Browns travel in Second Class, much of the action takes place in First. On Sunday 14 April, for instance, he includes the radio messages about icebergs received by the *Titanic*. While Thomas Andrews, the ship's designer, and Captain Edward Smith realised that the ship was badly damaged, 'this damage would always be speculated upon but never fully known' (ibid, 9). After the arrival of the *Carpathia* in New York, he includes extracts from local newspapers and from interviews that Edith and Elizabeth Brown gave. He also summarises the official American and British Inquiries (ibid, 84–7, 92–4).

For reasons that remain unclear, Haisman exaggerates the social standing of Thomas Brown (ibid, 24, 31, 43). Moreover, the book is full of reconstructed conversations, many of which Edith Brown could not have experienced, even if she had been able to remember them (ibid, 25, 28, 33). Certainly the least likely of the conversations is the one that the Browns allegedly had onboard the ship with Captain Edward Smith (ibid, 57). The account of Thomas Andrews and Smith discussing the damage to the ship (when of course none of the Browns was present) is reminiscent of similar scenes in the films 'A Night to Remember' (1958) and 'Titanic' (1997). Here and elsewhere, Haisman is clearly influenced by sensory imagery – both sights and sounds. For example, when the lifeboats are lowered, Fifth Officer Harold Lowe fires his pistol, saying to men trying to get in, 'try that again and it won't be warning shots I shall be firing, but shots aimed at anyone trying to get into this boat!' (Haisman 2009, 65). Possibly the fact that Edith Brown was 15 years old in 1912 makes her account more reliable that those of Goldsmith and Hart. However, as with Hart, some anecdotes are copied directly from Lawrence Beesley's earlier account, notably the tilting of the ship as it took on water, and the description of its sinking (ibid, 7, 42, 68; Beesley 1960, 30, 47). Edith Brown's story is similar to that of Eva Hart in that there are premonitions of the disaster, notably on the part of Thomas Brown, but as retold by her son it is much more self-conscious as an example of re-narrated heritage (Haisman 2009, 25, 28–9).

As with Goldsmith and Hart, David Haisman tries hard to recapture the outlook of his mother as an excited 15-year-old girl. Yet while Haisman reconstructs his mother's experiences quite imaginatively, the account is over-written. This is particularly true in the lengthy descriptions of the ship, with its 'plush' carpeting and 'beautiful' wood panelling, its 'crisp' white sheets, 'fluffy' white towels and 'neat' little curtains (ibid, 36). Even given quite extensive research, Haisman includes details not found in any other survivors' accounts. Another example is where he writes that, as dawn breaks on the lifeboats, 'there were bodies everywhere, some with life jackets on, floating silently in an upright position. Among the corpses were many women and children, frozen to death and still clinging to each other as they drifted amongst the ice' (ibid, 69–70). Instead, it is worth noting that very similar scenes are depicted in the film 'Titanic' (1997).

Conclusion

There are both similarities and differences between the three narratives examined in this chapter. Whereas Goldsmith's is an autobiographical account of the *Titanic* disaster, published after his death, that by Hart is really her life story, assisted by a friend, while the story of Brown is a posthumous biographical account narrated by her son. Each account is mediated by the author or editor, and each brings their own perspective to it. Of the three, it is Goldsmith's that seems the most naturalistic, and least stylised, and the Brown account the most sophisticated, in the way that it attempts to cut between different perspectives. Yet, conversely, it is the Brown account that appears the least reliable, notably with its reconstructed conversations that cannot be checked against any other sources. What motivated the authors to write and publish the accounts – a simple impulse to tell their own, or a family member's, story; a sense of injustice at the causes of the disaster or the treatment of survivors; or a desire to profit commercially, particularly following the success of the James Cameron film in 1997?

Yet the similarities between the three are also striking. Each is informed by additional research by the author. All three make claims to their subjects predicting the disaster to different extents, and each tends to repeat myths such as the *Titanic* being 'unsinkable'. Each seeks to recover the perspective of their subject as a child, with varying degrees of success, and each follows a non-linear structure, either jumping ahead to the disaster and its consequences, or going back in time to describe some earlier experience. Each makes a claim to offer an original account while at the same time clearly borrowing from earlier accounts. And all three are examples, less of displaced heritage, than of re-narrated heritage.

Why draw on sources that are so problematic, particularly those written many years after the disaster in question? It is when we compare the three texts to a real letter, written by the 11-year-old Eileen Lenox-Conyngham to her nursery maid, Louisa Sterling, from onboard the *Titanic* on 10 April 1912, that their flaws as historical sources become most apparent (Lenox-Conyngham 1912). Eileen's letter recounts images and experiences through the eyes of a child at the time, rather than reconstructed by an adult many years later. However, the three accounts are similar because, despite the challenges that they present to the historian, they do still offer vivid perspectives as eye-witness accounts, or the retelling of stories from an eye-witness. It is the occasional but revealing human detail of these stories that provide a window on to the worlds from which these people came. Thus if the three texts considered here have any value as historical sources, it must be less in terms of any supposed empirical accuracy and more in revealing how three different authors or narrators have set about reconstructing experiences, creating in the process a complex blend of memory, experience, research and myth. In that respect 'creative non-fiction', or re-narrated heritage, seems a suitable description of them.

Overall, rather than a loss of self or loss of place, *Titanic* survivor narratives in general reveal the steady accretion of heritage over time, from the publication of eye-witness stories immediately after the disaster in 1912, through the memoirs of crew members published in the 1930s, to the emergence of the most recent accounts (often written with the help of children and grandchildren). The *Titanic* has become a site of memories. In turn, the survivor accounts have become an archaeology of memories, where heritage is not displaced or lost but is continually rediscovered, reconstituted and transmitted in response to an apparently insatiable worldwide demand for 'new' survivor narratives.

Bibliography and References

Bainbridge, B, 1997 (1996) *Every Man For Himself*, Abacus, London

Ballard, R D, 1987 *The Discovery of the Titanic*, Hodder and Stoughton, London

Beesley, L, 1960 (1912) The Loss of the S S Titanic: Its Story and Its Lessons, in *The Story of the Titanic: As Told by its Survivors* (ed J Winocour), Dover Publications, New York, 1–109

Denney, R C, and Hart, E M, 1994 *Shadow of the Titanic: A Survivor's Story*, Greenwich University Press, Dartford

Geller, J B, 1998 *Titanic: Women and Children First*, Patrick Stephens, Sparkford

Goldsmith, F J W, 1991 *Echoes in the Night: Memories of a Titanic Survivor*, Titanic Historical Society, Indian Orchard MA

Haisman, D, 2009 *Titanic: The Edith Brown Story*, Authorwise, Milton Keynes

Howells, R, 1999 *The Myth of the Titanic*, Macmillan, Basingstoke

Lenox-Conyngham, E, 1912, Letter to L Sterling, 10 April, Springhill, Moneymore, Northern Ireland

Lord, W, 2005 (1955) *A Night to Remember*, Henry Holt, New York

Welshman, J, 2010 *Churchill's Children: The Evacuee Experience in Wartime Britain*, Oxford University Press, Oxford

— 2012 *Titanic: The Last Night of a Small Town*, Oxford University Press, Oxford

14

Troubled 'Homecoming': Journey to a Foreign yet Familiar Land

Aron Mazel

Introduction

In August 2011, my daughter Nicola and I travelled to Lithuania to commemorate the memory of my paternal grandparents, Mashe and Mordechai Mazel, and other family members who were murdered on Saturday 23 August 1941, and to say kaddish[1] for them. On that day in 1941, the Nazis and their Lithuanian collaborators killed 7523 people in the Pajouste forest, around eight kilometres east of Panevėžys. Panevėžys experienced six recorded episodes of killing between 21 July and 23 August 1941 (Anon 2005, 145).[2] Altogether, the Nazis and Lithuanians killed 8837 people in Panevėžys, 99% of whom were Jews (see Table 14.1; Levinson (ed) 2006, 146–8). When I planned the trip it was unknown to me that the Pajouste forest was also the death site for my maternal relatives from the village of Ramygala, some 30 kilometres south of Panevėžys. The Jews of Ramygala had in fact been removed to the Panevėžys ghetto in mid-July 1941 (Yahadut Lita vol 4, cited in Levinson (ed) 2006, 329–31). This realisation added additional poignancy to the visit. I have since learned, however, that my maternal great-grandmother, Sarah Nochemovitz, was likely to have died in Ramygala because the 'elderly and sick were left in the Beit Midrash[3] that was set on fire and they were burned alive' (Rosin 1996).

As part of the journey we spent a few days in Vilnius, the capital of Lithuania, visiting museums and heritage centres, some of which focused on Jewish heritage. Visiting Lithuania, Panevėžys, Ramygala and Pajouste forest has stirred in me a series of emotional and academic responses about the Holocaust, its impact on my parents and my family, and the Lithuanian response to this fraught heritage. This short chapter explores various aspects of the visit in relation to public and private issues. Accordingly, it is both autobiographical and academic. Three issues are addressed: (i) the feeling of 'homecoming' that I experienced in Lithuania and especially Panevėžys, although it was my first visit; (ii) the emotions and thoughts that emerged during our walk down M Valančiaus Gatve, the street where my father grew up; and (iii) the representation of Jews in Vilnius' museums.

---

[1] Jewish prayer for the dead.
[2] This information was recorded in a 'Report on the mass murder in Lithuania of 1 December 1941 by the Commander-in-Chief of the SS Security Police and the Security Services in Kovno, SS-Standartenführer (SS-Colonel) Dr Karl Jäger'.
[3] House of Learning.

Table 14.1: The number of people killed in Panevėžys between 21 July and 23 August 1941

Date	Breakdown of people killed	Totals
21.7.1941	59 Jews 11 Jewesses 1 Lithuanian woman 1 Pole 9 Russian Communists 22 Lithuanian Communists	103
22.7.1941	1 Jew	1
28.7.1941	234 Jews 15 Jewesses 19 Russian Communists 20 Lithuanian Communists	288
4.8.1941	362 Jews 41 Jewesses 5 Russian Communists 14 Lithuanian Communists	422
11.8.1941	450 Jews 48 Jewesses 1 Russian Communist 1 Lithuanian Communist	500
23.8.1941	1312 Jews 4602 Jewesses 1609 Jewish children	7523
Total		**8837**

A SENSE OF BEING AT 'HOME'

> Did I dare say it? In some ways I felt very much at home in this land of bears and forebears.
>
> (Cassedy 2012, 70–1)

As I had anticipated, the journey was associated with considerable hurt and anger because of the devastation the Holocaust had inflicted on both my paternal and maternal families and the ongoing emotional and psychological suffering it has caused. However, the trip also presented the upwelling of a feeling of homecoming inside me, especially in Panevėžys. Despite being aware of the literature surrounding these issues I did not anticipate that it would touch me and I was not prepared for this feeling. This unexpected emotion was crystallised in Panevėžys when Gennady Kaufman, Chairman of the Jewish community, asked me to give a brief presentation, on my experience of visiting the town and the Pajouste forest and on my experience in the anti-apartheid struggle in South Africa, at a workshop on tolerance on 24 August 2011. This impromptu request compelled me to consider my relationship with Lithuania and especially Panevėžys. It became evident that I had a strong sense of 'return' to this 'foreign' town, which is c. 10,000 miles from Cape Town, where I grew up. Significantly, my daughter did not share this feeling, despite being acutely aware of our family's connections to these places.

FIG 14.1.   MY FATHER, MORRIS MAZEL, WHO IS SECOND FROM THE LEFT IN THE BACK ROW
WITH 'LANDSLITES' FROM PANEVĖŽYS. TO HIS LEFT IS MY GODFATHER, DAVE WITTEN.

I informed the gathering that although I was born in Cape Town and now live in the United
Kingdom, Panevėžys has always loomed large in my life. Most of my father's friends, who we
interacted with at family occasions such as batmitzvahs and barmitzvahs, were from Panevėžys
(see Fig 14.1). They were known as 'landslites', the Yiddish term for someone who hails from the
same place. Moreover, although my father did not spontaneously talk about his life in Panevėžys
he could be prompted to recall things, especially when viewing old photographs of football teams
and Zionist youth (see Fig 14.2). Later in his life my father attended the Ponevez[4] synagogue in
Cape Town where I believe he was the last living link to the town. It has been commented by
my partner of 35 years, Ann Macdonald, that in essence I was 'marinated' in a Jewish Panevėžys
reality; it is central to my identity and how I belong and function in the world. This is not
something my daughter experiences to the same extent.

According to Čiubrinskas (2006, cited in Dargufiytė 2010, 49, 50), increasing numbers of
Jews have being travelling to Eastern Europe 'to honour their ancestors, out of nostalgic reasons
or even "to be of use to one's own country"'. Moreover, Dargufiytė (2010, 50) states: 'Inevitably, these East European "homelands" are loaded with traumatic memory for the majority
of the Diaspora people, and this is always the case for the descendants of East European Jews'.
In a related comment, Aviv and Shneer (2005, 8) note that 'In a post-Holocaust world, many

4   One of the Yiddish spellings of Panevėžys.

FIG 14.2. THE MACCABI SOCCER TEAM IN PANEVĖŽYS IN THE LATE 1920s. MY FATHER,
MORRIS MAZEL, IS ON THE LEFT IN THE FRONT ROW AND HIS BROTHER, NATHAN MAZEL, IS ON
THE RIGHT IN THE BACK ROW.

American Jews [I believe this applies more broadly] came to see Eastern Europe no longer as the
real place from which to draw roots but one that they want to bear witness to. It is a land of
Jewish ghosts and lost cultures.' Drawing on these sentiments, Dargufiytė (2010, 52) maintains
that 'precisely this "nostalgic sacredness" is becoming the main motive which draws Jewish roots
tourists to their cultural and familial lands'.

Regarding other contexts, commentators offer various reasons as to why people travel to
the land of their birth or of their ancestors. For Scotland, Basu (2004) proposed three primary
reasons: homecoming, pilgrimage and quest, while Bohlin (2011, 290) comments that:

> Particularly in cases of 'ancestral' or 'roots' return (Stefansson 2004), when people return to a
> geographical place where their ancestors used to live, the idea of homecoming as an opportunity
> to reconnect with a valuable, shared past lies at the heart of accounts of homecoming, whether
> personal narratives, public discourses or state policy.

According to Timothy and Teye (2004), African-American trips to West Africa are a form of
appreciation of the cultural and family legacy of one's community, while Austin (1999) notes
that Africans in the Diaspora on trips to Africa view themselves as 'coming home'. The feeling of
'homecoming' and the reconnection with the land of their forefathers represents the essence of
their trip. The analogy is sometimes drawn between these trips to Africa and those undertaken by

Jews to Eastern Europe: 'just as "the dispersal of Africans taken from the continent to the New World has left its mark on the very tropes of heritage and identity" (Ebron 1999, 910), so did the Holocaust leave its mark on the very tropes of Jewish heritage and identity' (Dargufiytė 2010, 49–50). These insights are strongly connected to the notion of place identity: 'The construction of identity for or by people(s) through reference to place and/or the construction of identity for places through reference to their morphology, histories, cultures and inhabitants' (Whitehead *et al* 2012, 14).

Reflecting back on my experiences and the brief appraisal of the appropriate literature I suggest that the literature does not adequately treat the changing emotions that people may feel through the different stages of their journeys. The possibility of 'shifting feelings' needs to be given greater consideration. Furthermore, in the context of Jewish visits to Eastern Europe, reference has been made to 'nostalgic sacredness', which only partially resonates with my feelings: 'sacredness' yes, but not 'nostalgia'. There was nothing 'nostalgic' about the trip that was dominated by pain and sadness, and the unexpected feeling of 'homecoming'. These emotions were evident in the walk we did down the street my father grew up on, to which we turn next.

DOWN M VALANČIAUS GATVE – A FRAUGHT WALK

At subsequent stages of ghettoization, deportations, and eventually extermination, Jewish victims were gradually deprived of everything. (Stola 2007, 241)

In 1997, one of my brothers visited Panevėžys with his son; the first Mazel to enter the town since 1941. In preparation for the visit my father, who was in Cape Town, indicated to him, via a fax sent by our brother Abraham Mazel,[5] that his

grandfather owned a few houses on Vishkopevelensis Gatwe where he [ie my father] lived with his family until he left for SA. The grandfather had his own shul[6] on that street – apparently a second-hand car dealership on that spot now. If there is a 16th of February Street (Main Road of Ponevez then). Your grandmother had her grocery shop there.

In the same year, on 18 July, I interviewed my parents and my father commented: 'Yes, there were a lot of houses. My grandfather had a whole suite of houses' (Mazel 1997, *pers comm*).

Given the family's connection to M Valančiaus Gatve, my daughter Nicola and I walked the length of the street, some 360 metres between Respublikos Gatve and Naujamiesčio Gatve, on the morning of the 70th anniversary of the massacre. It appeared to us that many of the houses pre-dated 1941, while the above-mentioned shul was still standing (see Fig 14.3). Regarding the ownership of houses, an interesting insight has been offered by Yoffee (1998), who referred to it as 'Ruta's Rule' after his guide/interpreter from Vilnius University. Jewish houses customarily had street-facing main entrances, mostly with an equal number of windows on either side in order to facilitate the public visiting the shop usually located within the house. In contrast, Christian houses normally had side entrances to facilitate access to vegetable gardens and/or animal pens

---

5   The content of the fax is reproduced here exactly as in the original (including punctuation, etc).
6   A Yiddish word for synagogue.

Fig 14.3.   It is believed that my great-grandfather owned this shul on M Valančiaus Gatve.

and barns. If correct, this 'rule' suggests that there are still houses standing on M Valančiaus Gatve that retain the architectural character of their Jewish owners. As we walked, we considered that perhaps one or more of the houses on the street might have belonged to Abram Abelski, my great-grandfather. Moreover, we speculated whether my father was raised in one of the houses and whether there might still be items belonging to my family in one or more of them. These were unsettling thoughts.

As we walked down the street people emerged from a few houses and watched us. I believe they were aware that we were not there as conventional 'tourists'. This is not a tourist area of the town and we were walking slowly, with cameras, stopping intermittently and clearly discussing the houses. Furthermore, it is very possible that they would have identified us as being of Jewish origin. We did not avert our eyes, but watched back. Thus ensued an unexpected passive aggressive 'standoff' with some of the occupants of M Valančiaus Gatve.

Questions of ownership and restitution were uppermost in our thoughts during this walk. Contextualising the theft of properties, Dean (2007, 28) commented:

> there is no doubt that the seizure of Jewish property in Europe was primarily a state-directed process linked closely to the development of the Holocaust. However, the widespread participation of the local population as beneficiaries from Jewish property served to spread the

complicity, and therefore also acceptance of German measures against the Jews, beyond the smaller circle of immediate perpetrators. In this way the Nazis and their collaborators were able to mobilize society in support of Nazi racial policies to a greater extent than the spread of racial antisemitism alone would have permitted.

In Panevėžys, the seizure of Jewish properties and belongings was likely to have occurred between early July 1941, when the Jews were ordered to move into the ghetto, and 23 August 1941, when the final massacre occurred.

In Lithuania, restitution arrangements have focused (i) on communal Jewish properties and (ii) on Jews who survived the war (ie fewer than 5%). Regarding the former, the 'Law On Restoration of the Rights of Ownership of Citizens to Existing Real Property (enacted in 1997 and amended in 2002), provided that former property owners, and certain heirs, were eligible to recover their confiscated property, so long as claimants were Lithuanian resident citizens' (World Jewish Restitution Organization n.d.). In terms of the latter, the Law on Goodwill Compensation for the Immovable Property of Jewish Religious Communities of the Republic of Lithuania, passed in 2011, has meant that the Lithuanian government will give about US $50 million to its Jewish community during the next decade in compensation for buildings expropriated by previous totalitarian regimes, and that c. US $1.1 million shall be paid 'to persons of Jewish nationality who resided in Lithuania during the Second World War and suffered from the totalitarian regimes during the occupations' (World Jewish Congress 2012). People eligible to receive these payments need to show that they meet the required criteria by holding a document confirming that they lived in the current territory of Lithuania during World War II or were forced to leave following the outbreak of war.

Neither of these laws considers the rights of the families of Lithuanian Jews, such as mine, who were dispossessed of their properties and other material possessions and savings during the Holocaust. Are our claims now buried and forgotten? Should the residents of M Valančiaus Gatve continue to live in our houses and possibly use the belongings of my murdered family without even acknowledging how these came into their possession? Should the survivors from families decimated by murder be compensated for their losses? Or, does there need to be a process of recognition of the theft that occurred and for people to 'move on', on the understanding according to Feldman (2007) that although

> there are cases in which profits from stolen assets have been taken into account in compensation cases, the notion that the sins of the parents shall be visited upon the children from generation to generation is grist for the mill of those who argue that the unlimited quest for justice can lead to new injustice and that it is legitimate to demand that compensation be brought to a reasonable termination. (Feldman 2007, 260)

These are problematic issues that require greater acknowledgment and more in-depth discussion. Museums represent one of the places in which public debates surrounding these issues could be aired but, as we shall see in the next section, the museum community in Lithuania is struggling to come to terms with the slaughter of the Jews that occurred in 1941.

MUSEUMS AND THE POLITICS OF ABSENCE

> History can also be told in a way that denies the past, manipulates the truth and deliberately misleads. In an environment where culture and identity is highly contested, exclusion from the canon of the established notion of history can be interpreted as a deliberate act of suppression.
>
> (Crooke 2005, 135)

In Vilnius we visited several museums, but I wish to focus on only three: the Vilna Gaon Jewish State Museum (hereafter Jewish Museum), the Museum of the Victims of Genocide (hereafter Genocide Museum), and the National Lithuanian Museum (NLM). While the Jewish Museum concerns the historical and cultural heritage of Lithuanian Jewry with a large emphasis on the Holocaust, the mission of the Genocide Museum, situated in the old Headquarters of the KGB, is to 'collect, keep and present historic documents about forms of physical and spiritual genocide against the Lithuanian people, and the ways and the extent of the resistance against the Soviet regime' (Kuodytė n.d., 3). The NLM was defined, via legislation in 2004, as a 'national budget enterprise which collects, stores, researches, conserves, restores and popularizes the values of Lithuanian archaeology, history and ethnic culture' (Rindzevičiūtė 2011, 540).

Investigation of these museums reveals a response to the Holocaust and the history of Jewry in Lithuania that can be characterised by 'denial' and 'suppression' as reflected in the above-mentioned quote from Crooke (2005). While the Jewish Museum addresses these topics, it was striking that at the other museums there is a paucity of Holocaust material in the Genocide Museum and a complete absence of Jewish history and culture in the NLM. The scant reference to the Holocaust in the Genocide Museum has been emphasised by Mark (2010, 283). According to him, while the displays begin in 1940, when the building was occupied by the Russians, they overlook the German occupation between 1941 and 1944 when the building housed the Vilnius Extraordinary Detachment, which exterminated Jews, and the use of it by the Gestapo to interrogate and torture Jews, Poles and Communists. Moreover, Rindzevičiūtė (2013) notes that the Genocide Museum only actually acknowledged the Holocaust in its displays 18 years after it opened.

In a somewhat defensive response to Mark's (2010) observations about the Genocide Museum displays, Rindzevičiūtė (2013, 63) has argued that 'It has become a cliché to argue that Lithuanian public sector organisations, particularly museums, emphasise the terrible legacy of communist crimes and that they tend to forget – and even actively avoid making public – information about the killings of Lithuania's Jews'. According to Rindzevičiūtė (2013), the Jewish Museum is one of the ways in which Lithuanian museums have marked the Holocaust. While this might be the case, it is nonetheless undeniable that even with the belated inclusion of Holocaust material in the Genocide Museum, the topic still remains markedly under-represented, conveniently eclipsed by the Communist oppression to mask culpability in the Holocaust. Trying to understand the avoidance of the Holocaust, Rindzevičiūtė (2013) has argued that a 'kind of compartmentalisation of histories emerged as the Holocaust and communist terror were housed in two different organisations', which according to one of her interviewees was informed by the 'administrative logic' of the state's desire to avoid duplications.

The 'compartmentalisation' apparent in the Genocide and Jewish museums is even more pronounced in the NLM, which totally ignores 700 years of Jewish history (Greenbaum 1995), a group that comprised c. 7% of the country's population prior to World War II. This writing of people out of history is not unique to the NLM. There are many examples internationally of museums doing likewise. At a personal level, the NLM resonated with my South African experiences where, for example, the apartheid museums of Natal celebrated the white colonial culture and heroics with almost no recognition of precolonial African history. Analysing this phenomenon, Wright and Mazel (1991, 65) argued that ignoring African people in the colonial history of Natal obviated the need to recognise the region's precolonial history:

> To admit the existence of precolonial history is to admit the existence of precolonial African population, and to raise a host of uncomfortable questions about what happened to it in the colonial period. Better, then, to exclude African people from history altogether, and, since their existence can hardly be denied, push them off into separate ethnic rooms and into separate museums.

The parallels with the NLM are salient: ignoring the Jews in the NLM avoids addressing uncomfortable questions in the national museum about what happened to the Jews of Lithuania during the Holocaust. This is easier to do than to begin to address a series of difficult and troubling issues, including the role that ethnic Lithuanians played in the Jewish genocide.

Conclusion: the Burden of 1941

Visiting Lithuania raised a number of personal issues that have public resonance. I am not sure what troubled me most: encountering passive aggression on my father's street; the questions of ownership and restitution; the apartheid-like separation of museum topics; or the uncanny feeling of being at home in a country I had not previously visited and which my father left in 1929 and my mother in 1935. These phenomena all connect to the tragic events of 1941 when many of my family perished, with thousands of other Jews, at the hands of the Nazis and their Lithuanian collaborators. The avoidance evident in the Genocide Museum and NLM suggests that the events of 1941 still bear heavily on Lithuanian society, especially since its independence from Soviet rule in 1992. Sužiedėlis (2001, n.p.) wrote:

> [the] only way for Lithuanians to lighten the load of the difficult history of 1941 is to embrace it. However artfully presented, the strategies of denial and evasion, the finger-pointing and righteous indignation directed at the Other, serve only to further weigh society down. … Recognizing a historic burden is not the same as accepting collective guilt. No honest person argues that Lithuanians are a nation of criminals, or that today's Lithuanians are responsible for what happened in 1941 (any more than contemporary Americans are responsible for slavery). But the legacies of such crimes, the historical burdens, remain. As a general proposition, attempts to evade, deny, minimize or misrepresent historical offenses are unsuccessful in the long run.

It would appear that there have been increasing attempts to confront this past during the last decade (see for example Vitureau 2013); however, if my experiences are anything to go by, there is still a way to go, especially in terms of reaching ordinary Lithuanians such as the residents of M Valančiaus Gatve, the street on which my father grew up.

BIBLIOGRAPHY AND REFERENCES

Anon, 2005 *Materials on the Memorial to the Murdered Jews of Europe*, Nicolaishe Verlagbuchhandlung GmbH, Berlin

Austin, N K, 1999 Tourism and the transatlantic slave trade: Some issues and reflections, in *The Political Economy of Tourism Development in Africa* (ed P U C Dieke), Cognizant, New York, 208–16

Aviv, C, and Shneer, D, 2005 *New Jews: The End of the Jewish Diaspora*, New York University Press, New York

Basu, P, 2004 Route Metaphors of 'Roots-Tourism' in The Scottish Highland Diaspora, in *Reframing Pilgrimage: Cultures in Motion* (eds S Coleman and J Eade), Routledge, London, 150–74

Bohlin, A, 2011 Idioms of Return: Homecoming and Heritage in the Rebuilding of Protea Village, Cape Town, *African Studies* 70 (2), 284–301

Cassedy, E, 2012 *We are here: Memories of the Lithuanian Holocaust*, University of Nebraska Press, Lincoln NE

Crooke, E, 2005 Dealing with the Past: Museums and Heritage in Northern Ireland and Cape Town, South Africa, *International Journal of Heritage Studies* 11 (2), 131–42

Dargufiytë, Z, 2010 Jewish Roots Tourism and the (Re)Creation of Litvak Identity, *LCC Liberal Arts Studies, Volume III: Responses to Cultural Homogeny: Engagement, Resistance, or Passivity*, 47–53

Dean, M, 2007 The Seizure of Jewish Property in Europe: Comparative Aspects of Nazi Methods and Local Responses, in *Robbery and Restitution: The Conflict over Jewish Property in Europe* (eds M Dean, C Goschler, and P Ther), Berghahn Books, New York, 21–32

Ebron, P A, 1999 Tourists as Pilgrims: Commercial Fashioning of Transatlantic Politics, *American Ethnologist* 26, 910–32

Feldman, G D, 2007 Concluding Remarks, in *Robbery and Restitution: The Conflict over Jewish Property in Europe* (eds M Dean, C Goschler, and P Ther), Berghahn Books, New York, 259–68

Greenbaum, M, 1995 *The Jews of Lithuania: a history of a remarkable community, 1316–1945*, Gefen Publishing House, Jerusalem

Kuodytë, D, n.d. Introduction, in *The Museum of Genocide Victims: A Guide to the Exhibitions*, The Museum of Genocide Victims, Vilnius

Levinson, J, 2006 The Shoah and the Theory of the Two Genocides, in *The Shoah (Holocaust) in Lithuania* (ed J Levinson), The Vilna Gaon Jewish State Museum, Vilnius, 322–53

Levinson, J (ed), 2006 *The Shoah (Holocaust) in Lithuania*, The Vilna Gaon Jewish State Museum, Vilnius

Mark, J, 2010 What Remains? Anti-Communism, Forensic Archaeology, and the Retelling of the National Past in Lithuania and Romania, *Past & Present* 206, 276–300

Mazel, M, 1997 Personal communication (interview between the author and his parents, Morris and Lily Mazel), 18 July, Cape Town

Rindzevičiūtė, E, 2011 National Museums in Lithuania: A Story of State Building (1855–2010), in *Building National Museums in Europe 1750–2010*, conference proceedings from EuNaMus, European National Museums: Identity Politics, the Uses of the Past and the European Citizen, Bologna 28–30 April 2011 (eds P Aronsson and G Elgenius), Linköping University Electronic Press, Linköping, EuNaMus Report No 1, 521–52

— 2013 Institutional Entrepreneurs of a Difficult Past: The Organisation of Knowledge Regimes in Post-Soviet Lithuanian Museums. European Studies: An Interdisciplinary Series in European Culture, History and Politics, *European Cultural Memory Post-89*, 63–96

Rosin, J, 1996 Ramygala, in *Encyclopedia of Jewish Communities in Lithuania* (ed D Levin), Yad Vashem, Jerusalem, available from: http://www.jewishgen.org/yizkor/pinkas_lita/lit_00640.html [16 November 2013]

Stefansson, A, 2004 Homecomings to the Future: From Diasporic Mythographies to Social Projects of Return, in *Homecomings: Unsettling Paths of Return* (eds F Markowitz and A Stefansson), University Press of America, Lanham MD, 2–20

Stola, D, 2007 The Polish Debate on the Holocaust and the Restitution of Property, in *Robbery and Restitution: The Conflict over Jewish Property in Europe* (eds M Dean, C Goschler, and P Ther), Berghahn Books, New York, 240–58

Sužiedėlis, S, 2001 The Burden of 1941, *Lituanus: Lithuanian Quarterly Journal of Arts and Sciences* 47 (4)

Timothy, D J, and Teye, V B, 2004 American children of the African diaspora: journeys to the motherland, in *Tourism, Diasporas and Space* (eds T Coles and T Dallen), Routledge, London, 111–23

Vitureau, M, 2013 *Restoring the Nazi-decimated heritage of Lithuanian Jews* [online], available from: http://artdaily.com/news/65188/Restoring-the-Nazi-decimated-heritage-of-Lithuanian-Jews#.UojWdY3GI7C [6 November 2013]

Whitehead, C, Eckersley, S, and Mason, R, 2012 *Placing Migration in European Museums: Theoretical, Contextual and Methodological Foundations*, MeLA Books, Milan

World Jewish Congress, 2012 *Application phase for compensation payments to Lithuanian Holocaust survivors to begin in January* [online], available from: http://www.worldjewishcongress.org/en/news/12747/application_phase_for_compensation_payments_to_lithuanian_holocaust_survivors_to_begin_in_january [25 October 2013]

World Jewish Restitution Organization, n.d. *WJRO Lithuania Operations* [online], available from: http://www.wjro.org.il/Web/Operations/Lithuania/Default.aspx [25 October 2013]

Yoffee, B, 1998 *My recent Jewish Heritage Roots Tour to Lithuania and Belarus: Moving and inspiring* [online], available from: http://www.litvaksig.org/index.php/litvaksig-online-journal/my-recent-jewish-heritage-roots-tour-to-lithuania-and-belarus-?task=article [19 May 2013]

Wright, J B, and Mazel, A D, 1991 Controlling the past in the museums of Natal and KwaZulu, *Critical Arts* 5 (3), 59–78

Displaced Heritage:
Lived Realities, Local Experiences

15

Humiliation Heritage in China: Discourse, Affectual Governance and Displaced Heritage at Tiananmen Square

Andrew Law

Introduction: Humiliation Discourse

In recent years, scholars and commentators (such as Gries 2004; Broudehoux 2004; Callahan 2010; and Wang 2012) on China's political history and national identities have noted the ever increasing rise in a complex set of narratives that proposes that from the Opium Wars of the 1840s to the Japanese invasion of the Mainland between 1937 and 1945, the Chinese nation and the Chinese people have been subject to a series of historic humiliations. However, writers (such as Wang 2012) have also argued that alongside this narrative of decline is a connected discourse that suggests that with the formation of the People's Republic of China (PRC) in 1949, the country has slowly redeemed itself. Connected to these two preceding discourses, a third has emerged, which suggests that since the formation of the PRC – yet within a historic context of humiliation, suffering, hardship and struggle – the country has been rejuvenated, obtaining growing prosperity and international respect (Gries 2004; Broudehoux 2004; Callahan 2010; and Wang 2012). I suggest in this chapter that this 'humiliation discourse' can be regarded as a form of heritage that captures the essence of displacement and loss.

Humiliation Theories

Gries (2004, 50), in his work on Chinese nationalism and international relations, noted that humiliation history can be understood as the resurfacing (from the 1990s) of 'long-suppressed memories of past suffering'. Broudehoux (2004) noted that such a history (and the fate of heritage associated with it) serves an important role in the production of new forms of nationalism, the legitimation of the Chinese state and of the Chinese Communist Party (CCP). Callahan (2010) and Wang (2012) have provided more in-depth studies into the origins of humiliation discourse and its role in contemporary Chinese culture.

Callahan (2010) explores the role of humiliation history from the early 1910s and 1920s to the present, suggesting that humiliation discourse has a history that existed long before the formation of the PRC. He states that 'More than a decade before Chiang [Kai-shek] came to power, an education professor designed a patriotic curriculum in 1915 around the principles of "national studies, national humiliation and hard work"' and warns that 'it would be a mistake to conclude that patriotic education [and humiliation history] is merely propaganda that is instrumentally used by the party elite to manipulate the people…' (Callahan 2010, 19). However, although tentative about the position of humiliation history within contemporary CCP propaganda, Callahan concedes, nevertheless, that 'Beijing's current positive/negative patriotic education campaign is so successful because it builds upon a structure of feeling that actually precedes this particular propaganda policy and predates the PRC' (Callahan 2010, 19).

Wang (2012) goes slightly further, asking 'how have… Chinese leaders used history and memory to reshape national identity so as to strengthen their legitimacy for ruling China after the end of the Cold War?' (Wang 2012, 14). To answer this question, Wang investigates China's education system, its popular culture and the public media; importantly for Wang, humiliation history and Chinese nationalism were not a feature of the Mao years[1] and only became a feature of CCP propaganda after the Tiananmen crackdown of 1989 (Wang 2012, 94). Importantly, Wang contends that humiliation history became increasingly an essential component of patriotic education after 1989:

> In the new textbooks approved after 1992, the official Maoist victor narrative was superseded by a new victimization narrative that blamed the west for China's suffering. 'China as victor' was slowly replaced by 'China as victim' in nationalist discourse. This change of narrative is found in the official documents, history text books and popular culture. (Wang 2012, 102)

Mediating Humiliation History: The Role of Heritage

As Broudehoux (2004), Callahan (2010) and Wang (2012) have reported, one of the central mechanisms by which humiliation history is mediated to the public is through the *medium of heritage*. This 'humiliation heritage' can take many forms, including museums, monuments, ruins and historical public displays. Wang contends that:

> Today, the Chinese people are living in a forest of monuments, all of which are used to represent the past to its citizens through museums, historic sites, and public sculptures. Although people all over the world cherish their own memorials, the special effort made by the Chinese government since 1991 to construct memory sites and use them for ideological re-education is unparalleled. (Wang 2012, 104)

Wang explores the CCP's Ministry of Civil Affair's patriotic-heritage campaign; he notes 100 patriotic-education sites selected by the CCP. These include 40 memory sites that allude to China's past conflicts or war with foreign countries, 24 civil war sites that represent the wars between the CCP and the KMT; 21 sites described as wonders of Chinese civilisation (including the Great Wall) and 15 other sites that pay homage to China's heroes. For Wang, what is crucial about these spaces is that they all reify narratives of humiliation and associated discourses of struggle and national rejuvenation.

Broudehoux (2004) and Callahan (2010) focus on the symbolism surrounding a few key heritage spaces. Both discuss the Yuanmingyuan gardens in Beijing; Callahan states that what is really important about the gardens is their commercial branding through posters, calendars and t-shirts; humiliation history and national identity can now be consumed by visitors and tourists.

---

[1] Indeed, as Wang suggests, Mao thought that nationalism and patriotism itself was narrow and detrimental to the CCP's international political aims (Wang 2012, 94).

Callahan also notes that the ruins are often used as a site of patriotic performance; indeed, as Callahan points out, 'The party-state certainly uses [the gardens] to stage commemorative events on National Humiliation Days: the first national defence education day in 2001, the return of Hong Kong in 1997, and the 130th anniversary of the looting [of the gardens]… in 1990' (Callahan 2010, 88; my words in brackets).

Importantly, then, these heritages not only reproduce humiliation history, they legitimise the party state. Indeed, as Wang suggests, humiliation history and heritage allows the CCP to construct itself as the 'bearer of China's struggle for national independence' (Wang 2012, 100). In sum, humiliation history, reified and concretised through humiliation heritage heroises the CCP and justifies its continued existence even when dissenting voices offer other alternative modes of governance.

Problematic issues

Gries and Callahan agree that the construction of humiliation history cannot simply be reduced to a contemporary strategy of the state. Like Gries and Callahan then, I also contend that humiliation history should be read as a series of complex and understandable traumas that predate the PRC. Nevertheless, the contemporary administration has utilised the traumatic power of humiliation history and heritage (and the feelings of Chinese people) to their own advantage. Indeed, as we have seen, Broudehoux and Wang position humiliation history and humiliation heritage as a key patriotic strategy and a source of party-state legitimacy.

In what follows, I concur with Wang's contention that the CCP have utilised history and heritage in this way, but suggest that contemporary analyses could go much further. Thus, while Broudehoux and Wang connect humiliation history and heritage with state control, additional analysis could explore the actual techniques or strategies of governance used to implement and reify these discourses. What is missing in the current literature is a more concentrated analysis of the art of governance and an analysis of the way in which political, historical and heritage space is utilised in tactical ways to convey particular messages and symbols.

The following analysis explores the concept of governmentality and particularly the idea of affective governance (Thrift 2004; Jansson and Lagerkvist 2009; Grusin 2010; Staheli 2011) to understand the contemporary reification and production of humiliation discourse. Specifically, I suggest that through affectual media (I will explain this term in more detail, below), humiliation history and associated narratives of struggle and rejuvenation are critical to a governmental presentation of identity.

These ideas are explored by investigating the role of a particular political assemblage – an assemblage of media, theatrical and heritage agents – that came together in the construction of the celebrations leading up to the 60th Anniversary of the PRC (1 October 2009). This anniversary was a key moment in Chinese governmental history and in the reproduction of humiliation discourse; it was also critical to a very affectual presentation of self, one that embedded the CCP within a particular narrative of the nation. In effect, the anniversary helped the government to be seen as heroic.

This construction of a heroic governmental identity was also enabled through a particular political geography. The encoded messages of the parades, galas, exhibitions and socialist heritage (including theatrical and musical portrayals of the history of the PRC) were reinforced by their symbolic location at Tiananmen Square in the nation's capital. While Tiananmen has often been associated with a history of political symbolism, through the lens of the Anniversary the Square was reinvented to serve as a stage for the reification of affectual humiliation narratives; with its assemblage of monuments, halls, friezes and symbols, the CCP could further reinforce its status as a heroic political actor.

GOVERNMENTALITY, AFFECTUAL GOVERNANCE AND AFFECTUAL, ATMOSPHERIC MEDIA

Governmentality, which has its roots in Foucauldian ideas, refers to the art of government, including the practices of government and the effects on the people governed (Foucault 1991; Burchell *et al* 1991; Rose 1996; 1999; Jeffreys 2009; Bröckling *et al* 2011; Walters 2012). However, contemporary re-evaluations of governmentality have often explored novel and more complex tactics of governance. Reflecting the development of non-representational and affectual theories in the human sciences, writers such as Thrift (2004; 2007), Grusin (2010) and Staheli (2011) have all hinted (whether directly or indirectly) at the role of affectual discourses in the production of new complex political strategies of power.

The concept of affectual government is related to ideas of emotion:

> Affects are social but they are pre-individual and non-significatory flows. Affects circulate through social and psychic layers of meaning, without being meaningful themselves (Ahmed, 2004). In contrast emotions – such as anger, greed and anxiety – are individualised and normalised social constructions. (Staheli 2011, 276)

Simply speaking, where emotions refer to socially constructed 'moods' associated within the body, affects are trans-human flows of pre-cognitive registers that construct and infiltrate subjects and bodies. That is, affects refer to the often imperceptible (and non-representational) moods constructed between people (or bodies); the most obvious example of this can be seen in crowds; thus for instance at a political protest the mood of a crowd can turn nasty or triumphant as a result of minor imperceptible fluctuations of mood or feeling.

However, as Thrift (2004; 2007), Pile (2010), Grusin (2010) and Staheli (2011) have all suggested, affects can also be manipulated on broader levels; indeed, in a discussion of media and political representation, Massumi and Thrift have discussed the role that 'confidence' played in the legitimation and justification of President Ronald Reagan during his tenure in the US. As Massumi (2002) and Thrift (2004) point out, Reagan's legitimacy was created through his ability to produce an 'air of confidence' and by the media's ability to accurately highlight this emotional performance (Massumi quoted in Thrift 2004, 65–6). But, as well as subjects, Thrift also points to the role of space and particularly urban space in the legitimation of political discourses and/or administrations; as Thrift states:

> Increasingly, urban spaces and times are being designed to invoke affective response according to practical and theoretical knowledges that have been derived from and coded by a host of sources. It could be claimed that this has always been the case – from monuments to triumphal processions, from theatrical arenas to mass body displays. (Thrift 2004, 68)

Informed by these ideas, recently writers have discussed the historical and contemporary role of affectual governance in contemporary China (Dutton 2008; 2009; Jansson and Lagerkvist 2009). For example, Jansson and Lagerkvist (2009) have argued that the city of Shanghai has been constructed in politico-emotive ways by urban agents, planners and policymakers; specifically through media, film and actual staged events (such as the Shanghai EXPO of 2010), Jansson and Lagerkvist suggest that citizens, visitors and investors are encouraged to gaze in admiration at the city and its buildings as a site of awe and futurity. Affectual geographies work, through urban atmospheres, bus tours, glittering lights, phantasmagoria and the actual production of urban landscapes themselves: with buildings planned, designed and built in meaningful and cinematic relations to one another (Jansson and Lagerkvist 2009).

But while Jansson and Lagerkvist (2009) focus on local political agents, arguably it can be suggested that the same affectual technologies of power have also been employed by the central government; specifically through the affectual discourse of humiliation history and heritage, it is arguable that the CCP have constructed imaginaries of themselves through staged galas, exhibitions and performances. These ideas are discussed here by focusing on museums, exhibitions, galas, the media and displays of political pageantry that came together to reify and reinforce humiliation history, prior to the anniversary of the PRC on 1 October 2009.

Moreover, as well as legitimising the CCP, the importance of these exhibitions and performances is that they also valorise the party as a hero of the state and the people; thus through affectual discourse the CCP constructs itself as the only government of the Chinese people.

Furthermore, as I am trying to suggest here, the political assemblage of exhibitions and performances can all be read as acts of *performative heritage*. Indeed, where humiliation discourse refers to affectual social and cultural memory, heritage is the *actualisation* of humiliation history and affect; it is the materialisation of humiliation discourse/affect that allows it to be recognised by viewer and subjects as something real (as opposed to something purely intangible). In this way, humiliation discourse/affect and heritage are inseparable entities.

PREPARATIONS FOR THE 60TH ANNIVERSARY OF THE PRC, 1 OCTOBER 2009

Several important galas, exhibitions and performances were staged prior to the event. Before the actual day, a gala, the 'Road to Rejuvenation', debuted on 23 September 2009 at the Hall of the People. The *China Daily* reported that the gala would be composed of a 2.5-hour-long performance, and would take the viewer through a timeline divided into five parts including '1840 to 1921, when China was invaded by western countries and people lived miserable lives… [to the year] 2009, characterised by a booming economy, a flourishing culture and a harmonious society' (*China Daily* 2009a).

Importantly as the journalist of this article observed, 'all five segments are presented in the most romantic, symbolic and artistic fashion' and, commenting on the first segment, observes that:

> The heavy and mournful tune rises above the sharp sounds of shattering glass, as the foreign powers attack the gardens, loot precious relics and kill mercilessly, till the officials bow and scrape to sign unjust treaties, ceding territory and paying indemnities. Then as one grief-stricken man reads out a poem, hundreds of young men perform a powerful but heart-breaking dance that portrays the Chinese people's awakening and struggle. (*China Daily* 2009a)

Another major feature of the festivities leading up to the anniversary was the exhibition *The Road of Rejuvenation*, opened in October 2007 at the Military Museum of the Chinese People's Revolution in Beijing. The exhibition was rehoused at the National Museum of China on

25 September 2009; along with the gala, as Johnson (2011) and Pedroletti (2011) note, the opening of the exhibition should not be regarded as another historical addition to the dynastic antiquities and political paintings that are housed at the museum. These authors imply the exhibition was crucial because it enforces a very one-sided version of Chinese history. Pedroletti states the exhibition starts with a 'period of humiliation to which western colonial powers subjected China, highlighting the steps on the way to nationalist reawakening and modernisation' (Pedroletti 2011). However, rather than simply discussing the horrors of the colonial period, within these very emotional discourses the CCP is represented as the guiding force upon which the Chinese people can rely. The museums' website explains the exhibition in the following terms:

> *The Road of Rejuvenation* is one of the museum's permanent exhibitions that reflects the Opium War of 1840 onward, the consequential downfall into an abyss of semi-imperial and semi-feudal society, the protests of people of all social stratum who have suffered, and the many various attempts at national rejuvenation – particularly the Communist Party of China's fight for the people's liberation and independence of every ethnicity. The exhibition demonstrates the glorious but long course of achieving national happiness and prosperity and fully reveals how the people chose Marxism, the Communist Party of China, socialism, and the reform and opening-up policy…

> …Today, Chinese civilization already stands tall in the East. With the bright prospects of the Great Revival already before us, the dreams and pursuits of Chinese sons and daughters will surely be achieved. (National Museum of China n.d.)

A cursory scan of Chinese newspapers reviewing the exhibition at the time of its production also reveals a further reinforcement of these discourses; in papers such as the *China Daily*, writers such as Chen (2009) and Zhu (2009) position the exhibition within an affectual narrative of China's past and future. For instance, Zhu comments on the exhibition as having encapsulated the entire spirit and zeitgeist of the anniversary moment; the exhibition is described as beguiling in its propensity to take its viewers back '170 years, when the Qing Dynasty (1644–1911) began declining due to isolation, backwardness, corruption and invasions by Western powers' (Zhu 2009, 1). Zhu suggests the major force of the exhibition lies in its optimistic overtones; thus despite historic moments of suffering, the exhibition also displays:

> [the] numerous trials and explorations along the road to national salvation and rejuvenation. In addition, viewers can better understand the ceaseless struggle to win freedom, democracy and independence at the beginning of the 20th century, and the twisting trajectory of socialist construction in the early years of New China. (Zhu 2009, 1)

Implicit within all these affectual narratives, then, is not a simple act of state remembrance, but a monologue that locates the state as a hero of the people; a hero that helped guide 'the people' to their 'freedom'.

THE 60TH ANNIVERSARY OF THE PRC (ON 1 OCTOBER 2009): TIANANMEN SQUARE AS A PERFORMATIVE STAGE

Lee (2011) noted that Tiananmen Square was the political and symbolic site for the construction of the 60th Anniversary. He emphasised that what was particularly striking about the parades at the Square was the 'unprecedented display of advanced weaponry, signalling a shift from… "militarised socialism" to "spectator-sport militarism"' (Lee 2011, 419). But while weaponry was an important feature of the display, Wang commented that:

> To begin the ceremony, China's elite color guard began marching in perfect lockstep exactly 169 strides from the centre of Tiananmen to the flag pole to raise the national colors. Each of these literal steps was meant to represent one year since the beginning of the Opium War in 1840. (Wang 2012, 6)

What is particularly interesting about this march was that the nation's history was being defined in very teleological terms; rather than a procession symbolising China's long history, the 169 strides constructed here symbolise a very modern story of the nation that is only 169 years old. But if the parades and marches represented a certain emotional symbolism, it is arguable that the speeches of Hu Jintao, the General Secretary of the Party, were the most striking of all in their positioning of the meaning and symbolisms of the day. Analysing his speech, Hu made references to '100 years of bloody struggle', a struggle that was finally reversed with the great victory of the 'Chinese revolution' (Hu 2009). Rather than describing only suffering and humiliation, Hu showed significant confidence in the rejuvenation of the nation:

> With the hard work and wisdom of all ethnic groups of the country, the Chinese people have joined hands to overcome great hardships and have made great contributions that have been recognized by the world and proved our perseverance and endurance. (Hu 2009)

Importantly, what was central to Hu's speech was a further affectual story of *unity in suffering*; thus in the statement above, Hu portrays an image of the Chinese people as unified despite the hardships they have faced. His speech calls not only for historical reflection, but also for a language of emotional unity and possibly even insularity in its celebration of a subjugated people. It is similarly evident that within these discourses the central administration is also embedded in this language of emotional homogeneity; indeed, it is noticeable that the premier concluded his anniversary speech with the following rallying slogans:

> Long live the Great People's Republic of China
> Long live the Great Communist Party of China
> Long live the Chinese people
>
> (Hu 2009)

What is particularly effective (and affective) about these final statements is the way that the People's Republic, the Communist Party and the Chinese people become symbolically interchangeable; and thus also the story of suffering, humiliation and hardship are connected to a sub-narrative (both explicit and implicit) that the central administration has always helped 'the people' over the hurdles of suffering to their rightful destiny of independence. Within this discourse, the more difficult histories surrounding the administration (and indeed the Square itself) become overshadowed by an imaginary of the government as invaluable and indistinguishable from 'the people' themselves.

CONCLUSION: HUMILIATION DISCOURSE AND HERITAGE AS A TOOL OF AFFECTUAL GOVERNANCE

Humiliation history and humiliation heritage in contemporary China is not merely a matter of historiography; it is also a tool of governmental representation. Moreover humiliation history/heritage is actually a tool of passionate or affectual governance. Quite simply, humiliation history/heritage draws people in emotionally; it is affective in that it binds subjects to a national story and more importantly to an idea of a national heritage. It fills subjects with feelings of sadness, humiliation, struggle and hope for rejuvenation and the future. Most importantly of all, humiliation history/heritage and the associated discourses of struggle and rejuvenation have also allowed the CCP to construct itself as the hero of China's history; by positioning the party within a monolithic historical narrative, the CCP allows itself to be understood as a saviour who has liberated Chinese people from a century of humiliation. While of course the CCP can certainly be read as an agent involved in the production of a new modern Chinese state, its self-aggrandisement is entirely contestable. Moreover, this presentation of party-identity and history also displaces the more complicated and regrettable histories that haunt the party-state.

This historical image of the party-state has been reinforced over time and at particular moments. Focusing on the 60th Anniversary of the PRC, it is suggested that a number of theatrical, museological, media and journalistic agencies came together prior to and during the anniversary to construct and reproduce humiliation discourse, humiliation heritage and state narratives of legitimacy and power. These agencies also reinforced their discourses through the symbolic political geography of Tiananmen Square, which acted as a performative stage through which affectual discourses of governmental identity could be reproduced. By examining the symbolic use of the Square, it is suggested that this site remains crucial to the production of passionate narratives and a heroic vision of the CCP.

BIBLIOGRAPHY AND REFERENCES

Bröckling, U, Krasmann, S, and Lemk, T, 2011 *Governmentality: Current Issues and Future Challenges*, Routledge, London

Broudehoux, A M, 2004 *The Making and Selling of Post-Mao Beijing*, Routledge, London

Burchell, G, Gordon, C, and Miller, P (eds), 1991 *The Foucault Effect: Studies in Governmentality*, University of Chicago Press, Chicago

Callahan, W A, 2010 *China, the Pessoptimist Nation*, Oxford University Press, Oxford

Chen, J, 2009 Show highlighting days of old embarks down new road, *China Daily*, 23 September, available from: http://www.chinadaily.com.cn/cityguide/2009-09/23/content_8725479.htm [3 December 2013]

China Daily, 2009a Much to shout about, *China Daily*, 1 September, available from: http://www.chinadaily.com.cn/showbiz/2009-09/01/content_8640230.htm [3 December 2013]

— 2009b China confident of rejuvenation: Hu, *China Daily*, 1 October, available from: http://www.china-daily.com.cn/60th/2009-10/01/content_8759106.htm [3 December 2013]

Dutton, M, 2008 Passionately Governmental: Maoism and the Structured Intensities of Revolutionary Governmentality, *Postcolonial Studies* 11 (1), 99–112

— 2009 Passionately Governmental: Maoism and the Structured Intensities of Revolutionary Governmentality, in *China's Governmentalities, Governing Change, Changing Government* (ed E Jeffreys), Routledge, London and New York

Foucault, M, 1991 'Governmentality', trans Rosi Braidotti and revised by Colin Gordon, in *The Foucault Effect: Studies in Governmentality* (eds G Burchell, C Gordon, and Peter Miller), University of Chicago Press, Chicago, 87–104

Gries, P H, 2004 *China's New Nationalism, Pride, Politics and Diplomacy*, University of California Press, Berkeley

Grusin, R, 2010 *Premediation: Affect and Mediality after 9/11*, Palgrave Macmillan, Basingstoke

Hu, J, 2009 Speech at Tiananmen Square (for the 60th Anniversary of the People's Republic of China), 'President delivers speech at the ceremony', translated by *China Daily*, available from: http://www.chinadaily.com.cn/video/2009-10/01/content_8759569.htm [3 December 2013]

Jansson, A, and Lagerkvist, A, 2009 The Future Gaze: City Panoramas as Politico-Emotive Geographies, *Journal of Visual Culture* 8 (1), 29–53

Jeffreys, E, 2009 *China's Governmentalities: Governing Change, Changing Government*, Routledge, London and New York

Johnson, I, 2011 At China's New Museum, History Toes Party Line, *The New York Times*, 3 April, available from: http://www.nytimes.com/2011/04/04/world/asia/04museum.html?pagewanted=all&_r=0 [3 December 2013]

Lee, H, 2011 The Charisma of Power and the Military Sublime in Tiananmen Square, *The Journal of Asian Studies* 70 (2), May, 397–424

Massumi, B, 2002 *Parables for the Virtual: Movement, Affect, Sensation*, Duke University Press, Durham NC

National Museum of China, n.d. *The Road of Rejuvenation* [online], available from: http://en.chnmuseum.cn/english/tabid/520/Default.aspx?ExhibitionLanguageID=83 [3 December 2013]

Pedroletti, B, 2011 National museum lauds patriotic China, *Guardian Weekly*, 10 May, available from: http://www.theguardian.com/culture/2011/may/10/museums-architecture [3 December 2013]

Pile, S, 2010 Emotions and affect in recent human geography, *Transactions of the Institute of British Geographers* 35, 5–20

Rose, N, 1996 *Inventing Our Selves*, Cambridge University Press, Cambridge

— 1999 *Powers of Freedom: Reframing political thought*, Cambridge University Press, Cambridge

Staheli, U, 2011 Decentering the Economy: Governmentality Studies and Beyond, in *Governmentality: Current Issues and Future Challenges* (eds U Bröckling, S Krasmann, and T Lemk), Routledge, London

Thrift, N, 2004 Intensities of Feeling: Towards a Spatial Politics of Affect, *Geografiska Annaler: Series B, Human Geography* 86 (1), 57–78

— 2007 *Non-Representational Theory, Space, Politics, Affect*, Routledge, London and New York

Walters, W, 2012 *Governmentality: Critical encounters*, Routledge, London

Wang, Z, 2012 *Never Forget National Humiliation*, Columbia University Press, New York

Zhu, L, 2009 The twists and turns of a new road into the past, *China Daily*, 1 October, available from: http://www.chinadaily.com.cn/cndy/2009-10/01/content_8758430.htm [3 December 2013]

16

Revitalising Blackfoot Heritage and Addressing Residential School Trauma

Bryony Onciul

The residential school era (1831–1996) is a traumatic and hidden part of Canadian history that has only recently begun to be addressed by government and recognised by the public. For former pupils, known as survivors, there has been long-term intergenerational suffering, worsened by previous public denial. As the official government apology in 2008 stated:

> Two primary objectives of the residential schools system were to remove and isolate children from the influence of their homes, families, traditions and cultures, and to assimilate them into the dominant culture. These objectives were based on the assumption Aboriginal cultures and spiritual beliefs were inferior and unequal. Indeed, some sought, as it was infamously said, 'to kill the Indian in the child'. (Parliament of Canada 2008)

Rather than providing closure on this period, the apology has raised awareness and helped to reveal more about the era. The residential school system is within living memory and the cycles of trauma are still playing out within First Nations communities, affecting new generations who did not attend these schools. The government policies that established the residential schools are yet to be repealed, and current debates over First Nations educational reform highlight the relevance and influence of this history today.

This chapter considers the impact upon Blackfoot communities and some of the strategies the communities are using to address this difficult history and revitalise traditional culture and community pride. The Old Sun residential school on Siksika reservation is an example of a site of trauma that is now being reused as a place for healing and cultural renewal. While the residential school system disrupted traditional Blackfoot learning systems, it did not destroy them and the Blackfoot, along with many First Nations, continue to fight for their right to teach their culture both within their schools and through traditional ceremonies and sacred societies.

Blackfoot Education

The Blackfoot community consists of four Nations situated within traditional Blackfoot territory which extended from the North Saskatchewan River in Canada to the Yellowstone River in America and from the Eastern slopes of the Rocky Mountains to the Great Sand Hills in Saskatchewan (BGC 2001, 4). Three of the Blackfoot Nations are located in Southern Alberta, Canada, and are known as Kainai (Blood), Piikani (Peigan) and Siksika (Blackfoot); one Blackfoot Nation is located in Northern Montana, USA, known as Amsskaapipikani (Blackfeet).

Blackfoot history is long and rich and the people still live in the landscape where their creation stories are based. The Blackfoot have intricate traditional models for lifelong learning through participation in sacred societies, led by Elders who are the knowledge keepers in the community.

In 1877, the Blackfoot signed Treaty 7 with the British Crown at Blackfoot Crossing (see Fig 16.1). As part of the peace treaty, the Blackfoot ceded a proportion of their land in exchange for reservations, rations, agricultural assistance and education. As stated in Treaty 7: 'Her Majesty agrees to pay the salary of such teachers to instruct the children of said Indians as to Her Government of Canada may seem advisable, when said Indians are settled on their Reserves and shall desire teachers' (AANDC 1877).

The *Indian Act* (1876) came into law a year earlier and enabled the federal government to establish Indian residential schools and 'enter into agreements with religious organizations for the support and maintenance of children who are being educated in schools operated by those organizations' (*Indian Act* 1876; AANDC 2010). Treaty payments to students went directly to the schools and the Act required children to attend from age 7 to 16, although this could be extended at the discretion of the Minister (*Indian Act* 1876). A truant officer enforced the law, imprisoning parents who did not comply with the legislation and returning absent children to school 'using as much force as the circumstances require' (ibid). This legislation marked the start of the Blackfoot communities' experience of residential schools.

The Treaty and *Indian Act* disrupted and suspended traditional Blackfoot learning processes, mainly by removing the children from the community. The 1895 amendment to Section 114 of the *Indian Act* further disrupted traditional educational processes by prohibiting ceremonies, including the Blackfoot Sundance, which were vital to the intergenerational teaching and renewal

FIG 16.1.   BLACKFOOT CROSSING, LOCATION OF THE SIGNING OF TREATY 7 IN 1877.

of Blackfoot culture. An additional amendment in 1914 prevented First Nations people from appearing in traditional dress, which aimed to further curtail the maintenance and celebration of First Nations cultures.

THE RESIDENTIAL SCHOOL SYSTEM

The residential school system was a systematic effort to assimilate Indigenous children into Canadian culture, designed to end the practice of Indigenous cultures and languages within Canada. As Narcisse Blood, Kainai Elder and survivor of the residential school system, explains: 'The residential school [...] was a very, very concerted effort that was well resourced to destroy a knowledge base, because knowledge was transferred through the family' (Blood 2013, *pers comm*).

In 1887 the first Prime Minister of Canada, John A Macdonald, stated that: 'the great aim of our legislation has been to do away with the tribal system and assimilate the Indian people in all respects with the inhabitants of the Dominion as speedily as they are fit to change' (quoted in Milloy 1999, 6). According to the Minister of Indian Affairs, Frank Oliver in 1908, education would 'elevate the Indian from his condition of savagery' and make 'him a self-supporting member of the State, and eventually a citizen of good standing' (Milloy 1999, 3).

In practice the system was not only effective at disrupting Indigenous languages and cultural practices but also the lives of the students who attended. Residential schools were chronically underfunded and overcrowded, 'undermining the health of students' (TRC 2012a, 30). Children lived in poor conditions, on insufficient diets and were overworked in an effort to make the schools self-supporting, while receiving sub-standard education (TRC 2012a, 30; Milloy 1999). As 'late as 1950, according to an Indian Affairs study, over 40 per cent of the teaching staff had no professional training' (LHF 2013).

> The purpose of residential schooling was to assimilate Aboriginal children into mainstream Canadian society by disconnecting them from their families and communities and severing all ties with languages, customs and beliefs. To this end, children in residential schools were taught shame and rejection for everything about their heritage, including their ancestors, their families and, especially, their spiritual traditions. (Chansonneuve 2005, 5)

Indigenous communities have critiqued residential schools since their inception; however complaints were largely ignored until 1990 when the Grand Chief of the Assembly of Manitoba Chiefs, Phil Fontaine, spoke about his own experiences of sexual abuse at a residential school (TRC 2012a, 41). After this, survivors of these schools increasingly began to talk publicly about their experiences, which included physical, mental and sexual abuse; malnutrition and 'forced eating of rotten food'; corporal punishment; 'bondage and confinement'; forced labour; overcrowding; disease; high death rates and medical experimentation (LHF 2013). 'Generations of Aboriginal people today recall memories of trauma, neglect, shame, and poverty. Those traumatized by their experiences in the residential school have suffered pervasive loss: loss of identity, loss of family, loss of language, loss of culture' (ibid).

Although some residential school survivors recall happy memories of their childhood, the systematic destruction of Indigenous culture and the neglect, abuse and death of children in the care of the residential school system under government control is widely acknowledged as a shameful part of Canadian history. The system has been described as a national crime

(Bryce 1922; Milloy 1999), as genocide (Churchill 2004) and as the Canadian Holocaust (Truth Commission into Genocide in Canada 2001).

OLD SUN RESIDENTIAL SCHOOL

The Old Sun residential school on the Siksika Reservation is an important example of the trauma Blackfoot children experienced and how these sites of suffering have been reinvented today as places of healing.

The school was opened by Anglican Missionary Archdeacon Tims in 1886 with the enrolment of 15 pupils (GSA 2008a). Old Sun frequently featured in government reports. In 1907 the Bryce Report condemned the school, stating that the 'wells [were] all bad', the dormitories were crowded, and 'only 3 pupils [were] not under care of physician [sic]. Most are tuberculized [sic]. So many were sick at hospital that their classes were interrupted' (Bryce 1907). The government was forced to close the school for extended periods between 1907 and 1911 owing to 'rampant disease outbreaks […] and recurring unsanitary conditions' (GSA 2008a). During this time: 'disease outbreaks at the Old Sun School affected most of the student body, causing many deaths, quarantines and several temporary school closings ordered by the government' (ibid).

In a follow-up study in 1909 two Blackfoot schools, Old Sun and Peigan, were recorded as having death rates of 47% (Milloy 1999, 92). Corbett's survey of the residential school in 1920 and 1922 'found that little had changed' (ibid, 98).

> Such conditions had left their indelible and mortal mark on the children who Corbett found to be 'below par in health and appearance'. Seventy per cent of them were infected. They had 'enlarged lymphatic glands, many with scrofulous sores requiring prompt medical attention'. But it was the discovery that sixty per cent of the children had 'scabies or itch... in an aggravated form' that most upset Corbett, for this was unnecessary and a sign of gross neglect. (ibid, 99)

The conditions at Old Sun did not improve over time; by 1957 the situation was dire. A visiting medical doctor recorded that 'The children are dirty. The building is dirty, dingy and is actually going backwards rather than forwards' (ibid, 263).

In addition to the poor conditions and ill health, children were also subjected to physical, mental and sexual abuse, which was widespread in residential schools across Canada (TRC 2012a; BGC 2001, 76). Corporal punishment was the norm and at times was so severe that the Indian Department would attempt to intervene.

> In 1919, when a boy who ran away from the Anglican Old Sun School was shackled to his bed and beaten with a horsewhip until his back bled, Indian Commissioner W A Graham tried, without success, to have the principal fired. … The lack of support from Ottawa led Graham to complain that there was no point in reporting abuses since the department was too willing to accept whatever excuses the principals offered up. (TRC 2012a, 40)

The Blackfoot Elders acknowledge that 'sexual and physical abuse by staff and students was widespread' (BGC 2001, 76). While 'the official records of the residential school system make little reference to incidents of [sexual] abuse', the Truth and Reconciliation Commission has found that sexual abuse was known about by authorities from the early days of the residential school era (TRC 2012a, 42). Abuse was reported as early as 1868 and some perpetrators were fired; however, many were either tolerated or relocated to other schools (ibid, 42–3). 'Victims were often treated as liars or troublemakers. Students were taught to be quiet to protect themselves' (ibid, 45). Abuse created cycles of abuse, violence and trauma that have ongoing consequences for the communities affected (ibid, 44; Elias *et al* 2012).

The high death rates at Old Sun are part of a nationwide history of child mortality at residential schools that remains largely unaddressed:

> There are no clear records as to how many children died while attending residential schools, and the total may reach into the thousands… many children were buried in school cemeteries. In some cases, parents never were told what had become of their children. The memoirs of former residential school students are filled with remembrances of death and disease. (ibid, 30)

A *Missing Children and Unmarked Graves Project* is currently compiling a list of names, causes of death and possible grave sites (ibid, 17).

The Apology

On Wednesday 11 June 2008, the Prime Minister of Canada, Stephen Harper, officially apologised for the trauma caused by the residential school era.

> The government now recognizes that the consequences of the Indian Residential Schools policy were profoundly negative and that this policy has had a lasting and damaging impact on Aboriginal culture, heritage and language. … The burden of this experience has been on your shoulders for far too long. The burden is properly ours as a Government, and as a country. There is no place in Canada for the attitudes that inspired the Indian Residential Schools system to ever prevail again. You have been working on recovering from this experience for a long time and in a very real sense, we are now joining you on this journey. The Government of Canada sincerely apologizes and asks the forgiveness of the Aboriginal peoples of this country for failing them so profoundly. (Parliament of Canada 2008)

During the apology, one school – Old Sun – was mentioned by name by MP Duceppe for its staggering death rate of 47% (ibid).

The apology was preceded by the 2006 Indian Residential Schools Settlement Agreement that provided 'various measures for healing, reconciliation and redress, and accompanying funding commitments, including a Common Experience Payment' to former students and a Truth and Reconciliation Commission tasked with publicly addressing the hidden history of the residential school era (Senate Canada 2010, 1).

Transforming Old Sun

Efforts to address and heal from the residential school era are ongoing within First Nations communities in Canada. On the Siksika Reservation, Old Sun residential school has undergone a number of transformations since its infamous days of child abuse and mortality. In 1969 the government assumed control of the Old Sun residential school from the Anglican Church, then in 1971 Mount Royal College took over most of the building and created a Native Learning Centre for adult education, renaming the site Old Sun Community College (GSA 2008a). In 1978 the college became a separate institution run by the Siksika Nation (ibid). This was a radical change and one that has enabled the Old Sun residential school building to be reimagined as a place of Blackfoot learning.

Old Sun Community College provides 'programs and services tailored to meet the needs of the Siksika Nation and individuals while preserving the Siksika Way of Life' (Old Sun 2013b). 'We, at Old Sun Community College, have an important responsibility to enhance the sovereignty and Nationhood of Siksika […] it promotes the lifelong learning of Siksika culture, language, history, and knowledge to our future generations' (ibid). Reviving the Blackfoot language, which was banned in residential schools, is vital to the process of maintaining Blackfoot culture and heritage as the language embodies traditional Blackfoot concepts.

In 1967 Old Sun also became the home of the Siksika Museum. The museum moved to a new building in 2007 and was renamed Blackfoot Crossing Historical Park (see Fig 16.2). Blackfoot Crossing presents the traumatic history of the residential school era in its display entitled 'Survival Tipi' (see Onciul 2014 for further discussion). It also promotes and maintains the Siksika language, culture and traditions. Reviving the language, as Jack Royal, Director of Blackfoot Crossing, explains 'is a huge undertaking… our language is in a critical stage… time is of the essence, so that is why… that is one of our priorities' (Royal 2008, *pers comm*).

FIG 16.2.   BLACKFOOT CROSSING HISTORICAL PARK.

While the efforts to reuse Old Sun for the positive processes of reviving and teaching Blackfoot culture and language are very important, Old Sun College recognises the need to directly address the trauma embodied in the history of the building. Plans are underway to build a Healing Lodge in the shape of a sweat lodge on the old school site, to directly 'address the horrific past of the boarding school era' and create a place for survivors to come and mourn (Old Sun 2013a).

> We cannot just paint over the walls with new paint without addressing its past; nor can we seal up the dark coal room in the Boiler room just because it makes us feel uncomfortable. In order for us as a people and individuals to move forward, the past needs to be addressed. That past, of multiple abuses that our family and relatives suffered continues to affect our people through depression, low self-esteem, inability to express and feel love, vulnerability towards re-victimization, alcohol or drug abuse, suicide, poor parenting skills and low academic achievement to name a few. We need to address the abuse that was paired with education and separate the two issues. Unless we address the root of the dysfunction that plagues our people we will continue to battle with its fruits. (Old Sun 2013a)

The task is daunting, as they explain:

> The Healing Centre will have to address the brain washing and mind control that went on by the authorities in charge and the negative policies and procedures that were implemented at the boarding school; the witnessing and receiving of physical abuse, sexual abuse, sexual harassment, prostitution, and child pornography to name a few; also, the pain and rage of the residential school victims and their children. (ibid)

The change in use of this building, from a place of cultural suppression to one that celebrates Blackfoot culture and directly addresses the trauma of this period, is extremely significant – particularly because the survivors of these schools are still living members of the community and the effects of the era continue to be felt within families and the wider community.

Traditional Teaching

Despite the disruption caused by the residential school era and the banning of sacred ceremonial practices in 1895, the Blackfoot Nations have maintained their culture and are reviving traditional practices. A key element to this cultural renaissance is the return of sacred cultural material that frequently ended up in museums during the residential school period. In 2000 the Blackfoot successfully changed the law in Alberta with the help of staff from the Glenbow Museum and Royal Alberta Museum. The *First Nations Sacred Ceremonial Objects Repatriation Act* came into force in 2004 and enabled these museums to return sacred material to the Blackfoot. This was the first enactment of its kind in Canada. Repatriation has supported the renewal of sacred societies and ceremonies in which traditional teaching can once again take place. These developments are extremely important in maintaining living Blackfoot culture.

Both traditional life and reconnection with the land can be a powerful healing tool for survivors, such as Elder Narcisse Blood. 'I am sitting here, I could have been dead. A lot of my fellow students have died from these residential schools. So it was our ways [traditional culture] that we hung on to that enabled people like myself and others to be able to experience who we are'

(Blood 2013, *pers comm*). The survival of both the Blackfoot people and their culture demonstrates the failure of the residential school policies of assimilation and the incredible resilience of the people.

RECENT REVELATIONS

Efforts to address the trauma of the residential school era and revive traditional culture and languages have been taking place on a local level, such as the changes made to Old Sun, and at a national level, through the work of bodies such as the Truth and Reconciliation Commission. However, these attempts to reveal and heal from the period have been challenged by recent findings of medical experimentation in residential schools.

In July 2013, historian Ian Mosby's research on the Canadian government's nutritional experiments on First Nations communities and residential school children shocked the Canadian press. Blackfoot children at St Paul's and Blood residential schools were part of the experiments Mosby exposed. While children at the Blood residential school received 'Vitamin B flour' supplements, St Paul's was the 'control' site and no changes were made to the children's diets, 'despite the fact that the initial investigation had found that students were being fed poor quality, unappetizing food that provided inadequate intakes of vitamins A, B, and C as well as iron and iodine' (Mosby 2013, 162).

Further evidence of ear experimentations were brought to light by CBC News in August 2013, and have led survivors to question the Indian Residential School Settlement Agreement reached in court in 2007 (Porter 2013). Survivor Richard Green argued that 'the legal process placed an unfair burden on survivors to recall what happened when they were children, while the government withheld documents like the ones obtained by CBC News' (ibid).

No new apology has been offered. In January 2013 the Truth and Reconciliation Commission took the government to court to order them to supply the relevant archival documents needed for their review of the residential school era (CBC News 2013a; Fontaine v Canada 2013). Justice Murray Sinclair, Chair of the Commission, publicly announced in July 2013 that they 'haven't seen the documents start to flow yet' and stated their concern that the government may not have 'sufficient resources and time to be able to get them to us before our mandate as a commission expires on July 1, 2014' (ibid). The AFN Regional Chief, Bill Erasmus, argued that 'the federal documents show that the government either doesn't know what's in its own records or that there may be an effort to actually suppress information' (ibid). 'We believe that what's already been exposed represents only a fraction of the full, true and tragic history of the residential schools. There are no doubt more revelations buried in the archives' (Erasmus quoted in CBC News 2013a).

EDUCATION POLICY

In addition to the exposure of medical abuses, the proposed *First Nations Education Act*, announced in October 2013, was felt by some First Nations to echo the attitudes of the residential schools era that the 2008 apology insisted would never again prevail. The Act proposes annual inspections of First Nations schools, with schools who fail to meet government standards being taken into administration by the government (AANDC 2013, 23–4). Grand Chief Derek Nepinak of the Assembly of Manitoba Chiefs stated that 'Mr Harper's new legislation is a trip back down residential school processes and the elimination of Indian control over Indian education' (CBC News 2013b). While First Nations schools have been allowed to deliver First Nations cultural content and employ First

Nations teachers, they are still run by federal government under the *Indian Act* and are under-funded, poorly maintained and continue to fail students (AFN 2012; Statistics Canada 2006).

On 7 February 2014 at the Kainai High School on the Blood Reserve, Prime Minister Stephen Harper announced 'an historic agreement between the Government of Canada and the Assembly of First Nations to reform the First Nations K-12 education system through the [renamed] *First Nations Control of First Nations Education Act*' (Prime Minister of Canada 2014). Harper stated that the Act will 'finally repeal the provisions of the Indian Act that established residential schools' (Graveland 2014). Chief Shawn Atleo of the Assembly of First Nations said 'today is a victory for First Nations leaders and citizens who have for decades, indeed since the first generation of residential school survivors, called for First Nations control of First Nations education' (AFN 2014). Despite the changes made to the Act, many First Nations people are still critical of the proposal in its present form and do not feel adequate consultation has taken place (CBC News 2014). The Kainai Nation rejected the original proposed Act and Kainai Chief, Charles Weaselhead, stated 'we agreed to host this national announcement, but in no way endorse the proposed legislation in its present form. However, we are open to continued dialogue and building relationships' (Blood Tribe 2014).

CONCLUSION

The story is constant in that it's only beginning to be told. It is suppressed today what happened, and yet back when it was being legislated it was honoured. You know, it's right there, the literature, the paper trail is there. ... What would motivate people? Some of the answers are in the way we dehumanise people so that we can destroy them. (Blood 2013, *pers comm*)

As new abuses from the residential school era come to light and current education policy continues to be debated, it is clear that the efforts for truth and reconciliation are ongoing. The work of the Truth and Reconciliation Commission is part of a process of decolonisation that continues to struggle against colonial laws and institutionalised racism in mainstream society.

Despite continuing racial inequality in Canada, the Blackfoot are successfully reviving their culture, traditions and language, both within and beyond the classroom. Traditional models of intergenerational learning and reconnection with Blackfoot culture can support healing for residential school survivors and the community as a whole. Sacred items are a vital part of living Blackfoot culture and the ongoing repatriation of this material from museums, such as Glenbow, supports these processes of healing and decolonisation. The work of the Truth and Reconciliation Commission in Canada is making important steps in addressing the residential school era. The depth and breadth of the formerly hidden history of the residential school era will require further research to fully understand the trauma, the ongoing consequences and ways to heal, for both the First Nations communities and Canadian society. How these findings are then disseminated, publicly received and acted upon will be the test of the apology.

A NOTE ON DATES

The Mohawk Institute in Brantford, Ontario, was the 'oldest continually operated Anglican residential school in Canada'; established in 1828 it began to take boys as boarders in 1831, extending this to include girls in 1834 (GSA 2008b). This and Mount Elgin school in Munceytown,

Ontario, founded in 1850, were funded by the Canadian government after its establishment in 1867 (TRC 2012b, 5). 1883 marks the beginning of the residential schools as a system, with the founding of three new schools in the Prairies by the Canadian government: Qu'Appelle, High River, and Battleford Schools (TRC 2012b, 6).

BIBLIOGRAPHY AND REFERENCES

AANDC, 1877 *Treaty Texts – Treaty and Supplementary Treaty No 7: Copy of Treaty and Supplementary Treaty No 7 between Her Majesty the Queen and the Blackfeet and Other Indian Tribes, at the Blackfoot Crossing of Bow River and Fort Macleod (1966, Reprinted from the Edition of 1877)* [online], available from: http://www.aadnc-aandc.gc.ca/eng/1100100028793/1100100028803 [15 April 2014]

— 2010 *Backgrounder – Changes to the Indian Act Affecting Indian Residential Schools* [online], available from: http://www.aadnc-aandc.gc.ca/eng/1100100015573/1100100015574 [12 September 2013]

— 2013 *Working Together for First Nations Students. A Proposal for a Bill on First Nations Education, October 2013* [online], available from: http://www.aadnc-aandc.gc.ca/DAM/DAM-INTER-HQ-EDU/STAGING/texte-text/proposal_1382467600170_eng.pdf [30 October 2013]

AFN, 2012 *Chiefs Assembly on Education: Federal Funding for First Nations Schools* [online], available from: http://www.afn.ca/uploads/files/events/fact_sheet-ccoe-8.pdf [16 October 2013]

— 2014 *Assembly of First Nations National Chief Agrees on a New Approach to Achieve First Nations Control of First Nations Education* [online], 7 February, available from: http://www.afn.ca/index.php/en/news-media/latest-news/assembly-of-first-nations-national-chief-agrees-on-a-new-approach-to-a [4 April 2014]

BGC (Blackfoot Gallery Committee), 2001 *Nitsitapiisinni: The Story of the Blackfoot People*, Key Porter Books, Toronto

Blood, N, 2013 Personal communication (interview with the author), 21 August, Fort McLeod

Blood Tribe, 2014 *Blood Tribe to Host the Prime Minister's Announcement for Action on First Nations Education* [online], 6 February, available from: http://www.bloodtribe.org/content/communiqu%C3%A9-blood-tribe-chief-and-council-february-6-2014 [5 March 2014]

Bryce, P H, 1907 *Report on Indian Schools of Manitoba and the North-West Territories*, Peel 3024, Government Printing Bureau, Ottawa

— 1922 *The Story of a National Crime*, James Hope & Sons, Limited, Ottawa, available from: http://www.archive.org/stream/storyofnationalc00brycuoft/storyofnationalc00brycuoft_djvu.txt [26 February 2014]

CBC News, 2013a *Aboriginal children used in medical tests, commissioner says* [online], 31 July, available from: http://www.cbc.ca/news/politics/aboriginal-children-used-in-medical-tests-commissioner-says-1.1318150 [12 September 2013]

— 2013b *Manitoba Chief slams First Nations education reform proposal* [online], 23 October, available from: http://www.cbc.ca/news/canada/manitoba/manitoba-chief-slams-first-nations-education-reform-proposal-1.2187509 [24 October 2013]

CBC News, 2014 *First Nations to get more control over education, Ottawa says* [online], 7 February, available from: http://www.cbc.ca/news/canada/calgary/first-nations-to-get-more-control-over-education-ottawa-says-1.2527266 [5 March 2014]

Chansonneuve, D, 2005 *Reclaiming connections: Understanding residential school trauma among Aboriginal people*, Aboriginal Healing Foundation, Ottawa

Churchill, W, 2004 *Kill the Indian, Save the Man: The Genocidal Impact of American Indian Residential Schools*, City Lights Books, San Francisco

Elias, B, Mignone, J, Hall, M, Hong, S P, Hart, L, and Sareen, J, 2012 Trauma and suicide behaviour histories among a Canadian indigenous population: An empirical exploration of the potential role of Canada's residential school system, *Social Science & Medicine* 74 (10), 1560–9

Fontaine v Canada, 2013 ONSC 684, available from: http://s3.documentcloud.org/documents/563338/ fontaine-v-canada-2013-onsc-684.pdf [26 February 2014]

Graveland, B, 2014 Harper unveils retooled First Nations education plan [online], *The Canadian Press*, available from: http://globalnews.ca/news/1132666/new-aboriginal-education-plan-coming-friday/ [5 March 2014]

GSA, 2008a *Anglican Church of Canada: Mission and Justice Relationships: Old Sun School – Gleichen, AB* [online], available from: http://www.anglican.ca/relationships/trc/histories/old-sun [2 September 2013]

— 2008b *Anglican Church of Canada: Mission and Justice Relationships: The Mohawk Institute – Brantford, ON* [online], available from: http://www.anglican.ca/relationships/histories/mohawk-institute [14 October 2014]

Indian Act, 1876 Statutes of Canada 1876, 39 Vict., c.18

LHF, 2013 *About Residential Schools* [online], available from: http://www.legacyofhope.ca/about-residential-schools/conditions-mistreatment [10 September 2013]

Milloy, J S, 1999 *A National Crime: The Canadian Government and the Residential School System 1879 to 1986*, University of Manitoba Press, Winnipeg

Mosby, I, 2013 Administering Colonial Science: Nutrition Research and Human Biomedical Experimentation in Aboriginal Communities and Residential Schools, 1942–52, *Histoire sociale/Social History* 46 (1), 145–72

Old Sun, 2013a *Future Healing Lodge* [online], available from: http://www.oldsuncollege.net/calendar-of-events/future-healing-lodge.html [1 October 2013]

— 2013b *Vision and Mission* [online], available from: http://www.oldsuncollege.net/discover-old-sun/vission-and-mission.html [1 October 2013]

Onciul, B, 2014 Telling Hard Truths and the Process of Decolonising Indigenous Representations in Canadian Museums, in *Challenging History in the Museum* (eds J Kidd, S Cairns, A Drago, A Ryall, and M Stearn), Ashgate, Farnham

Parliament of Canada, 2008 39th *Parliament, 2nd Session, Edited Hansard: Number 110, June 11* [online], available from: http://www.parl.gc.ca/HousePublications/Publication.aspx?DocId=3568890&Language= E&Mode=1&Parl=39&Ses=2 [20 June 2008]

Porter, J, 2013 Ear experiments done on kids at Kenora residential school, *CBC News* [online], 8 August, available from: http://www.cbc.ca/news/canada/thunder-bay/story/2013/08/08/tby-documents-show-kenora-residential-school-ear-experiments.html [19 August 2013]

Prime Minister of Canada, 2014 *PM announces an historic agreement with the Assembly of First Nations to reform the First Nations education system* [online], 7 February, available from: http://pm.gc.ca/eng/news/ 2014/02/07/pm-announces-historic-agreement-assembly-first-nations-reform-first-nations#sthash.SEEED vq0.dpuf [4 March 2014]

Royal, J, 2013 Personal communication (interview with the author), 25 August, Blackfoot Crossing Historical Park

Senate Canada, 2010 *The Journey Ahead: Report on Progress Since the Government of Canada's Apology to Former Students of Indian Residential Schools* [online], available from: http://www.afn.ca/uploads/files/the-journey-ahead.pdf [18 September 2013]

Statistics Canada, 2006 *Educational Portrait of Canada, 2006 Census*, Catalogue no 97-560-X

TRC, 2012a *Truth and Reconciliation Commission of Canada: Interim Report* [online], TRCC, Winnipeg, available from: http://www.tunngavik.com/files/2012/02/trc-interimreport-feb2012-1.pdf [1 October 2013]

— 2012b *Canada, Aboriginal Peoples, and Residential Schools: They Came for the Children* [online], available from: http://www.myrobust.com/websites/trcinstitution/File/2039_T&R_eng_web%5B1%5D.pdf [14 October 2014]

Truth Commission into Genocide in Canada, 2001 *Hidden from History: The Canadian Holocaust. The Untold Story of the Genocide of Aboriginal Peoples by Church and State in Canada* [online], available from: http://canadiangenocide.nativeweb.org/genocide.pdf [12 January 2012]

17

Reading Local Responses to Large Dams in South-east Turkey

Sarah Elliott

Introduction

Development interventions to transform natural resources often involve displacement in some form (WCD 2000, 102); in the case of large dams and their related infrastructure, this has led to people's physical dislocation, livelihood deprivation and loss of access to cultural resources and heritage as river basins around the world are significantly, and often irreversibly, altered. The adverse socio-cultural impacts for the often involuntarily and coercively displaced – significant globally in magnitude, extent and complexity – warrant detailed deliberation when decisions are made to construct large dams, and yet they are still often not acknowledged in the planning process and can remain unrecognised during project operations (ibid, 98). In south-east Turkey, where ambitious hydro-development of the Tigris and Euphrates basins is ongoing under the Southeastern Anatolia Project (Turkish acronym GAP), fundamental omissions regarding cultural heritage impacts have coalesced around the relationship between affected communities and their cultural resources as local-scale concerns and interests are inadequately understood, often misprized and largely excluded from programmes that collapse cultural heritage aspects into limited archaeological salvage work.

The Ilisu Hydroelectric Dam, currently under construction on the River Tigris and scheduled for completion in 2014, is a controversial example of the persistent failure of developers and state officials to adequately consult with local people on their impacted cultural resources. In the case of perhaps the most important of hundreds of cultural and historical sites at risk of inundation by the Ilisu reservoir – Hasankeyf – residents have not officially[1] been 'asked to explain which sites form a central core to their identities or how the loss or preservation of certain sites will influence their cultural well-being... [or] to assess the impact on their way of life by the loss of their heritage landscape' (Schmidt 2000, 14), leaving environmental impact assessment reports (eg Hydro Concepts *et al* 2005) and cultural heritage action plans (eg Ilisu Consortium 2006) devoid of meaningful significance assessment data. Understanding significance, particularly as it pertains to social value – a concept that 'embraces the qualities for which a place has become a focus of spiritual, political, national or other cultural sentiment to a majority or minority group' (Australia ICOMOS 1999, 12) – is often central to arguments for conserving a place and its liminal positioning for Hasankeyf requires urgent critical attention.

The literature on broader cultural heritage management for the Ilisu project indicates an array of problems and procedural failings underlain by structural, institutional and political factors

---

[1] There has been a recent, but limited, study by the Turkish NGO Doğa Derneği. See Işıklı *et al* 2012.

(Hildyard *et al* 2000; ICOMOS 2000; Özdoğan 2000; Ronayne 2002 and 2008), but the issues can be effectively interrogated and nuanced through local responses to the dam. Utilising data collected from a range of respondents in Hasankeyf in the summer of 2005,[2] this chapter explores how some responses – specifically evidence of muted rallying and mobilisation against threats to cultural and natural heritage – can complicate our understanding of the significance of heritage to affected communities, while others reveal the potential devastation of displacement as local people ascribe multiple typologies of value to their heritage resources.

HASANKEYF: CAPTURING SOCIO-CULTURAL VALUE

A full articulation of heritage values should form a central reference point for conservation decisions about cultural infrastructure in large dam impacted zones[3] and frame our understanding of the importance and consequences of heritage loss and displacement. However, taking an anthropological perspective that commends attempts to understand the full range of values and value processes attached to heritage requires a level of democratic agency and participatory stimulation not established for the Ilisu Dam project. The reasons for this arguably reside in the entrenched traditional attitudes of some heritage professionals tasked with the 'rescue' of Hasankeyf who hold normative, art historical views of significance, but are almost certainly more deeply located in the political and cultural geography of the south-east. This extremely difficult broader context of 'exceptionality, resistance and contestation' (Watts 2009, 2) that defines the region, rooted in war and the state's historical and ongoing cultural repression of ethnic Kurds, constrains contemporary, postmodern imperatives towards pluralism and denies the polyvalence of heritage. Communities have been disenabled and disempowered as their stories, objects and places are not valued.

Capturing socio-cultural value ascribed to cultural resources by local people – a demographic of predominately ethnic Kurds and Arabs[4] – in Hasankeyf and using it to better articulate displacement issues in the south-east is fraught with conceptual and practical difficulties. The subjectivity and contingency (socially and spatially constructed) of heritage values, with variance in epistemology, modes of expression and in how a specific type of heritage is appraised by different stakeholders, together with the constant changing of values, present difficulties in establishing a framework or typology of values (Mason 2002, 9) and the exigencies of researching communities experiencing fear, trauma and lack of freedom of expression are challenging. Attempts, however, have been made elsewhere to question what constitutes value and how frameworks can be constructed (see de la Torre 2002; Gibson and Pendlebury 2009) and these offer

---

2   Primary data was gathered by the author through in-depth, hybrid unstructured/semi-structured interviews with 25 residents of Hasankeyf together with local officials (Headman, Mayor, *Kaymakam* or District Head), historians and the Director of Archaeological Excavations for Hasankeyf. It formed the basis of AHRC-funded PhD work conducted at ICCHS, Newcastle University, from 2003 to 2007.

3   The World Bank (1994, 6), a major actor in development processes whose guidelines represent an oft-used international gauge for impact assessment, notes that environmental assessment 'should cover cultural heritage values of both major and minor significance as they may be subject to different types of impacts within the same project'.

4   In Hasankeyf the population of Kurds at the 1997 census was estimated at 60% of the 5624 total (ECGD 1999, 44) while its Arab and other ethnic communities accounted for 40%.

a point of departure for this discussion. Methodological frameworks that recognise the research as 'sensitive', potentially posing a threat to those who are or have been involved in it and thus encompassing the topic, the consequences and the situation (Lee 1993, 4), also provided the tools to meaningfully engage with respondents.

HASANKEYF: ASCRIBING VALUE

Historical value

The ancient city of Hasankeyf, with remains of mausolea, religious structures, monumental gates, water systems, stairways, palaces and private buildings extant at all levels (see Fig 17.1), bearing testament to a history dating back 2000 years (although its earliest communities probably settled in the late Assyrian and Urartu periods, c. 7th century BC) and featuring work of the highest aesthetic quality (see Meinecke 1996; Arık 2003; Gabriel 1940), is recognised by travellers and researchers through past centuries (eg Iosaphat Barbaro, Samuel Guyer, Friedrich Sarre and Gertrude Bell), archaeologists, ICOMOS Turkey[5] and the Turkish state[6] for its historical and unique artistic value.

Local people who live in and among its ruins also find in Hasankeyf the capacity to convey, embody and stimulate a relation, or reaction, to the past and ascribe historical value to the site. The very early settlement of the place has created a powerful dynamic for respondents, prompting liberal use of the descriptor 'cradle of civilisations', with many commenting on the fact that Hasankeyf has been capital to a multiplicity of faiths and ethnicities. This plurality seems to provide additional cultural value for respondents as they use it to produce a landscape sustained by nostalgic narratives of multicultural tolerance. Here, different religious buildings standing side-by-side serve as material evidence of the social memory of past harmony:

> When the Yezidi[7] come here and visit their grandparents' houses, or when a Christian comes here and says 'my ancestors lived here', or when a Muslim comes here and visits the tombs, I see many ethnicities and religions here together and this shows me that Hasankeyf is a world city. Also, you can see churches and mosques together, next to each other.
>
> (H15 2005, *pers comm*)

Although in reality the landscape of Hasankeyf and the social memory produced through it is not innocent, this pervasive notion is salient for communities attempting to recover from ethnic strife. The historical value of ancient sites more generally in recovery processes, particularly for the Kurdish milieu whose history in the region was subject to 'cleansing' by the Turkish state, has also been noted by Ronayne (2002, 64–5):

---

5   ICOMOS Turkey has promoted Hasankeyf's inclusion in the World Heritage List (see Ahunbay 1999).
6   The town was awarded complete, class 1 archaeological protection by the Turkish government in 1978 (decision no A-1105) under which the town, in its entirety, should be protected against negative impacts (see Arık 2001, 797).
7   The Yezidi are a Kurdish-speaking people who adhere to a religion linked to Zoroastrianism and Islamic Sufism.

FIG 17.1.   THE REMAINS OF AN ARTUKID MADRASSAH NEAR THE ONION-DOMED TOMB OF THE
AKKOYUNLU PRINCE ZEYNEL BEY AT HASANKEYF.

> One way in which ancient sites do matter for present day communities is in recovering from
> this repression of their cultural rights and untangling their place in the complex history of
> ethnicity, movements of tribes, ancient empires and the struggles against them, household,
> village and regional relations over millennia in the Upper Tigris.

Further historical value has accrued for many respondents through Hasankeyf's association
with particular events and individuals, most notably relating to the Muslim conquests and the
commanders Khālid ibn al-Walīd (known as the Drawn Sword of God), Ṣaḥābī ʿAbd Allāh
ʿYukanna' (a companion of the Prophet Muhammad) and Imām Muḥammad b. ʿAbd Allāh
aṭ-Ṭaiyār (Imam Abdullah, grandson of the Prophet Muhammad's uncle). The latter's tomb (see
Fig 17.2) forms part of the sacred geography of the area and is central to ascriptions of religious
and spiritual value.

Religious and spiritual value

Local people feel a profound connection to Hasankeyf through its many religious sites, one
respondent even venturing that 'every stone here is holy' (H23 2005, *pers comm*). The sites are
shrines and feature in the everyday lives of the community, highly valued not only as loci of
worship, but as places of new life and healing, steeped in mystical tradition and even miracle.

FIG 17.2. THE MAUSOLEUM OF WHAT ARE BELIEVED TO BE THE REMAINS OF THE ISLAMIC MARTYR IMAM ABDULLAH.

Several respondents relayed the power of these sites, some through anecdote and others through folkloric narrative. The Imam Abdullah tomb, privileged by many because 'Die Bedeutung des Wallfahrtsortes wird durch den Glauben ersichtlich, daß der siebenmalige Besuch der Grabstätte der für jeden Muslim verpflichtend vorgeschriebenen Wallfahrt nach Mekka gleichwertig sei'[8] (Väth 1992, 66) and attracting around 30,000 pilgrims annually (Hildyard *et al* 2000, 27), generated powerful responses locally. One couple pointed out the paradigmatic event connected with prayer at the tomb – begetting children: 'The Imam Abdullah tomb is important. After visiting and praying at the tomb, these children were born [indicates several offspring]. We accept that they are the gift of Allah' (H17 and H18 2005, *pers comm*). They also described the mausoleum as a site of ritual: 'We go every year to sacrifice a goat and have ceremonies' (ibid).

The resting place of another Muslim commander, known locally as Aslan Baba (Lion Father), is valued as a site of healing. Respondents attested to the highly therapeutic properties of the water there for sufferers of rheumatism and arthritis (H17, H18 and H25 2005, *pers comm*).

[8] 'The importance of the shrine is seen through the belief that visiting the tomb seven times is equivalent for every Muslim to the obligatory pilgrimage to Mecca.'

They also identified the shrine of Sheik Sevinç with restorative power, its water being particularly efficacious for those with stomach complaints.

The presence of these sites, and other important religious touchstones ranged around the locale, seems to amplify a proximity to God for local people, and this creates a particular dynamic with the impending dam. Armstrong (2002, 193) contends that when people are threatened, devotion to sacred places becomes problematic, for 'when they visit their shrines, they not only believe they are meeting God there, they also have an encounter with their deepest selves'. This suggests that if indeed these shrines are bound up with a sense of self, then as the sites are threatened, so the very selves of the people are in jeopardy.

Aesthetic value

> Whenever I go up onto the cliffs and see Hasankeyf under my feet, I feel like I own the entire world. (H10 2005, *pers comm*)

Comment from this respondent suggests an aesthetic dynamic to her appreciation of Hasankeyf. As well as an implicit understanding of the centuries-long interweaving of world civilisations evidenced in the fabric of the surviving structures and of her place within this continuum, her remarks allude to the appeal of exceptional natural beauty and the aesthetic qualities of grandeur and viewability (see Fig 17.3). Aesthetic value was constructed and ascribed using sight by many

FIG 17.3.   THE SWEEPING DRAMA OF THE LANDSCAPE AT HASANKEYF. IN THE BACKGROUND, THE HIGH ROCKY TERRAIN SURROUNDED BY DEEP GORGES CUT BY THE RIVER TIGRIS IS THE RESPONDENT'S VANTAGE POINT.

respondents in an enthusiastic articulation of the beauty of place. In local eyes, historical, cultural and natural landscapes mesh with delightful consequences: 'Hasankeyf will be like a paradise [if it is saved], because the history and geography combine in a beautiful way' (H24 2005, *pers comm*). Indeed, the descriptor 'beautiful' was often coupled to the city: 'Our city is very beautiful. We love our city' (H17 2005, *pers comm*) and 'We would like this beautiful city to become the heritage of the world' (H9 2005, *pers comm*). This coupling did not suggest a hierarchy of beauty in the elements of place, but more a perception of the exquisiteness of the whole. It is interesting to note here how incongruously this view sits with salvage plans that conceive the rescue of individual architectural elements through removal and relocation (see Ilisu Consortium 2006). Aesthetics is often observed in isolation from local communities, but revealed in a social context an interconnection with social value could be evidenced through respondents' connection to place.

Social (recreational) value

Another layer of local valuing features those aspects of Hasankeyf that enable and facilitate social connections, specifically recreational activities that 'do not necessarily capitalize directly on the historical values of the site but, rather, on the public-space, shared-space qualities' (Mason 2002, 12). Cultural proclivities mean that leisure participation is gender differentiated in Hasankeyf and swimming is thus recreationally valued by men. Many respondents expressed their great pleasure in the Tigris not only as a setting for swimming, but also for fishing, activities borne out by observation. The riparian spaces, however, are valued by the women of the town as sites for their primary recreational pursuit: picnicking. At certain times of the year whole streets of female neighbours join forces to enjoy lunch *al fresco* on the Tigris banks or up on the castle (H7 2005, *pers comm*).

'Use' and 'non-use' value

Although not socio-cultural valuing *per se* (there is overlap), respondents also attributed both 'use' value (pertaining to the goods or services emanating from material culture that are market tradeable and priceable), recognising the potential of the site to generate revenue through tourism, and 'non-use' economic value, apperceiving the public-good qualities of heritage. This respondent ascribes the latter as he expresses 'bequest motivations accruing from the desire to conserve cultural goods for future generations' and 'altruistic feelings associated with the knowledge that other people may enjoy cultural heritage' (Mourato and Mazzanti 2002, 51):

> As we were born in Hasankeyf, the thousands of years of heritage here is more valuable than a dam [with a life] of 40 years. So we would like to preserve the remains of the heritage for the next generations. ... Hasankeyf is not the heritage of Turkey, but the heritage of the world.
>
> (H9 2005, *pers comm*)

LOCAL RESPONSES TO THREAT

Despite multiple attributions of value, at the time of this research there appeared to be a duality of local scale response to the threats facing Hasankeyf. In the town itself, highly active rallying and mobilising contingents against the dam – most notably Mayor Abdulvahap Kusen and Çorban Ahmet, a former shepherd and leader of the nascent Surviving Hasankeyf Association – vied with the passivity of more muted, quiescent respondents. However, meta-narratives contained within the responses of the historical procrustean tendencies of the state towards ethnic minorities, 'militarised water' and economic marginalisation contribute to an understanding of this passivity.

The bleak acquiescence of many respondents towards the fate of Hasankeyf can be ascribed in part by the unforgiving everyday realities of poverty, producing feelings of powerlessness and fatalism. The notion of fatalism as theological necessity (it is God's will) is afforded particular emphasis in West Asian society as a result of repressive political and economic systems that have effectively immobilised communities (Khoury 2001, 23) and featured heavily in local responses. As 'a psychological mechanism that reconciles people with their harsh reality and justifies their inability to control and shape their destinies' (ibid), it has particular resonance:

> We have limited power. The whole community is against this idea [the Ilisu Dam] but we have limited power. They [the Surviving Hasankeyf Association] will do whatever they can, but the *last word is Allah's word. … Only Allah stops this project.* (H10 2005, *pers comm*)

> People are poor here and they are not able to protest. (H25 2005, *pers comm*)

Passivity in response to the dam is enmeshed in politics. Increasingly elsewhere, 'governments have imposed a security framework over water to insure the priority of sovereign rights to manage and use strategic resources. Defining water as a national security resource militarizes the water commons, and people who protest the consequences of dams can be charged with crimes against the state' (Johnston 2012, 310). GAP is a key example of the militarisation of the water commons (see ibid), and opposition to the dam is considered anti-state, misconstrued by the authorities as evidence of sympathy for, or belonging to, the Kurdistan Workers' Party (Turkish acronym PKK) (Yildiz 2001; Hildyard *et al* 2000). Many respondents were quick to locate anti-dam action within the regional dynamic; this young man hints at the consequences:[9]

> This region is different. It's not like a place in Turkey on the west side. If you undertake certain types of behaviour here, you will be potentially guilty because you are living in a different region. So you have to be more careful about your behaviour. If you do do something, you can easily be punished. (H25 2005, *pers comm*)

Planning blight, a phenomenon caused by delays between the decision to build a dam and construction (manifest for Hasankeyfians in decades of underdevelopment and a lack of welfare investment), together with its correlatives of psychological stress and outmigration, have also impacted community action:

> *H9*: The people here have made no attempt to stop the Ilisu Dam. As a community, there is no action. … Of course, if you want to do something, society has to have great power. But the period is very long. The discussions [about Ilisu] started before I was born. Now I am 60 years

---

[9] Accusation of PKK membership is punishable by up to 15 years' imprisonment (Hildyard *et al* 2000, 18), and recent penal narratives of the region extended to physical abuse and torture.

old and have finished my life and they still haven't come to a decision. The whole of the society is disturbed because of this. They want a decision. *Author*: If you get a decision, do you think this will prompt action by the community? *H9*: Yes. (H9 2005, *pers comm*)

The local people cannot really do anything because there are not many Hasankeyfians here now.
(H13 2005, *pers comm*)

Local people's ability to act is further stymied by a lack of knowledge. Information on impacts, as Ronayne (2002, 11) has noted, is central to an affected community's ability to effectively defend its cultural and social rights, and its non-dissemination further shaped responses to the dam.

CONCLUSION

Although only briefly sketched here, an indication of the range of socio-cultural values ascribed by the inhabitants of Hasankeyf to their cultural heritage has been demonstrated. They form an important component of the significance continuum, with other aspects, such as socio-cultural cognitions of place (part of a person's place identity), further underpinning valorisation processes. In Hasankeyf these were expressed through a sense of belongingness, as physical, social and autobiographical 'insideness'[10] emerged through respondent discourse (see Elliott 2007). Mobilising and rallying against threats to cultural resources is also a normative, conspicuous indicator of value, often presaging positive heritage interventions. However, while international, national, regional and local Mayoral action in opposition to Ilisu has been intense and to a degree successful,[11] the silence of the inhabitants themselves arguably compromises their own suite of values and concomitant position when heritage decisions are made. Interrogating this passivity, however, has revealed nuances of context, as essential to deeply and seriously consider as the site itself.

Given that the displacement and loss of valued cultural heritage comprising Hasankeyf would deracinate its inhabitants, potentially entail a dislocation of place identity, further weaken an already assailed ethno-cultural identity, and evoke a constellation of social and psychological responses, including problems of nostalgia, disorientation and alienation (see Fullilove 1996), a more encompassing assessment of heritage values for Ilisu is critical to a just and more sustainable way forward.

---

[10] See Rowles' (1983) research in which he took aspects of affinity with surroundings and expressed them as notions of 'insideness'.

[11] In January 2013, a Turkish high court ordered an immediate cessation to the construction of the Ilisu Dam on the grounds that the government had not conducted the legally required environmental impact assessment. The intervention – ultimately a hollow victory for campaigners, as the project General Manager has claimed that 'the ruling does not even remotely have anything to do with stopping the project' (Dundar, quoted in Güsten 2013) – is one of many attempts by a broad international coalition of *inter alios* lawyers, archaeologists, architects, celebrities, NGOs and British parliamentarians to pressure the government to reconsider its plans for the dam and related infrastructure. Campaigning has also led to company (Skanska, Balfour Beatty, Impreglio) and export credits guarantee (British, German, Austrian, Swiss government) withdrawal.

BIBLIOGRAPHY AND REFERENCES

Ahunbay, Z, 1999 Dünya Kültür Mirasi Ölçütleri Açısından Hasankeyf ve Kurtarılma Olasıkları, *Mimarlık* 12/99 (290), 29–34

Arık, O, 2001 Hasankeyf Excavation and Salvage Project, in *Ilısu ve Karkamış Baraj Gölleri Altında Kalacak Arkeolojik ve Kültür Varlıklarını Projesi 1999 Yılı Çalışmaları* (eds N Tuna, J Velibeyoğlu, and J Özturk), METU, TAÇDAM, Ankara, 795–804

—— 2003 *Hasankeyf: Üç Dünyanın Buluştuğu Kent*, Türkiye Bankası, Istanbul

Armstrong, K, 2002 Jerusalem: The Problems and Responsibilities of Sacred Space, *Islam and Christian–Muslim Relations* 13 (2), 189–96

Australia ICOMOS, 1999 *The Burra Charter. The Australia ICOMOS Charter for Places of Cultural Significance*, ICOMOS Australia Incorporated, Burwood, VIC

de la Torre, M (ed), 2002 *Assessing the Values of Cultural Heritage*, The Getty Conservation Institute, Los Angeles

ECGD (Export Credits Guarantee Department), 1999 *Stakeholders' Attitudes to Involuntary Resettlement in the Context of the Ilisu Dam, Turkey*, HMSO, London

Elliott, S, 2007 Rescuing Hasankeyf: Ecomuseological Responses to Large Dams in Southeast Turkey, unpublished PhD thesis, Newcastle University

Fullilove, M T, 1996 Psychiatric Implications of Displacement: Contributions from the Psychology of Place, *American Journal of Psychiatry* 153, 1516–23

Gabriel, A, 1940 *Voyages Archéologiques dans la Turquie Orientale*, vol 1, Institute Français d'Archéologie de Stamboul and E de Boccard, Paris

Gibson, L, and Pendlebury, J (eds), 2009 *Valuing Historic Environments*, Ashgate, Farnham and Burlington VT

Güsten, S, 2013 Construction of Disputed Turkish Dam Continues, *New York Times* [online], 27 February, available from: http://www.nytimes.com/2013/02/28/world/middleeast/construction-of-disputed-turkish-dam-continues.html?pagewanted=all&_r=0 [22 July 2013]

H7, 2005 Personal communication (interview with the author), 28 August, Hasankeyf

H9, 2005 Personal communication (interview with the author), 24 August, Hasankeyf

H10, 2005 Personal communication (interview with the author), 24 August, Hasankeyf

H13, 2005 Personal communication (interview with the author), 25 August, Hasankeyf

H15, 2005 Personal communication (interview with the author), 25 August, Hasankeyf

H17, 2005 Personal communication (interview with the author), 31 August, Hasankeyf

H18, 2005 Personal communication (interview with the author), 31 August, Hasankeyf

H23, 2005 Personal communication (interview with the author), 31 August, Hasankeyf

H24, 2005 Personal communication (interview with the author), 31 August, Hasankeyf

H25, 2005 Personal communication (interview with the author), 2 September, Hasankeyf

Hildyard, N, Tricarico, A, Eberhardt, S, Drillisch, H, and Norlen, D, 2000 '*If the River were a Pen...*' *The Ilisu Dam, the World Commission on Dams and Export Credit Reform. The Final Report of a Fact-finding Mission to the Ilisu Dam Region 9-6 October 2000*, Kurdish Human Rights Project, London

Hydro Concepts, Hydro Québec International and Archéotec, 2005 *Ilisu Dam and HEPP: Environmental Impact Assessment Report*, Update July 2005, DSI, Ankara

ICOMOS, 2000 *Heritage at Risk: ICOMOS World Report on Monuments and Sites in Danger*, K G Saur, Munich

Ilisu Consortium, 2006 *Ilisu Dam and Hydroelectric Power Plant Cultural Heritage Action Plan*, Revised May 2006, DSI, Ankara

Işıklı, E, Engin, D, and Akçay, E, 2012 *Hasankeyf Araştırma Raporu* 14.03.2012, Doğa Derneği, Istanbul

Johnston, B R, 2012 Water, Culture, Power: Hydrodevelopment Dynamics, in *Water, Cultural Diversity, and Global Environmental Change: Emerging Trends, Sustainable Futures?* (eds B R Johnston, L Hiwasaki, I J Klaver, A Ramos Castillo, and V Strang), UNESCO, Paris, and Springer, Dordrecht, Heidleberg, London and New York, 295–318

Khoury, S, 2001 A Cultural Approach to Diabetes Therapy in the Middle East, *Diabetes Voice* 46 (1), 23–5

Lee, R M, 1993 *Doing Research on Sensitive Topics*, Sage, London

Mason, R, 2002 Assessing Values in Conservation Planning: Methodological Issues and Choices, in *Assessing the Values of Cultural Heritage* (ed M de la Torre), The Getty Conservation Institute, Los Angeles, 5–30

Meinecke, M, 1996 *Patterns of Stylistic Changes in Islamic Architecture. Local Traditions versus Migrating Artists*, New York University Press, New York and London

Mourato, S, and Mazzanti, M, 2002 Economic Valuation of Cultural Heritage: Evidence and Prospects, in *Assessing the Values of Cultural Heritage* (ed M de la Torre), The Getty Conservation Institute, Los Angeles, 51–73

Özdoğan, M, 2000 Cultural Heritage and Dam Projects in Turkey: An Overview, in *Dams and Cultural Heritage Management. Working Paper submitted to the World Commission on Dams* (eds S Brandt and F Hassan), World Commission on Dams, Cape Town, 58–61

Ronayne, M, 2002 *The Ilisu Dam. Displacement of Communities and the Destruction of Culture*, Kurdish Human Rights Project, London

— 2008 Committment, Objectivity and Accountability to Communities: Priorities for 21st Century Archaeology, *Conservation and Management of Archaeological Sites* 10 (4), 367–81

Rowles, G D, 1983 Place and Personal Identity in Old Age: Observations from Appalacia, *Journal of Environmental Psychology* 3 (4), 299–313

Schmidt, P, 2000 Human Rights, Culture and Dams: A New Global Perspective, in *Dams and Cultural Heritage Management. Working Paper submitted to the World Commission on Dams* (eds S Brandt and F Hassan), World Commission on Dams, Cape Town, 13–14

Väth, G, 1992 *Hasankeyf am Tigris: Stiller untergang einer Stadt in Kurdistan*, Verlag Königshausen and Neumann, Würzburg

Watts, N F, 2009 Re-Considering State–Society Dynamics in Turkey's Kurdish Southeast, *European Journal of Turkish Studies* [online], 10, available from: http://ejts.revues.org/index4196.html [18 January 2012]

WCD (World Commission on Dams), 2000 *Dams and Development: A New Framework for Decision-Making*, Earthscan Publications Ltd, London and Sterling VA

World Bank, 1994 Environmental Assessment Sourcebook, Update 8 Cultural Heritage, in *Environmental Assessment*, World Bank Environment Department, Washington DC

Yildiz, K, 2001 *The Human Rights Impact of the Ilisu Dam* [online], presentation delivered to Criticizing Globalization, Public Eye on Davos 2001, the Berne Declaration, available from: http://www.evb.ch/en/p25001649.html [12 May 2006]

18

Placing the Flood Recovery Process

Rebecca Whittle, Will Medd, Maggie Mort,
Hugh Deeming, Marion Walker, Clare Twigger-Ross,
Gordon Walker and Nigel Watson

Contrarily, and without apparent irony, the preferred story in a natural disaster is one of good news: miraculous rescues and escapes; acts of heroism and bravery; selfless rescue workers from Rotherham; sniffer dogs from Barking; saintly surgeons from Surbiton. As the hope of more wide-eyed victims being plucked from the grave diminishes, as the disaster medics wrap up their kit and go, so too do the 24-hour rolling news teams. This is very expensive stuff, and nobody has the budget or the audience for the grim, dull depression of resurrection. (Gill 2010)

The writer A A Gill's heartbreaking portrait of the aftermath of the Haitian earthquake of January 2010, in which up to 230,000 people died and more than 1 million were made homeless, raises the question of what it means to recover from a disaster. In the immediate aftermath of an event like Haiti, there is inevitably a focus on physical action and progress – the rubble is moved, survivors treated, fed and clothed – all of which corresponds to what emergency planners term the 'response' phase of the emergency. And yet what Gill is hinting at, beautifully captured in the notion of *resurrection*, amounts to something else entirely: the idea of recovery as a spiritual, physical and emotional process that is deeply hard to achieve and much less visible than the usual metrics applied to such situations would suggest (How many people are back in their homes? How many businesses are open for trading? What aspects of key infrastructure have been reopened?).

This chapter argues that, if we want to understand the recovery process then it is essential to think about just exactly what it is that is being recovered. Our case study is a qualitative, longitudinal study of people's recovery from the floods of June 2007 in Kingston-upon-Hull, UK, in which over 8600 households were affected and one man died (Coulthard *et al* 2007). The aim of the research was to discover what the long-term disaster recovery process was like for people as they struggled to get their lives and homes back on track. The project, which is described extensively elsewhere (Whittle *et al* 2010), used in-depth qualitative methods that had been previously used to investigate people's recovery from the 2001 Foot and Mouth Disease (FMD) disaster in Cumbria (Mort *et al* 2004). This consisted of initial interviews, weekly diaries and quarterly group discussions with 44 participants over an 18-month period.

We begin the chapter by exploring some of the ways in which the recovery process has been described in the policy and research literature before moving on to the experiences of the Hull residents who show that the physical process of repairs to the home and the surrounding built environment go hand in hand with a broader process of recovering social life – including a sense of home and 'normality'.

UNDERSTANDING RECOVERY

One of the most prominent policy accounts of recovery to emerge in the aftermath of the 2007 floods was that presented in the Pitt Review, the UK government's review of the 2007 floods (The Cabinet Office 2008). Clearly presented in graphical format (see Fig 18.1), recovery is presented as a steady process of improvement – a smooth curve, which rises steadily until a previous level of 'normality' is perceived to be attained – or better still, an improvement to a state of 'regeneration'.

The Pitt Review describes what we might hope to be an ideal scenario for recovery, as other government documents acknowledge that recovery may not be so simple. For example, the National Recovery Guidance recovery plan guidance template admits that recovery 'usually takes years rather than months to complete as it seeks to address the enduring human, physical, environmental, and economic consequences of emergencies' (HM Government 2007, 3).

The research community has also emphasised the complexities of recovery in ways that challenge the picture presented by the Pitt Review. For example, Kai Erikson (1976) shows that, particularly in disadvantaged communities, a pre-existing disaster already exists in the form of the poverty and inequality experienced by residents. Consequently, the idea of 'recovery' as a straightforward bricks and mortar exercise that begins immediately after the disaster is already called into question. The recovery process itself has also been shown to be infinitely more complex than the idea of a smoothly rising upwards curve would suggest. Indeed, our work in Hull showed that recovery was experienced as an unpredictable series of 'ups' and 'downs', more akin to a game of Snakes and Ladders than a steady process of improvement (Whittle *et al* 2010; Deeming *et al* 2012). Crucially, we also discovered that recovery was, in many cases, experienced as more of

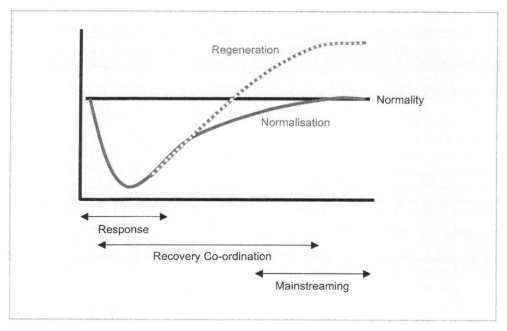

FIG 18.1.   RECOVERING FROM AN EMERGENCY (DIAGRAM OF FLOOD RECOVERY PROCESS).

a trauma than the flood event itself and that, during the recovery process, it was impossible to separate the physical work involved in recovering the home from the emotional and mental effort required to manage the repairs and create and maintain social and family life in a new setting (Whittle *et al* 2012; Sims *et al* 2009).

In this chapter, however, we take this process one step further to explore what it is that people are recovering towards. Is there an end point to recovery and, if so, what does it look like? How will we know when people are there? This is by no means a simple question to answer. In the Pitt Review graph (The Cabinet Office 2008, 398), the state of 'recovery' appears straightforward: it seems that the minimum we should aspire to is a return to 'normality'. Once again, however, research has shown this process to be more complex than is presented here. In the first instance there is the argument that, if we are interested in resilience, then returning to 'normality' is not necessarily a good thing if it simply reproduces – or worse still, exacerbates – people's pre-existing vulnerability to disasters (Manyena 2011). Rather, we should be looking to adaptation or perhaps even transformation – a more fundamental shift in our way of being, which offers us a better future than carrying on with an, albeit more strongly engineered, version of 'normal' (Whittle *et al* 2010; Medd and Marvin 2005).

Indeed, the contradictions encompassed in the idea of a 'return' to a pre-existing state are also clearly highlighted by Wisner (Wisner *et al* 2004), who suggests that recovery must instead be understood as something relative and contingent:

> The terminology associated with disaster recovery is biased towards optimism. The key words – 'recovery', 're-establish', 'reconstruction', 'restoration' and 'rehabilitation' – are all prefixed with 're', indicating a return to the pre-existing situation. A more realistic view challenges the assumption that such recovery will actually be achieved. Instead, the more pessimistic argument suggests there will be uncertainty, unforeseen events and even the reproduction of vulnerability. A rather depressing implication… is that in some cases the most vulnerable households and individuals do not recover. (Wisner *et al* 2004, 357)

We do not, of course, wish to imply that the participants that we worked with in Hull have not necessarily recovered (although, sadly, that may have been the case for some). Neither do we want to imply that recovering a sense of normality is a misguided or impossible goal. On the contrary, many participants longed to 'get back to normal'. However, what we wish to show is that this 'normality' is not easily captured: rather, it takes much work to recreate it piece by piece because – and this is the key point – it can never be the same 'reality' as that which existed before the flood (Convery *et al* 2005). As one local resident tellingly stated in describing the recovery process from the 2001 FMD outbreak in Cumbria, 'There was no normality: normality had gone' (Mort *et al* 2004, 49). This chapter explores the work involved in finding and rebuilding a sense of normality within the immediate home and family.

Rebuilding a Sense of Home

The experiences of the Hull diarists show that the recovery process does not end when the physical repairs are finished. Living with a *house* that has just 'arrived' and where everything is new, rather than a *home* that has been built up gradually and where particular objects have memories and meanings attached to them, can be very difficult psychologically. It is as if the home – which grows, as it does, with its inhabitants – becomes a reference point for individuals and families, leading to a stamp of individuality that makes it unique and special to the people concerned. This is what Erikson talks about when he describes how physical damage to the home is also experienced as damage to the very 'furniture of self' (Erikson 1976). In this extended extract from a group discussion (below), the diarists describe the physical, mental and emotional work that goes on in order to rebuild the home and, crucially, recover its meaning over time.

Amy:	You've heard on radio interviews and things like that, people ringing up and talking on the shows that were going on throughout the year, saying people have been flooded, you get this nice new home and you get possessions.
Isobel:	'Aren't you lucky you get a nice new home?' – I blow when someone says that to me.
Amy:	I had a nice home before it started, I didn't need any of this.
Isobel:	Exactly.
Amy:	It doesn't feel like home though does it?
Jan:	A home is something you build up gradually.
	[they talk over each other]
Amy:	Yes you do it gradually don't you?
Isobel:	You have to choose and you don't know what you want do you?
Amy:	I went into a furniture shop four times when I was looking at suites because I could not face walking round looking at them. So a lot of people must have gone through that.
Abby:	It's the shopping, every day it's like looking on the internet, just trying to replace things.
Amy:	You've also said about it that it's not a home, yes we had a nice home before and it was ours and we'd worked for it. And we've got downstairs replaced, we've got things replaced upstairs, we've got a brand new kitchen, which I'm sure walking in would say this is absolutely gorgeous. But it's not ours; we've not built it up like we did before. It's arrived.
Leanne:	Well mine is beginning to feel like home now, I do feel like I'm back at home. But I don't value it the same. I don't have the same sense of value and I feel very, very insecure. Now when you are in your home you should feel secure and content, and I don't have those feelings…
Abby:	We are not valuing it.
Amy:	It's not just that, it's the thought is in the back of your mind, is it going to happen again?
Isobel:	Exactly.

(Group discussion, 24 April 2008)

Here, then, we see how the home (as opposed to just a house) has many meanings for people – for example as a place of family heritage, safety, security and belonging. We also see how residents must work at recreating these feelings long after the physical work has been completed; almost as if the home has become a stranger with which one must become reacquainted. However, the physical form of the home is by no means incidental to this process. The endless shopping and decision-making to replace things, which Amy and Abby discuss here, was a process that was characterised on the one hand by a bewildering array of choice: because everything down-stairs was lost to the flood water there was no point of reference when looking to replace other items. On the other hand, however, participants also experienced a lack of choice: as all the houses across Hull were being renovated at the same time, selection was limited to the styles and colours considered 'fashionable' and in stock at the time. This was sometimes compounded when particular builders only dealt with particular suppliers, meaning residents were restricted to the ranges available in those companies' brochures. Consequently, many people ended up with houses that looked very similar, as Melanie reflected in her diary:

> We have started to pick our colour schemes for all the rooms, the only problem is everyone's house we have been in that has been flooded has got the same colour scheme browns and creams! Either we all have good taste or bad taste! (Melanie, resident, diary, 4 February 2008)

The challenge of adapting to new things seemed to be particularly acute for older people. In addition to the long memories of family life in their homes, there was also a loss of personal heritage associated with the fact that the look and style of their homes could not be recreated when all that was available from the shops was very 'modern' in design. For example, Sophie described how her elderly mother-in-law would often sit in her bedroom because this was the one part of the house that remained as it had done before the floods. By contrast, the downstairs of her house, which was new and modern, felt alien to her.

> The only thing that hasn't changed is her bedroom, so you'll often find her living in her bedroom because she feels comfortable and safe in there. Her house has gone from being quite old fashioned but how she really loved it, to now being a plastered wall, cream plaster, modern TV, modern cabinets, because you couldn't replace what she had, and she hates it.
>
> (Sophie, resident, interview, 6 February 2008)

Consequently, Sophie's mother-in-law was having to contend with the emotional, physical and mental challenges of recovering to a 'home' that looked and felt very different.

REBUILDING ROUTINES AND SKILLS

It was not only the décor, fixtures and fittings that people missed; particular household items, such as baking trays or recipe books handed down through generations were mourned as part of a link to life pre-flood that could never be recaptured (Harada 2000). For example, Betty really missed being able to cook and bake:

> I was always baking, I mean I've always cooked and baked. I had about 12 recipe books, good ones as well. And every one of them went. If I wanted to do something fancy, I can't remember, I've no books left, I've nothing left, it's all gone. ... We were just talking about baking for Christmas, I said, 'We'll have a job making some mince pies because I haven't go no trays'.
>
> (Betty, resident, interview, 19 December 2007)

Of course, other recipe books could be bought, but this is not the point – Betty *knew* her recipes – should she want a particular one, she would know exactly where to look and how to make it, as she had done many times before. In this sense it is not just the home that has changed, but the kinds of things that one is able to do confidently and easily within that space.

This point was also illustrated by the case of Sophie's mother-in-law who had lost many of her independent living skills after living with her son and daughter-in-law for an extended period while her own house was repaired. This made it very difficult for her to recover a 'normal' life back home:

> Every single thing in her life has changed… maybe somebody who was in their thirties or their forties, it's a case of 'yes it's been horrendous but we've got new things now, well let's just start a new life'. But when you are older – she got new things and she has no idea – she was only comfortable in her own home because she was comfortable with turning the TV on, the microwave, the oven – she knew in her head. She could control her memory loss because everything was where it had always been, it's not a new thing. Whereas now she cannot turn the TV on and we'll have her ringing up saying, 'I don't know which remote to press. … I can't remember which drawers, where do I put these?' (Sophie, resident, interview, 6 February 2008)

Once again, therefore, we see how it is impossible to separate the physical, emotional and mental aspects of recovery, since particular objects are a crucial part of the activities and practices that go towards making a house a home (Harada 2000).

Special Events

The ongoing work of recovery was also apparent during key occasions for participants and their families, since these events would often prompt reflection on what such times could or would have been like before the floods happened. In this way, people would sometimes reflect on the success – or otherwise – with which they had been able to recover a sense of 'normal' family life. For example, Sally described Christmas back in her home as 'lovely' and 'as it should be', while Laura relished the fact that she was actually able to find and put up all her usual decorations (they had been packed away in storage while she was in the rented house). However, for Leanne, things still did not feel right – so much so, that she decided to go away for Christmas with her husband and dogs:

> I just didn't feel I could do Christmas; I still didn't have that homely, loving, exciting feeling that you get when you are at home and you put your tree up and you do all this. We did put some decorations up and some of the neighbours made an effort and some didn't. Some didn't bother with anything; they just couldn't bring themselves to do it. We did put some Christmas lights up and a tree up but minimal. And then we went off on Christmas Eve and came back on the 29th and we had a totally unusual, very strange Christmas, it was a very strange experience.
> (Leanne, resident, group discussion, 12 February 2009)

This was Leanne's second Christmas back in her home, and yet this quotation shows how she is still struggling to feel the same about her property as the emotional bond that made it a 'home', not just a house, is missing. Once again, therefore, we can see that recovery does not have a clear end point and that instead of things going back as they were, residents may have to work to create a new version of 'normality' involving new ways of relating to their homes and families (Convery *et al* 2008).

For some diarists, there was a clear sense of loss associated with the fact that particular occasions were not able to live up to expectations of how they could or should have been. This sense of loss was particularly acute in the case of occasions that were not able to be repeated or recaptured in any way. For example, Melanie, who gave birth to her second child while living in a rented house, felt that she lost those early days with her new son. She wrote in her diary:

> Feeling very weepy at the moment, just want to cry all the time. ... Right now I just feel like I have failed both my sons and should be able to provide a home for them instead of just a house. ... Whenever I look at my new baby I cry just because he is so lovely and also because he deserves to be in his home with all his toys and his nursery which is all ready for him.
>
> (Melanie, resident, diary, 10 December 2007)

Again, this is not to argue that recovery does not happen; rather, that the 'thing' that is recovered is not the same as it was before. Indeed, many diarists accepted this difference, either consciously or unconsciously. Emily, who was in her eighties, had remarried following the death of her first husband. She was able to see a more positive side to the flooding as the repairs process meant that she and her new husband were able to redesign their home together. This was in contrast to the pre-flood situation where the house they shared had been designed by Emily and her first husband. Her story is a clear example of where something positive was extracted from the flooding. Equally, some younger diarists reported that the floods had acted as an incentive for them to change or improve their homes in some way. For example, some chose to pay a bit extra to get additional improvements made – such as a new bathroom – while they had disruption from the builders anyway. Such positive changes in no way compensated for the overall stress, anxiety and disruption resulting from the floods, but these examples do illustrate some of the more positive ways in which people responded to an otherwise very difficult set of circumstances by choosing to recreate a 'new normal'.

RECOVERY REVISITED

This chapter has focused on the question of what it means to recover from a disaster such as the 2007 floods. The diary methodology was helpful here as it reveals the ways in which the recovery process continues when a person moves back home. We have shown that a key issue for recovery is the question of what is 'normal'. Many diarists experienced a shift in how they felt about 'home' and everyday life, including a lack of security for fear of future flood, a sense in which their home was not the place that it was before (it both looked and felt different) and changes that may have occurred within the family. There is not, therefore, a clear end point to recovery. Instead, what we have found is a process of rebuilding family life that involves adjusting to a new sense of home and a changed sense of the future, since there are aspects of everyday life that may be fundamentally changed, whether for better or for worse – a finding consistent with other disaster research (Convery *et al* 2008). Recovery, then, is about readjustment and adaptation rather than a return, as such.

BIBLIOGRAPHY AND REFERENCES

The Cabinet Office, 2008 *The Pitt Review: lessons learned from the 2007 floods*, The Cabinet Office, London

Convery, I, Bailey, C, Mort, M, and Baxter, J, 2005 Death in the wrong place? Emotional geographies of the UK 2001 foot and mouth disease epidemic, *Journal of Rural Studies* 22, 99–109

Convery, I, Mort, M, Baxter, J, and Bailey, C, 2008 *Animal disease and human trauma: Emotional geographies of disaster*, Palgrave Macmillan, Basingstoke

Coulthard, T, Frostick, L, Hardcastle, H, Jones, K, Rogers, D, Scott, M, and Bankoff, G, 2007 *The June 2007 floods in Hull: final report by the Independent Review Body*, Independent Review Body, Kingston-upon-Hull, UK

Deeming, H, Whittle, R, and Medd, W, 2012 *Flood Snakes and Ladders* [online], Lancaster Environment Centre, Centre for Sustainable Water Management, Lancaster University, available from: http://www.lec.lancs.ac.uk/cswm/Hull%20Floods%20Project/HFP_%20FSL.php [11 December 2012]

Erikson, K, 1976 *Everything in its path: destruction of community in the Buffalo Creek flood*, Simon & Schuster, New York

Gill, A A, 2010 Raising Haiti: postcards from the edge, *The Sunday Times Magazine*, 20 April

Harada, T, 2000 Space, materials, and the 'social': in the aftermath of a disaster, *Environment and Planning D: Society and Space* 18, 205–12

HM Government, 2007 *National Recovery Guidance: recovery plan guidance template* [online], HM Government, London, available from: https://www.gov.uk/government/uploads/system/uploads/attachment_data/file/85962/recovery-plan-guidance-template.doc [29 January 2014]

Manyena, S B, 2011 Disaster resilience: a bounce back or bounce forward ability? *Local Environment* 16, 417–24

Medd, W, and Marvin, S, 2005 From the Politics of Urgency to the Governance of Preparedness: A Research Agenda on Urban Vulnerability, *Journal of Contingencies and Crisis Management* 13, 44–9

Mort, M, Convery, I, Bailey, C, and Baxter, J, 2004 *The Health and Social Consequences of the 2001 Foot and Mouth Disease Epidemic in North Cumbria*, Institute for Health Research, Lancaster University

Sims, R, Medd, W, Mort, M, and Twigger-Ross, C, 2009 When a 'Home' Becomes a 'House': Care and Caring in the Flood Recovery Process, *Space and Culture* 12, 303–16

Whittle, R, Medd, W, Deeming, H, Kashefi, E, Mort, M, Twigger-Ross, C, Walker, G, and Watson, N, 2010 *After the Rain – learning the lessons from flood recovery in Hull, Final Project Report*, 'Flood, Vulnerability and Urban Resilience: a real-time study of local recovery following the floods of June 2007 in Hull', Lancaster University, Lancaster

Whittle, R, Walker, M, Medd, W, and Mort, M, 2012 Flood of emotions: emotional work and long-term disaster recovery, *Emotion, Space and Society* 5, 60–9

Wisner, B, Blaikie, P, Cannon, T, and Davis, I, 2004 *At Risk: Natural hazards, people's vulnerability and disasters*, Routledge, London

19

Village Heritage and Resilience in Damaging Floods and Debris Flows, Kullu Valley, Indian Himalaya

Richard Johnson, Esther Edwards, James Gardner
and Brij Mohan

Introduction

Heritage is shaped, and reshaped, by the impacts of natural hazard events that are common in mountains. This chapter examines the heritage–resilience relationship through villages in the Phojal Nalla catchment, in the Kullu Valley of the Indian Himalaya (see Fig 19.1) in the context of a 1994 flood event. The Valley is rich in many forms of heritage, including vernacular architecture, material culture, and custom and religion.

Current definitions of heritage are complex and disputed; nonetheless, understandings are drawn from Smith (2006), Sorenson and Carman (2009) and Harrison (2013), where heritage may comprise: (1) objects, places and societal practices; (2) tangible (ie objects) and intangible (ie socio-cultural practices) contributors; (3) officially recognised and unofficial contributors; and (4) an entity that is intertwined and continually created. Of these, intangible heritage is particularly relevant, since it sees people as unofficial heritage creators through their thoughts, knowledge and approaches to life. This type of cultural heritage may arise in response to actual loss and perceived risk, enabling the past to inform understanding of, and engagement with, the present and the future. Intangible heritage can take many forms: art, dance, food, language, music, oral histories, religion and virtual media.

It may seem somewhat unusual to be discussing heritage and resilience together in relation to flood events, but heritage is an important resource that can contribute significantly to strengthening community resilience. The definition and utility of the resilience concept has been heavily debated (Cutter *et al* 2008; Brown 2014). In the context of communities subject to hazards, it is frequently discussed in association with societal vulnerability (a potential for harm). Resilience is often seen as a positive condition, opposite yet interlinked to vulnerability – a negative condition. Cutter *et al* (2008, 599) define resilience as:

> the ability of a social system to respond and recover from disasters and includes those inherent conditions that allow the system to absorb impacts and cope with an event, as well as post-event, adaptive processes that facilitate the ability of the social system to re-organise, change, and learn in response to a threat.

To improve community resilience, an understanding of contributing factors is necessary. Capital is key in developing and maintaining resilience where economic, social and environmental resources facilitate coping (during event) and adaptation (after event). Strong capital may include: (1) economic – diversified employment, wealth and robust infrastructure; (2) social – active community networks, female empowerment, accessible education and health provision, low levels of corruption and strong governance; and (3) environmental – biodiversity, local energy, water and food supplies, sustainable environmental management and an array of hazard mitigation and preparedness measures (Gardner and Dekens 2007; Cutter *et al* 2008; Collins 2009; Wilson 2012). Implicit here is the ability for people to learn before, during and following a hazard event; the application of this learning enables diversity and self-organisation, which are important facets of a resilient community (Gardner and Dekens 2007). Intangible heritages provide a means for observation, experiences and memories to be shared and repeated, facilitating intergenerational learning, thus focusing the link between heritage and resilience through learning. Maintaining heritage, and in particular the traditional knowledge accumulated over centuries, is therefore likely to result in increased community resilience, helping communities to better withstand natural hazard events such as earthquakes, floods or slope instability processes.

The Context: Kullu Valley and Phojal Nalla Catchment

Kullu Valley

The Kullu Valley, within the Kullu District of the State of Himachal Pradesh, is the uppermost 90km of the Beas River. Though narrow (2–3km), the area of physical and social influence extends into tributary catchments, of which Phojal Nalla is one (see Fig 19.1). Upstream of the Pandoh Dam, the Beas River catchment covers 5278km² and rises from 890 to 6632m ASL (above sea level). The primary water sources are rainfall (65%) and snow and glacial melt contributions (35%) (Kumar *et al* 2007). Particularly important to understanding the hazardous setting is the summer monsoon season. This is characterised by multiple rainfall episodes (including long-duration and high-intensity), which influence slope instability and stream flow. Tectonically, the area lies within the Pir Panjal Himalaya and is seismically active with ongoing uplift (Valdiya 2002). In contrast, active erosion processes are evidenced by valley incision, high fluvial sediment yields and a variety of glacial, paraglacial, colluvial and alluvial landforms (Owen *et al* 1996; Sah and Mazari 2007). The valley-side fans are particularly prominent and of importance to the livelihood of the Kullu Valley (Gardner 2002).

The natural and social contexts enable understanding of the regional hazard, vulnerability and risk relationship. Hazards that impact the area include: earthquakes (Chandel and Brar 2010), floods and debris flows (Gardner 2002; Sah and Mazari 2007), slope instability (Sah and Mazari 1998), snow avalanches, wildfires and crop infestations. Hazard event impacts vary in size, ranging from regional-scale disaster (eg 1905 Kangra earthquake) to more localised and frequent floods, debris flows and slope instability. Of the latter, the 1894 Phojal Nalla landslide-dam outburst flood and debris flow event has been one of the most significant in terms of loss of life among people (indigenous and migrant populations) and livestock (Gardner 2002). The vulnerability of migrant populations to floods and debris flows in the Kullu Valley continues to be substantial (Sah and Mazari 2007).

The social, economic and political history of the Kullu Valley is complex (see Harcourt 1972; Agnew 1899). These contexts, alongside present conditions, influence hazard, vulnerability and risk. Strong village-based governance and land-use administration were present prior to British administration (1846–1947); as a consequence of British rule rural settlement and commercial

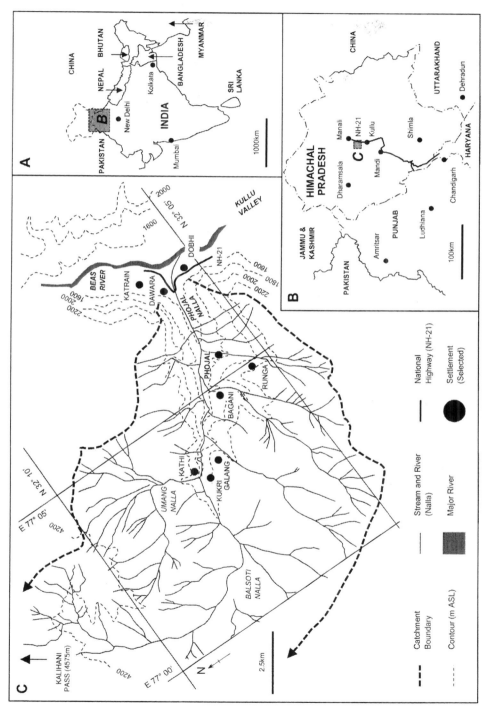

FIG 19.1. THE GEOGRAPHICAL LOCATION OF THE PHOJAL NALLA CATCHMENT IN THE KULLU VALLEY. (A) INDIA AND SURROUNDING NATIONS, (B) HIMACHAL PRADESH STATE AND ADJOINING ADMINISTRATIVE AREAS, (C) MID AND LOWER EXTENTS OF THE PHOJAL NALLA CATCHMENT (ADAPTED FROM HPSEB 1998).

extractive land uses expanded. The Kullu District has a resident population of 440,000 of whom 90% are rural dwelling (Census of India 2011). Literacy is relatively high (80%), though the gender gap is 18% (female lower). Religion is dominated (96%) by Hinduism and a strong attachment to village deities (Dhumal and Ahuja 2012). The Kullu Valley also has a significant semi-permanent/transient population related to transhumance, trade, economic migration, pilgrimage and tourism. Of these, the growth of tourism has been exponential from c. 10,000 in 1964 to c. 1 million in 2004 (Singh 2008).

At the state level, economic activity comprises: agriculture and horticulture (most important), tourism, transportation, small industries (hydro-electric power – HEP), local commerce and administration (Balokhra 2007). Baldi (2012) suggests a trend of economic diversification in which the role of agriculture and allied industries is declining, whereas industrial and services sectors inclusive of tourism are growing.

This combination of population dynamics and socio-economic change has elevated the vulnerability and risk in the region: settlements have grown beyond traditional village arrangements; roads have expanded and land uses are increasingly commercialised (Gardner 2002). As will be discussed, such changes have impacted on the reproduction of heritage within the region.

Phojal Nalla catchment

The catchment (see Fig 19.1C and 19.2) upstream of the NH-21 Dobhi Bridge ranges from 1485m to 5100m ASL, over 130.8km^2 (HPSEB 1998; BEHEPD 2009). Between Phojal and Dawara, the Phojal Nalla is a steep gradient bedrock/boulder bed river where floodplain segments contain palaeochannels and terraced boulder deposits. The channel corridor is heavily vegetated, locally cultivated, and has local channelisation and irrigation offtakes. Current construction works relate to hydro-electric power generation and transmission.

Mean annual rainfall at nearby Katrain (1962–2009) is 1124 ± 248mm (1σ); and the monsoon months July (160 ± 82mm, 1σ) and August (156 ± 78mm, 1σ) are the wettest (Jangra and Singh 2011). Flows are perennial, and discharge records at Dobhi Bridge (1967–1983) range 0.63 to 41.27 m^3 s^{-1} (BEHEPD 2009). However, palaeo-deposits and 19th-century flood accounts (Forbes 1911; Chetwode 1972) all suggest larger flow magnitudes; for example the disastrous event of 1894 (Gardner 2002).

People reside in small settlements up-catchment (Phojal c. 550 people). More sizeable are Dobhi and Dawara, where migrant populations have settled alongside the Phojal Nalla (see Fig 19.2D). The indigenous population are an 'ecosystem people', traditionally reliant upon their environment for much of their food, fuel and resources, but are also dependent on a wage economy (Berkes *et al* 2000).

METHODS: AN INTERDISCIPLINARY APPROACH

An interdisciplinary approach was undertaken to describe environmental characteristics and the heritage–resilience condition. The methods include: (1) walkover observation and GPS mapping of hillslope and channel attributes; (2) semi-structured interviews (n=28) with villagers and officials were conducted in Hindi and translated to English; and (3) archive searches in India and the UK.

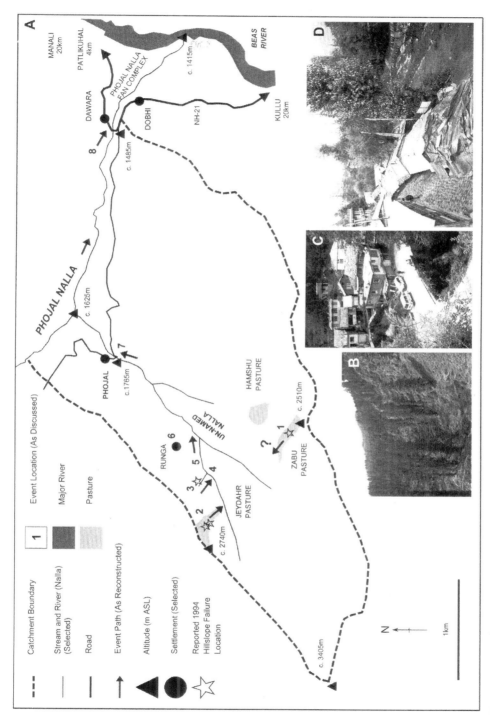

FIG 19.2. 1994 EVENT PATH BETWEEN JEYDAHR/ZABU TO THE BEAS RIVER. (A) EVENT PATH BETWEEN JEYDAHR/ZABU TO THE BEAS RIVER. (A) EVENT SUB-CATCHMENT, (B) FORESTED SLOPE AND STREAM BED [EVENT LOCATION 5], (C) ROAD BRIDGE AND STREAM IN PHOJAL VILLAGE [EVENT LOCATION 7], (D) INFORMAL DWELLINGS ON THE LEFT BANK OF THE PHOJAL NALLA DOWNSTREAM OF THE NH-21 AT DAWARA [EVENT LOCATION 8]. (ALL OCTOBER 2013).

THE 1994 FLOOD: A REVISED ACCOUNT OF A NATURAL DISASTER EVENT

Existing knowledge of the 1994 event

Gardner and Dekens (2007) outline a debris flow in the 'lower reaches and confluence with the Beas River' in association with a 'cloudburst', impacting properties adjacent to the channel. Kuniyal *et al* (2004) suggest Phojal experienced two cloudbursts in 1994, resulting in 11 fatalities and damage to 21 temporary stalls, 5 water mills, 3 vehicles and 4–5ha of agricultural land. The Dartmouth Flood Observatory Archive (Brakenridge 2014) details a storm duration of three days (8–10 August 1994). Using new data, a revised synthesis is offered.

Timing and meteorological causation

Event timings are determined from interview (Table 19.1a) and documentary data (see Fig 19.3). A Runga farmer notes heavy rain for 2 hours, with event activity on 9 August 1994 at 22:00 IST (Indian Standard Time). In Phojal the Pradhan (Village Head) suggests heavy rainfall, with flooding at c. 02:00–03:00 IST following two days of prolonged rainfall; the shopkeeper indicates a storm duration of c. 3 hours. Fig 19.3 suggests event occurrence at 02:30 IST on 10 August 1994.

Physical impacts

The event path is reconstructed from interviews (Table 19.1a and 19.1b) and field reconnaissance (see Figs 19.2 and 19.4), beginning in the unnamed Nalla sub-catchment, south of Phojal Nalla, then continuing 2.8km to Dawara and then 1.6km down the fan to the Beas River confluence. From the Jeydahr pasture (c. 2740m ASL) to the Beas River confluence (1415m) is c. 8.1km.

Hillslope instability

The Runga farmer suggests three sites of instability: on Fig 19.2A at locations 1 (Zabu), 2 (Jeydahr) and 3 (stream side slope). The Jeydahr failures were notable, resulting in the loss of livestock, verified at the time by government veterinarians, whose records were subsequently lost in the September 1995 flood at Patlikuhal (Sah and Mazari 1998; 2007).

Fig 19.4 indicates two 1994 failure locations at Jeydahr: 1' and 2'. At 2' (2620m ASL) a scar remains (c. 20m × 5–8m; 31°; 160m³); sediment transfer to the nearby stream (c. 275m) is plausible.

Slope-channel coupling

Accounts detail the transfer of materials from hillslopes to channels (Table 19.1a). The Runga farmer provides details for locations 1 and 3–4 (see Fig 19.2A). At location 1 the destruction of trees occurred and sediment from location 3 fell over a cliff into the stream at location 4 (see Fig 19.2A and B). In the aftermath, villagers planted trees to stabilise deposits (see Fig 19.2A, location 5).

Sediment-water flow characteristics

Correct identification of flow processes (eg fluvial or debris flow) is important for hazard and risk management; accordingly interviews (Table 19.1a) explored this. The Runga farmer reports ground-shaking (see Fig 19.2A, location 6), known to occur with debris flows (Arattano 2000) and an associated constant noise, which would be atypical for debris flows as they have pulsed

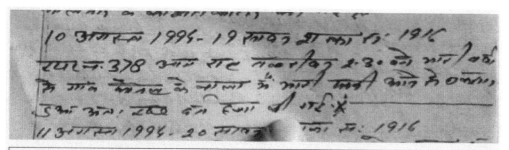

English Translation (Mohan, 2013):

*'10 August 1994, this is another date [Hindu calendar system date: 1916]…
Whole night around 2.30 AM there was a heavy rainfall and it was in village
Phojal and there are few other villages which are far away from here, and
later there was a lot of damage…'*

FIG 19.3. PHOJAL VILLAGE REGISTER ENTRY FOR 10 AUGUST 1994.

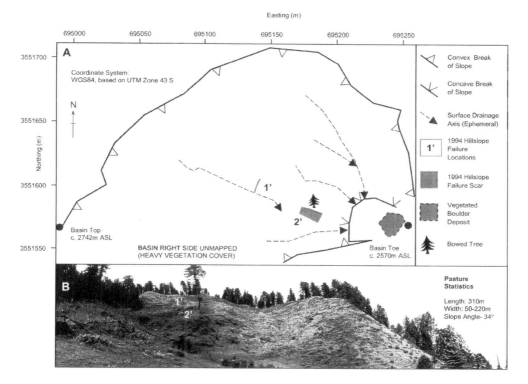

FIG 19.4. THE JEYDAHR PASTURE WITH 1994 HILLSLOPE FAILURE SCARS. (A) RECONNAISSANCE
MORPHOLOGICAL MAP OF THE FAILURE BASIN DERIVED FROM SMOOTHED GPS WAYPOINT DATA,
(B) PHOTO PANORAMA LOOKING UPSLOPE (20 OCTOBER 2013).

Table 19.1a: Interview accounts from April and October 2013: timing, meteorological causation and physical characteristics.

1994 Event Question Theme	Location of Interview	Interviewee (Generalised)	Responses (Via Interpretation of Mohan, April & October 2013 [Phojal and Downstream] & Singh, October 2013 [Landslide & Runga])
Event timing	Runga	Runga Farmer	• 'He is saying 9 August 1994...in the night time...ten-o'clock [In relation to slope failures]'
	Phojal Village	Pradhan	• 'That flood came in the month of August/September 1994...it came around, in between 2pm to 3pm, night'
Meteorological conditions	Runga	Runga Farmer	• 'Yeh, rain...there was too much rain there. Heavy rain...[lasted] 2 hours'
	Phojal Village	Grandmother	• 'This lady is saying in 1994 when the flash flood came...due to the heavy rainfall'
		Shopkeeper at Road Bridge	• '...before this flash flood there was another flash flood came which is 90km away place from here...Some relatives made alert, there is a flash flood, and you also be careful' • 'The water came with the cloudburst...the place where the cloudburst occurred in 2km away from here... That cloudburst was mainly for regular 3 hours'
		Pradhan	• 'There was a heavy rain...there was a cloudburst' [In April 2013 also said heavy/regular rainfall 2 days prior to this]
Hillslope failures and impacts	Runga/Jeydahr	Runga Farmer	• 'He is saying one place is Jeydahr, from start from landslide there...' [Continued in conversation: Jeydahr was the largest site of hillslope instability in 1994] • 'This landslide...they are eating calves up there, that time some sheeps and cows also going [downhill animated] in this landslide' [Continued in conversation: Jeydahr was visited and photographed in 1994 by the Kullu 'Animal Hospital Department' to certify livestock fatalities] • In conversation identifies three locations of hillslope failures in this event: Zabu, Jeydahr, and stream basin east of Jeydahr (location 3); at Jeydahr assuredly pin-pointed two failure locations from the 1994 event on the pasture
	Phojal Village	Pradhan	• '...there is one village called Runga...and there was the landslide occur'
Slope-channel coupling	Runga/Jeydahr	Runga Farmer	• In conversation indicated mud, rock and tree mixes transferred downslope to channels: – From failure at location 3, falling over a cliff to the channel at location 4; indicated was muddy at time, but now re-vegetated; locals also replanted vegetation from seed in event debris at location 5 – Material from Zabu moved through the forest to the stream below, knocking down trees
	Phojal Village	Pradhan	• 'When there was a landslide and that landslide came in the Phojal Nalla [ie tributary of]... the whole landslide mixed with water and that landslide/mud came into the Phojal Nalla'

1994 Event Question Theme	Location of Interview	Interviewee (Generalised)	Responses (Via Interpretation of Mohan, April & October 2013 [Phojal and Downstream] & Singh, October 2013 [Landslide & Rungal])
Channel sediment-water flow process type	Runga	Runga Farmer	• *'He say like very big noise…and his house is like shaking…not big shake but little…one noise, one noise'* [location 6]
	Phojal Village	Grandmother	• *'It was a sudden flash water, nobody was able to understand what happened…'*
		Shopkeeper at Road Bridge	• *'It was mixed with lots of debris, is the type of water was little blackish and lot many stones, boulders, pebbles, everything, and big long trees also came with that flash water…the water came with cloudburst, it was not very thick, it was not very viscous, but it was little thin… This person has seen the water from here with his torch light…'* • *'…before the flash flood there was a swift [down channel] wind was there…And after 5 minutes they saw the water coming'*
		Pradhan	• *'They saw sound first…continuous… and a lot of light because many stones were striking with each other…they were producing a small kind of light…sparks…because there was a heavy rain, and so far there is no record of the [a] fire… the whole [?] was having little shaking'* • *'In that flash flood there mud, stones, sand, big boulders…it was muddy…it was very very furious and very fast'*

Table 19.1b: Interview accounts from April and October 2013: physical characteristics continued and immediate economic impacts.

1994 Event Question Theme	Location of Interview	Interviewee (Generalised)	Responses (Via Interpretation of Mohan, April & October 2013 [Phojal and Downstream] & Singh, October 2013 [Landslide & Rungal])
Channel water/ sediment yield and form dynamics	Runga	Runga Farmer	• Conveyance of material down to Phojal: *'Just straight away,* [but] *near the Phojal there is one space like…when the landslide is going down and everything is stopped there'*
	Phojal Village	Shopkeeper at Road Bridge	• *'…It was full of stream'* [subsequently pointed to bank top stage indicator – a coarse boulder deposit on the right bank margin downstream of the road bridge] • *'…earlier the flood area was very narrow. Now due to the flood it has become wider…'*
		Pradhan	• *'…It was muddy, it was continue for more two days. First it was heavy flash flood and later there was a continuous water supply was there…'* • *'*[River after the flood] *it has become bigger. A lot of debris…so it all deposit here…and still lying there in the same shape.'*
Village physical/ immediate economic impacts-Phojal	Phojal Village	Grandmother	• *'…many houses were there which were washed away and that time with that water…'* • *'Even a few vehicles were parked along the riverside – which were also washed away with that flash water…the flour mill it was also washed away… shops also; everything whatever was available in the shop, their things; everything was washed away…one big truck was also washed away in that'*
		Shopkeeper at Road Bridge	• *'…truck was blown away by that swift wind.'* The truck was found in the Beas River at Kullu 22 days later [Stated April 2013] *'my building all finished…the whole house was washed away by the fast flowing water'* [Later confirmed as a 3 story building, located on right bank margin of the stream down of the road bridge] • The left bank side of the stream suffered the greatest damage [Stated April 2013]
		Pradhan	• *'Property was totally lost and totally damaged…'* [In April 2013 suggested 8-12 properties lost] • *'…agricultural land which was washed away, their orchards…and that was the apple season…they fill the boxes of the apple to import to the other parts of the country, and they saw those apple boxes also were washed away with that water'* • *'Round about 20 hectares of you can say 20 acres of the land from top to end is washed away…'* • 100-150 metres of road lost preventing road access to other villages for 12-16 months [Stated April 2013]
		Bank Manager	• Indicated the bank financed the truck which got washed away
Village physical impacts- downstream in Phojal Nalla	Dawara	Shopkeeper Assistant	• *'…one truck and few houses were washed away that flood, that night'* [Uncertainty where this applies to?]
		Shopkeeper	• *'…that night there were 2 trucks standing there, parked there, and they were washed away'*

sediment transfer. In Phojal, the accounts of the shopkeeper and Pradhan are similarly incon-clusive. In support of fluvial flows, the Pradhan states it was 'muddy' (cf Pierson 2005), and the shopkeeper expresses a lower-viscosity mixture (cf Costa 1988). In contrast, potential indicators of a debris flow include the Pradhan's observation of furious and fast flow conditions (cf Pierson 2005), and the shopkeeper's observation of an air blast, which may reflect a rapidly approaching debris flow front.

Channel yield and form dynamics

No hydrological monitoring of Phojal Nalla was performed in 1994, so interview accounts were used instead (Table 19.1b). The shopkeeper indicates a bankfull flood stage near the Phojal village road bridge (see Fig 19.2C). The Pradhan details elevated flow and sediment loads following the event for in excess of two days. The consequences were in-channel deposits (Runga farmer and Pradhan) and increased channel size (shopkeeper and Pradhan).

Physical impacts: Phojal Village, Phojal Nalla, Dawara

Interview data (Table 19.1b) reveal the breadth of physical and immediate economic impacts. In Phojal (see Fig 19.2A, location 7), respondents indicate: loss of multi-storey buildings (residen-tial, commercial and flour mills); loss of vehicles including a truck; destruction of agricultural land and the loss of harvested apples; and severing of the local road. Downstream at Dawara, shopkeepers make reference to the loss of assets but do not confirm locations accurately.

Societal impacts in Phojal Village

Table 19.1c indicates how the community were both recipients and responders to the event, from health, long-term economic and socio-cultural perspectives.

1994

Human fatality accounts are inconsistent in Phojal, ranging from 7 (shopkeeper) to 20–30 (grandmother), and in Dawara, 5–6 people (shopkeeper). Victims came from both indigenous and migrant populations (Table 19.1c); among the latter many lived (illegally) along channel margins alike today at Dawara (see Fig 19.2D). Fatality accounts (eg grandmother) demonstrate the enduring trauma for those who survived but lost others.

During the response and initial recovery phases, the villagers helped each other (Table 19.1c), since there was limited government intervention (Pradhan) and no aid provided by the local bank (bank manager).

Post-1994

Economic impacts are also persistent, in respect of both housing and income generation. The permanent loss of some agricultural land in Phojal, and the temporary loss of road access for 12–16 months (Tables 19.1b and c) caused considerable anguish (eg Pradhan, shopkeeper). Berkes *et al* (2000) similarly found that Kullu Valley villagers give great weight to land areas producing cash crops, market access and roads in maintaining a sustainable lifestyle. The shopkeeper details his personal loss of property (Table 19.1b) and the subsequent construction of a temporary house (in which he still lives) (Table 19.1c). Twenty years on, people still perceive a lack of compensa-tion from the government for their losses. For example, the shopkeeper asserts that his house, shop and stock were worth millions but he was offered 5000 Rupees (around £50 in 2014). He

Table 19.1c: Interview accounts from October 2013: societal impacts.

1994 Event Question Theme	Location of Interview	Interviewee (Generalised)	Responses (Via Interpretation of Mohan, October 2013)
Human fatality	Phojal Village	Grandmother	• '…many people died…nobody survived from some of the families…and from here dead bodies were found more than 10km away from this area…around 20 or 30 people died in that' [She confirmed she personally knew many of them]
			• 'Some people were from far away and they died. Now nobody lives here…this place is a hell for them, because they have lost their families and they do not want to stay…so they have shifted from this area to somewhere else'
		Shepherdess	• 'never ever came any flood before 1994…is the saying by the old people, those who lived here … only in 1994 the flood came which has made a total destruction and many people have died in that…11 people died in that flash flood'
		Pradhan	• 'Ten people died' [He confirmed these were in Phojal village]
		Shopkeeper	• '…Seven people died'
	Dawara	Shopkeeper	• '…inside the truck there were 2 conductors…they were also died…and 3-4 Nepalese died in that' [Uncertainty where applies to?]
Emergency response & initial recovery	Phojal Village	Grandmother	• 'Many people came with the torch, some other things to see is there anybody survived in this flash flood or not…It was a horrible situation…'
		Pradhan	• 'They had help for each other at that time. But how much can they do? Still, we have seen some people walking here and there in search of food…'
			• '…the Acting Chief Minister of the state came here… People ask from him the help but nobody has listen, and still people are waiting for the aid, waiting for the help…'

Post-1994 Event Question Theme	Location of Interview	Interviewee (Generalised)	Responses (Via Interpretation of Mohan, October 2013)
Long-term economic impacts-compensation	Phojal Village	Grandmother	• '…nothing is given by the government as compensation. So, some people have lost everything, only they survive themself'
		Shopkeeper	• '…has purchased a little piece of land here, which is not very costly and he has made this temporary residence over here because everything is [lost] and he is not having plenty of money left…waiting for the compensation from the government…not paid anything to him so far…owned everything, millions of property, but only offered 5000 Rupees' • '…big problem of repairing road, government has done nothing, even if money has come… not reached to the people'
		Pradham	• '…people are still waiting for the compensation…[Losses were] in millions, compensation in thousands'
Socio-cultural issues	Phojal Village	Shopkeeper & wife	• '…his daughter has grown up now…but there is a problem…he is not having anything left behind [money], the problem is facing the marriage of his elder daughter because he is not going to get a proper bride [ie groom] for his daughter'
Psychological issues	Phojal Village	Shopkeeper & wife	• '…very sad after that incident because everything has finished…temporary house is not in good condition – they don't have much money to repair it' • 'they feel secure, but they still scared of, because what they saw… When there is a rainy season… there is terror in their mind'
		Pradhan	• '…we are all scared because at least how much land is left, that should not be washed away with that water now. And whenever there is rain…they have a kind of fear in their mind, in their heart…and they pray to God'
		Grandmother	• 'If in future the flood will come they are scared of, because they have seen the destruction here, and they are still scared of and they pray to God that there should not be that kind of tragedy in the future…if the flood will come there is no chance for us to run…will leave everything in the hands of God'

also implies that the compensation was top-sliced by the local government. Whereas it seems that the indigenous population rebuilt their lives in the local community, there has been out-migration of Nepalese, resulting in a cultural and economic loss to Phojal. Although migrant communities do exist nearby in Dawara and Dobhi, they remain vulnerable. For example, in Dawara one interviewee said they are aware of the risks, and in the event of a future flood would move away and expect to receive no government help.

The psychological and cultural impacts of the 1994 event continue (Table 19.1c). All inter-viewees expressed fear of future flooding, especially when heavy rains occur. Several people talked of helplessness in the face of future events and place themselves in the hands of God. Intergen-erational impacts have also been reported – for example, the shopkeeper and his wife say that their loss of wealth, and therefore status in society, means that their well-educated daughter cannot marry a 'man of consequence' because of her impoverished background; she is also now unwilling, or unable, to bring her town friends to her home village, and this is a cause of sorrow to her parents.

HERITAGE AND RESILIENCE IN THE PHOJAL NALLA CATCHMENT

The 1994 heritage–resilience condition

Building community resilience is predicated on the ability to learn, in part by recognising and exchanging collective indigenous knowledge (McEwen *et al* 2012). The foregoing data report this theme and are used to interpret the community heritage–resilience condition. Oral histo-ries of former events (1994, 1894) are numerous in the community, providing confirmation of intangible heritage. However, written accounts (other than the village record, see Fig 19.3) and photographs are not apparently in the public domain, so these aspects of heritage are possibly underutilised.

More positively, Table 19.1 indicates good collective awareness of event temporal sequence, spatial patterns, hydro-geomorphological attributes and impacts. Appreciations include: (1) season-ality and event timings; (2) connections between hillslope and channel settings (downslope transfer of material and livestock); (3) approaching flow indicators (ground-shaking, noise, wind blast and sparks); and (4) physical impacts on the channel system and adjacent socio-economic assets.

Measures for coping and adaptation are also evident in Table 19.1. Coping strategies included: responses (evacuation of houses, search by torchlight and inspection of bank assets) and recovery (food sharing, community financial self-help, compensation claims and rebuilding of lives among the indigenous population). Adaptations included: (1) limited rebuilding and non-residential use of remaining structures along the Un-named Nalla in Phojal by the indigenous popula-tion; (2) new land ownership maps (2006-2007) clearly delimit the Un-named Nalla channel corridor, which may suggest *ad hoc* land-use planning; and (3) localised re-vegetation of 1994 event deposits near Runga by farmers to modify sediment source erodibility. The fate of the migrant population, however, seems less fortunate, with no apparent recovery in Phojal.

Developing capital for resilience in the Phojal Nalla catchment

Opportunities to develop capital to enhance resilience exist and are offered for consideration, recognising that further investigation is required to evaluate their necessity and feasibility. This is timely given the development pressures in the Kullu Valley.

Economic capital

Communities with diversified income and credit sources generally exhibit more resilience (Gardner and Dekens 2007). In the Phojal Nalla catchment, horticultural and agricultural activities dominate, supplemented by smaller-scale retail and taxi businesses. Financial hardship following 1994 (Table 19.1c) questions the extent of diversification. Further, the bank did not offer any aid and available credit was poorly utilised in 1994. This may present an opportunity for the creation of a community disaster aid fund, being a partnership between the local bank, local authorities and villagers. This would empower the community to be proactive rather than responsive, addressing the perceived lack of government compensation (Table 19.1c).

Reliable communication networks support emergency response and recovery (Pfefferbaum *et al* 2007). Phojal village currently has landline and mobile telephone coverage, but these may be at risk from natural hazards. In this regard, the Phojal Secondary School headmaster stated they did not have access to two-way radios, and more so did not perceive the need for this communication redundancy.

Roads enhance access to markets, health services, food, water and emergency response (Berkes *et al* 2000). Villages in the Phojal Nalla catchment are served by a single road (see Figs 19.1 and 19.2). The hydro-power development in the catchment (HPSEB 1998; BEHEPD 2009) may present an opportunity for road construction and upgrading. Villages could also stockpile materials and equipment to facilitate emergency repair of roads and bridges, reducing dependence and impacts of road loss as reported in 1994 (Table 19.1b and 19.1c).

Social capital

Education is key to building resilience (Frankenberg *et al* 2013). In Phojal village, discussions with the Pradhan and school headmaster reveal that there is no formal syllabus regarding hazards, although awareness is raised during morning assemblies when pupils are advised on behaviour during earthquakes. Emergency telephone numbers and evacuation route signs are displayed in the playground, although it was not clear whether emergency response training and rehearsals take place. There is also an absence of IT facilities and specific textbooks; having these would provide a means to better incorporate hazard and risk awareness into the curriculum.

Muttarak and Pothisiri (2013) suggest that improved education has wider intergenerational benefits, as information is exchanged among the community. However, knowledge is not unidirectional, as memories passed between generations create a sense of place, and this helps to reproduce heritage and maintain readiness (Norris *et al* 2008; McEwen *et al* 2012); such accounts exist in respect of the 1894 flood.

To expand social capital, opportunities may include: (1) citizen participation in community organisations (Pfefferbaum *et al* 2007; Norris *et al* 2008), which in Phojal would benefit from a new village hall (lost in 1994); (2) information boards and places of disaster memory; and (3) establishment of a community hazard warden team to record events, operate early warning sensors and liaise between stakeholders (McEwen *et al* 2012).

Environmental capital

Gardner and Dekens (2007) suggest indigenous knowledge has long influenced settlement locations in the Kullu Valley, and this hazard avoidance affords permanent populations some resilience. Exposure to natural hazards does remain in tributary valleys, and on river terraces and fans, and particularly afflicts vulnerable migrant populations, eg Dawara (see Fig 19.2D). In such locations, Gardner (2002) contends that detailed land-use zonation and management is limited; it follows that developing such measures may assist. However, more achievable in the short-term would be the assisted relocation of migrant populations away from Phojal Nalla, and measures to deter resettlement such as rewilding the channel corridor. While reducing exposure these measures do not eradicate hazard occurrence, so pre-event planning is also needed. Although a Kullu District Disaster Management Plan exists (Nanta 2011), integrated emergency response planning seems absent at village level; in Phojal the Pradhan indicated it was beyond their financial means. The Phojal School headmaster also indicated an absence of emergency kits; hence opportunities to improve self-sufficiency in the aftermath of disasters remain.

CONCLUSIONS

The loss of heritage may impact negatively on people, due to cultural significance, loss of learning, identity and connection to place, as well as more specific socio-environmental factors such as memory and materials connected to flood events. This chapter has examined the heritage–resilience condition of communities in the Phojal Nalla catchment in the context of a 1994 flood, using physical and social science methods. The objective is to contribute to understanding of the benefits of intangible heritage in helping people enhance resilience in the context of hazard and risk. Three keys points emerge:

(1)   The 9–10 August 1994 flood was triggered by convective rainfall, resulting in slope instability and channelised flows over c. 8km. This inflicted health, physical, economic and socio-cultural losses both at the time and in the aftermath; these span tangible (physical objects impacted) and intangible (people centric) heritages;

(2)   Oral accounts and village records demonstrate a heritage of learning from the 1994 and 1894 events. Some resilience is demonstrated in 1994 in terms of coping during the event, and in subsequent adaptations;

(3)   Opportunities to enhance heritage–resilience are offered in respect to economic, social, and environmental capital.

Research findings demonstrate the complexity and sensitivity of the social–physical environment in the Kullu Valley; many opportunities to extend knowledge and promote heritage–resilience remain, and these have added urgency in a region prone to a number of hazardous processes that is undergoing rapid socio-economic development.

BIBLIOGRAPHY AND REFERENCES

Agnew, P D, 1899 *Gazetteer of the Kangra District, Parts II to IV, Kulu, Lahul and Spiti*, Indus Publishing Company, New Delhi

Arattano, M, 2000 On Debris Flow Front Evolution Along a Torrent, *Physics and Chemistry of the Earth (B)* 25 (9), 733–40

Baldi, S, 2012 *Economic Survey of Himachal Pradesh 2012–2013. Economic and Statistics Department, Government of Himachal Pradesh* [online], available from: http://admis.hp.nic.in/himachal/economics/pdfs/EconSurveyEng2013_A1b.pdf [9 February 2014]

Balokhra, J M, 2007 *The Paradise Himachal Pradesh*, H G Publications, New Delhi

BEHEPD (Bhrigu Enterprises Hydro Electric Power Developers), 2009 *Kukri Small Hydroelectric Project, Modified Detailed Project Report*, NHP Consultants, Shimla-171004

Berkes, F, Gardner, J S, and Sinclair, A J, 2000 Comparative Aspects of Mountain Land Resources Management and Sustainability: Case Studies from India and Canada, *International Journal of Sustainable Development and World Ecology* 7, 375–90

Brakenridge, G R, 2014 *Global Active Archive of Large Flood Events, Dartmouth Flood Observatory, University of Colorado* [online], available from: http://floodobservatory.colorado.edu/Archives/index.html [31 January 2014]

Brown, K, 2014 Global Environmental Change 1: A Social Turn for Resilience? *Progress in Human Geography* 38 (1), 107–17

Census of India, 2011 *Provisional Population Totals, Figures at a Glance: Himachal Pradesh* [online], available from: http://www.censusindia.gov.in/2011-prov-results/paper2/data_files/punjab/2.Paper-2_pb_at_a_glance.pdf [8 February 2014]

Chandel, V B S, and Brar, K K, 2010 Seismicity and Vulnerability in Himalayas: the case of Himachal Pradesh, India, *Geomatics, Natural Hazards and Risk* 1 (1), 69–84

Chetwode, P, 1972 *Kulu The End of the Habitable World*, John Murray, London

Collins, A E, 2009 *Disaster and Development*, Routledge, Abingdon

Costa, J E, 1988 Rheologic, Geomorphic and Sedimentologic Differentiation of Water Floods, Hyperconcentrated Flows and Debris Flows, in *Flood Geomorphology* (eds V R Baker, R C Kochel, and P C Patton), Wiley Interscience Publications, New York, 113–22

Cutter, S L, Barnes, L, Berry, M, Burton, C, Evans, E, Tate, E, and Webb, J, 2008 A Place-based Model for Understanding Community Resilience to Natural Disasters, *Global Environmental Change* 18, 598–606

Dhumal, V K, and Ahuja, P, 2012 *Know Your State: Himachal Pradesh*, Arihant Publications (India) Limited, New Delhi

Forbes, M C, 1911 *To Kulu and Back*, Thacker Spink & Co, Simla

Frankenberg, E, Bondan, S, Cecep, S, Wayan, S, and Thomas, D, 2013 Education Vulnerability and Resilience after a Natural Disaster, *Ecology and Sociology* 18 (2), 16

Gardner, J S, 2002 Natural Hazards Risk in the Kullu District, Himachal Pradesh, India, *Geographical Review* 92 (2), 282–306

Gardner, J S, and Dekens, J, 2007 Mountain Hazards and the Resilience of Social-ecological Systems: Lessons Learned in India and Canada, *Natural Hazards* 41, 317–36

Harcourt, A F P, 1972 (1871) *The Himalayan Districts of Kooloo, Lahoul and Spiti*, Vivek Publishing House, Delhi

Harrison, R, 2013 *Heritage Critical Approaches*, Routledge, Abingdon

HPSEB (Himachal Pradesh State Electricity Board), 1998 *Fozal Small Hydro Electric Project, Project Report*

Jangra, S, and Singh, M, 2011 Analysis of Rainfall and Temperatures for Climatic Trend in Kullu Valley, *Mausam* 62 (1), 77–84

Kumar, V, Singh, P, and Singh, V, 2007 Snow and Glacier Melt Contribution in the Beas River at Pandoh Dam, Himachal Pradesh, India, *Hydrological Sciences Journal* 52 (2), 376–88

Kuniyal, J C, Vishvakarma, S C R, Badola, H K, and Jain, A P, 2004 Tourism in Kullu Valley: An Environmental Assessment, *HIMAVIKAS Publication* 15

McEwen, L, Krause, F, Hansen, J G, and Jones, O, 2012 Flood Histories, Flood Memories and Informal Flood Knowledge in the Development of Community Resilience to Future Flood Risk, paper presented at the BHS Eleventh National Symposium *Hydrology for a Changing World*, 9–11 July, Dundee

Muttarak, R, and Pothisiri, W, 2013 The Role of Education on Disaster Preparedness: Case Study of 2012 Indian Ocean earthquakes on Thailand's Andaman Coast, *Ecology and Sociology* 18 (4), 51

Nanta, B M, 2011 *Kullu District Disaster Management Plan*, District Disaster Management Authority Kullu District, Kullu

Norris, F H, Stevens, S P, Pfefferbaum, B, Wyche, K F, and Pfefferbaum, R L, 2008 Community Resilience as a Metaphor, Theory, Set of Capacities, and Strategy for Disaster Readiness, *American Journal of Community Psychology* 41, 127–50

Owen, L A, Derbyshire, E, Richardson, S, Benn, D I, Evans, D J A, and Mitchell, W A, 1996 The Quaternary Glacial History of Lahul Himalaya, Northern India, *Journal of Quaternary Science* 11 (1), 25–42

Pfefferbaum, B J, Reissman, D B, Pfefferbaum, R L, Klomp, R W, and Gurwitch, R H, 2007 Building Resilience to Mass Trauma Events, in *Handbook on Injury and Violence Prevention* (eds L S Doll, S E Bonzo, D A Sleet, and J A Mercy), Springer, New York, 347–58

Pierson, T C, 2005 Distinguishing Between Debris Flows and Floods from Field Evidence in Small Watersheds, *USGS Fact Sheet* 2004-3142

Sah, M P, and Mazari, R K, 1998 Anthropogenically Accelerated Mass Movement, Kulu Valley, Himachal Pradesh, India, *Geomorphology* 26, 123–38

— 2007 An Overview of the Geoenvironmental Status of the Kullu Valley, Himachal Pradesh, India, *Journal of Mountain Science* 4 (1), 3–23

Singh, S, 2008 Destination Development Dilemma-Case of Manali in Himachal Himalaya, *Tourism Management* 29, 1152–6

Smith, L, 2006 *Uses of Heritage*, Routledge, Abingdon

Sorensen, M L S, and Carman, J, 2009 Introduction: Making the Means Transparent: Reasons and Reflections, in *Heritage Studies Methods and Approaches* (eds M L S Sorensen and J Carman), Routledge, Abingdon, 3–10

Valdiya, K S, 2002 Emergence and Evolution of Himalaya: Reconstructing History in Light of Recent Studies, *Progress in Physical Geography* 26 (3), 360–99

Wilson, G A, 2012 *Community Resilience and Environmental Transitions*, Routledge, Abingdon

20

Cultural Heritage and Animal Disease: The Watchtree Memorial Stone

Josephine Baxter

In the far north-west of England are two immense Scottish stones. One is in a dark underpass in Carlisle where dim light catches the inscription on its polished surface: the vitriolic interdiction of a 16th-century archbishop and his mighty curse upon the Border Reivers. Put there to mark the Millennium, this seven-foot monolith was conceived by a Carlisle-born artist and carved in Galloway. The other stone stands some seven miles away to the west at a place now called Watchtree. This squat boulder, carried by ice across the Solway during the last glaciation, lay deep underground until it was unearthed during the Foot and Mouth epidemic of 2001. It is now a memorial and bears a small bronze plaque (see Fig 20.1). Both stones commemorate times of grief and destruction and are intended to use art and nature respectively to place those sorrows firmly in the past. Yet both continue to attract controversy and bitterness.

In May 2013, the warden of Watchtree Nature Reserve gave a sigh of resignation as he told us that there are still visitors who come to remember – more in anger than sorrow – the events of 2001. That was the year when Great Orton airfield, as Watchtree was then known, became a place of slaughter and burial for an estimated half a million animals. The warden told us that he had recently been confronted by an enraged woman who took issue with an information board that referred to animals being 'laid to rest': 'They were not laid to rest', she insisted, 'they were murdered!' Others come as far as the entrance and cannot go any further; their memories, or the unconscious feelings they arouse, are too powerful to overcome. He told us that these visitors tend to come in the spring, and that he recognises them immediately. There is something in the way that they approach the visitor centre. What is it he sees? Is it dread, anger, or a desire to make sense of the past? What brings one man to the memorial stone where he stands for over an hour, a book in his hand, performing a ritual that only he understands or gains comfort from?

How too do we account for the atavistic fear that surrounded the installation of the Bishop's Stone in the underpass at Tullie House Museum? After Christian activists insisted that it should be mitigated by a Christian blessing, a cross and a prayer were duly embossed on a metal door nearby. In succeeding years protesters blamed it for bringing misfortune of 'biblical proportions' to the city, claiming that it was full of 'spiritual violence', a 'lethal weapon' that should be exorcised, or utterly destroyed. When Carlisle was flooded in 2005 Graham Dow, then Bishop of Carlisle, asked Bishop Mario Conti of Glasgow to lift the curse and the future of the stone was debated in Carlisle's council chamber. Opinion was divided between those who felt truly cursed and those who were angered at Carlisle being ridiculed in the national press for what they saw as superstitious nonsense.

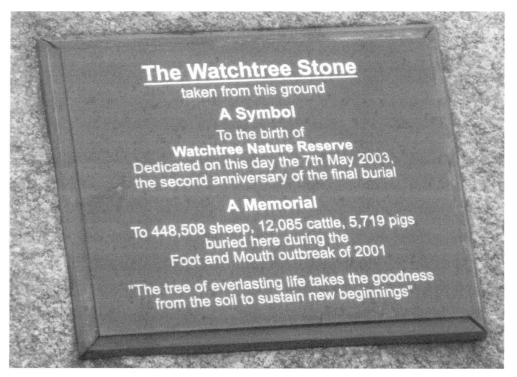

Fig 20.1.   Memorial plaque on the Watchtree Stone.

Each of these stones is a silent testament to the tensions and tolerance of the people of this region. This chapter attempts to trace that history, to understand how fear and trauma have recurred over the centuries, and to observe how Cumbrians have survived these travails. Watchtree weaves those threads in a way that is unique and yet typical of the region.

It is impossible simply to arrive at Watchtree. It is the opposite of a mirage, a place you know exists and yet cannot quite locate; it slips and slides and shapeshifts; it hides among the names and subdivisions of place and parish. As Great Orton airfield it became infamous in 2001, but you will not find its history among the records of Great Orton parish because it lies in the township of Wiggonby, in the parish of Aikton, in the old barony of Burgh by Sands. To find it you must swerve and feint along narrow lanes until you see its landmark windmills, only to find that the road curves away and even these giants disappear from view. Like the Solway villagers who fastened chains across their roads against their enemies, so Watchtree seems determined to slow us down and test our determination.

Watchtree, or perhaps Watchtrees, is the name of the farm that once stood on this plateau above the Solway. It had houses and cottages, barns and outbuildings, woods and pastures. That old Watchtree has vanished, obliterated over the course of 70 years in as thorough and brutal a manner as could be imagined, and replaced by a sanctuary of sorts, something rare and strange. It is a site of wilderness and control, decomposition and renewal, trauma and refuge. The contemporary Solway Plain is a wiry, windy version of the English country idyll with quiet lanes, clean air, healthy livestock and fields of crops. In order to comprehend it fully, however, and fathom its depths, we must first recognise that whichever direction we approach from we are travelling across ground that is steeped in blood.

The recurrence of places called Watch Hill on maps of Cumberland and Northumberland tells its own story, and the name Watchtree is believed to refer to two trees that grew on the farm, from whose branches incoming raiders from across the Solway could be seen. The early history of this region is one of almost continuous conflict. The decision of the Romans to draw a borderline from Wallsend to Bowness-on-Solway and mark it with an immense wall created an arbitrary locus for violent disorder. The people either side of it had much in common; indeed the northern baronies of Cumberland and Westmorland were sometimes under the rule of Scotland, sometimes England. Yet they fought and pillaged both across the border, and among their own countrymen, for centuries. After the Romans they were colonised by Angles, Norwegians and Danes, and mercilessly harried into submission by the Normans.

The late 13th century brought both natural and political disaster. In 1292 a hurricane blew for 24 hours. A fire raged through Carlisle destroying most of the city, and the wind forced the sea over Burgh and Rockliffe marshes, drowning immense numbers of cattle and sheep. The implacable Edward I made it his life's work to subdue the Scots and, together with his successors, inflicted 150 years of warfare on the Borders. Edward spent his last months in Cumberland and died, sick and weary, on the windswept Solway marshes at Burgh by Sands. His body was laid in the modest church at Burgh by Sands until it began its protracted journey to Westminster Abbey. As Ferguson says in his *History of Cumberland*:

> The fourteenth century was the most miserable… that the citizens of Carlisle and the men of Cumberland ever had to endure. … Strange and marked must have been the splendour and plenty in Carlisle during the visits of Edward I, II, and III, and the poverty of the country around: the citizens of Carlisle waxing fat on the wages of the soldiery and the money of the courtiers, while the wretched peasants round them starved. (Ferguson 1890)

The waths, or fords, across the Solway sands meant that the apparently isolated plain west of Carlisle was a route for incursion and its inhabitants were prey to the ravages and demands of passing militias. Carlisle's strategic position bridging the River Eden and commanding the only safe road south made it the seat of power and consequently the focus of siege and conflict. But when its castle proved impregnable its hinterland suffered grievous collateral damage as frustrated troops took their revenge on the surrounding countryside, plundering, killing and burning as they made their way home. The evidence of the threat and terror in which the people of the Solway Plain lived is still poignantly visible in its small, scattered churches. Usually built on high ground, their enormously thick walls speak of more than a determination to keep out the weather. They were built as places of refuge, with stout towers, narrow doors and slit windows. Standing in one of these churches while the wind roars at the door, it is easy to imagine the foetid, terrified press of people and livestock waiting for the tumult and uproar outside to pass.

When peace finally came to Cumberland, many of these churches fell into disrepair, to the dismay of church officials who came to inspect them and detected a general Godlessness among their flock. Was it Godlessness or did the churches fall into ruins because they were resonant with the horror and trauma of people huddled together in the dark in fear of their lives?

By the 16th century, reiving,[1] a culture of vendetta and raiding, had become the norm in the borders. It involved shifting allegiances of families and their supporters on both sides of the border in a system not unlike the Mafia or tribal warlords. Reiving gave us the words 'bereavement' and 'blackmail'[2] and, whatever romantic gloss they may have acquired over the intervening years, the Reivers were looters, kidnappers, murderers and extortionists. The ceaseless fear and uncertainty they inflicted across the border countryside was a miserable affliction for the whole population. By 1525 Archbishop Dunbar of Glasgow was so incensed by their pitiless criminality that he issued an interdiction to be read in all public places. He laments: 'that God-fearing men, women and children are innocently murdered, slain, burnt, harried, spoiled and robbed by day and night; their farms and lands laid waste while they are forced to flee, all this by the common traitors, reivers and thieves... who seem unafraid of either church or temporal laws' (Gavin (Dunbar) Archbishop of Glasgow 1525). What follows is perhaps the most comprehensive malediction ever committed to paper in which every possession, body part and activity of the reivers is cursed and biblical plagues wished upon them. A thousand words later he issues his interdiction, excommunicating them from all sacraments except baptism, and with a final flourish condemns them to the deep pit of hell 'till they forbear their blatant sins and rise from this terrible cursing'.

This diatribe seemed to have little or no effect on the reivers, but it was to haunt the people of Cumberland down the centuries to the present day. They could be forgiven for believing that they were cursed when, in the late 16th century, there were severe outbreaks of plague in Cumberland, together with poor weather and scant harvests. A contemporary inscription (see Hughes 1971) recorded that 1196 people died of plague in the Carlisle area, and the unusually large number of burials in Great Orton in August in 1597 may account for some of that number. Attempts were made to prevent or minimise infection by whatever means were available. The city fathers of Carlisle drew up 'necessary observations thought meet to be kept in this city' (ibid), and they bear an eerie similarity to the regulations of the Foot and Mouth outbreak of 2001. Infected houses were sealed off and arrangements made to provision any live inhabitants. There were strict orders for the manner and timing of burials, and men were paid generously to work as corpse-bearers or to clean the houses of the dead. At Penrith a large stone was sited at the edge of the town and its hollow centre filled with vinegar. Money was placed in it to pay for necessities, and only when the money was 'disinfected' in this way would traders leave their goods and take their payment, a practice that would seem familiar to those who dipped their boots in disinfectant and obsessively washed their cars in 2001.

We do not know how Watchtree fared through these times, but it is reasonable to assume that it would not have been immune to violence or sickness. Two miles away across the fields, Down Hall, moated and built on the site of a Roman fort, was razed by the Scots. After the unification of England and Scotland, the last sieges of the Civil War and the 1745 rebellion, the 19th century brought peace and comparative security at last.

The 1843 tithe map for the parish of Aikton depicts a parish vividly described by its place-names. Water Flosh, Quarry Gill and Shuttle Hall, the daunting Hardcake, Black Brow, Ugly Gill and the mischievous-sounding Lovely. For the most part it is striped with narrow rectangular fields, remnants of the old rig and furrow system. Their patterns remain there today on maps and aerial

---

[1]   Reiving: from Old English, meaning theft, plunder.
[2]   Blackmail: early version of protection money.

FIG 20.2.   THE BISHOP'S STONE, TULLIE HOUSE, CARLISLE.

photographs, along with the ancient tracks that ran between them out of the village of Great Orton. And in 1843 Watchtree has a substantial farmhouse and buildings, and fields far larger than those neighbouring it. The parishes of Aikton and Great Orton teemed with working men and women including blacksmiths and shoemakers, weavers and bleachers, dressmakers and victuallers, coopers and wallers, nailers and thatchers, as well as servants, agricultural labourers and farmers.

By 1851 the little township of Wiggonby had a population of 856, of whom 34 lived in Watchtree's cottages, but the farmhouse remains elusive and there is no listing on the census for the farmer or landowner. Despite its 230 acres and large house, Watchtree had a transient population of tenant farmers. The only family to stay longer than ten years were the Erringtons: in their 20s when they first came to Watchtree, they employed six agricultural labourers as well as four farm servants by the time they left in the 1870s. We do not know what made them move on, but a poor winter in 1879 followed by a severe outbreak of Foot and Mouth Disease in 1881–3 marked the start of a decade of hardship for farmers.

'Cattle Plague', 'Rinderpest' or 'Murain' was a recurring blight in the North of England. A school inspector at Aikton in 1864 said: 'During the late violence of the cattle plague the numbers of the scholars [fell] off considerably' (see www.stevebulman.f9.co.uk 2008). And in 1866 methods of control both for livestock and people again prefigure those applied in 2001. The *London Gazette* (1866, 581) of 2 February lists in detail the restrictions and prescriptions for the sale and movement of livestock, including the issuing of licences to drovers moving cattle along a highway. The recommended method of carcass disposal was burial 'as soon as possible', with a sufficient quantity of quicklime or other disinfectant and a covering of 'at least five feet of earth'. That year John Ostle, a farmer near Abbeytown, Cumberland, records in his diary that magistrates had stopped the movement of livestock. It appears that they had also adopted a policy of 'slaughter on suspicion' – as unpopular then as it was in 2001. Ostle wrote that Sir R Brisco and Captain James 'insist on destroying all the cattle where the disease breaks out whether they are healthy and weal [sic], or ailing or recovering', and that, unlike 'The Lord', who 'will have some compassion on us and leave some alive', Brisco and James 'will leave none alive' (Ostle 1866). There are other resonances across the years: a commission was set up to investigate 'the origin and nature of the cattle plague' and 'regulations which should be made to prevent the spreading of the disease and to avert any future outbreak' (Collinge 1984). In a precedent to the multiple reports of 2001, they produced three reports over two years, the first and third accompanied by dissenting reports while the second was agreed by all.

In 1899 Daniel Scott noted that when contemporary farmers found disease among their cattle 'they no longer pin their faith to the old-time observances' (Scott 1899). He describes the old custom of needfire, which he believed was last used in 1841. Needfire was an amalgam of rough science and folklore in which a fresh fire was kindled and livestock were driven through the smoke. The fire was passed from village to village, and was not to be extinguished until it had done its work. Scott tells of a farmer between Grasmere and Keswick who allowed the flame to go out and refused either to rekindle it or help it on its journey north. Although berated at the time, it seems his reluctance to take part in the rite extinguished the custom altogether.

The 20th century endured its own conflicts, and World War II saw a proliferation of airfields across the northern plain of Cumberland. Most are now ghosts on the map, visible only through aerial photography. The compulsory purchase of Watchtree for the Royal Air Force marked the end of its old identity. The orchards, woodlands and the two eponymous watch trees were felled, the house and buildings demolished and buried, either as hardcore under runways or to level out areas that needed to be raised. The lanes that led there from the village of Great Orton were cut off altogether, creating an impasse to the east that adds to Watchtree's sense of isolation. Great Orton airfield's beginnings were inauspicious, when the first landing was forced by engine failure, but it served its purpose until 1945 when it became the repository for a vast tonnage of unused ordnance. It closed in 1952 but the Air Ministry did not relinquish ownership until the early 1960s when the airfield was returned to farming and its runways became a favoured spot for learner drivers, boy-racers and microlight enthusiasts. The first wind turbines were built there in 1992, the ten originals being replaced by six larger ones in 1999; they are rarely still.

On 19 February 2001, Foot and Mouth Disease was discovered at an abattoir in Essex. 'A rare set of circumstances had determined that this would be one of the worst epidemics of FMD the world has ever seen. Numbers alone cannot capture the sense of what unfolded' (Anderson 2002, 20). Cumbria went on to be the worst affected county in Britain with almost 3000 farms undergoing partial or complete culls of their stock. On Sunday 11 March, Nick Brown, the Minister for Agriculture, stressed in a BBC interview that he was 'absolutely certain' that the disease was under control, but by then over 40,000 carcasses were rotting in farmyards around Cumbria awaiting disposal (ibid, 79) and public opinion was hardening against the use of pyres. The Ministry of Agriculture had turned first to fire to dispose of infected carcasses, but when this failed to keep pace with requirements and was believed by some to help spread the disease, mass burial was adopted to deal with the vast backlog of bodies.

Great Orton airfield was identified by the Environment Agency as a suitable site and the army organised the logistics to deal with the dead stock. Overnight on 25 March, trenches were opened and the following morning lorries carrying the decomposing carcasses of 7500 sheep arrived. Two days later the first consignments of live sheep were brought to be slaughtered and buried as part of what was hoped would be a 'firebreak' cull. At its peak the site took between 10,000 and 14,000 live animals and 10,000 carcasses a day. Some slaughtermen recoiled from the mechanistic killing of healthy stock, but others preferred the abattoir-like detachment of Great Orton to the emotionally fraught atmosphere of the farms.

For six weeks the residents of Great Orton, Wiggonby and surrounding villages had to close their windows against the stench of putrefaction and the roar of heavy lorries making their way along the narrow lanes to and from the airfield. The name Great Orton appeared on television every night beneath horrifying footage of slaughter and disposal. It is disturbing to recall the industrial scale and impact of the site, and in another time and place the population might have risen in anger at the violation of their environment, but these were not normal times and people were desperate to stop the spread of the disease. As the chairman of the parish council said: 'It had to be done.' Following the burials it was clear that the site would have to be monitored indefinitely. Systems were devised to separate surface water, ground water and leachate; lagoons were constructed and a massive wall of bentonite sunk 12 metres deep around the pits to ensure that there could be no leakage from the cells.

Rumours abounded about DEFRA's[3] plans for the airfield's future. DEFRA might retain ownership, but whatever possibilities London might have in mind, the people of Great Orton and Wiggonby were not prepared to be used as a future dumping ground for Britain's toxic waste, either organic or chemical. Opinions were sought, and the consensus was that a nature reserve would be a good use of the site. DEFRA's regional works officer recalled a crucial public meeting in Great Orton village hall:

> We got this feeling for a nature reserve. We were taken back by the depth of feeling, especially among the farmers. I said to my boss that night, 'We have had a profound meeting. These guys are serious and we have to take them seriously. They see this as a memorial.' (Ward 2002)

---

3 DEFRA: Department for Environment, Food and Rural Affairs. DEFRA was formed in June 2001 following the general election. It replaced MAFF: Ministry of Agriculture, Fisheries and Food. For a time this change added to feelings of uncertainty and suspicion in the areas most affected by Foot and Mouth Disease.

The chairman of the parish council canvassed the relevant government agencies but met with little response until, almost exactly a year after the outbreak of Foot and Mouth, he was summoned to the airfield and given the news that a nature reserve had been approved. Great Orton airfield became Watchtree once again and opened to the public in October 2004. It is now a place of conservation, education and work experience; where the award-winning social enterprise 'Watchtree Wheelers' sets people free from the dangers of traffic and the restrictions of their physical difficulties, to ride specially adapted bicycles.

There were times in 2001 when it seemed that Cumbria would never escape the oppression of Foot and Mouth. Perhaps those age-old feelings of helplessness and persecution account for the fury that surrounded the Bishop's Stone at Tullie House, and for the man who stands beside that other stone, at Watchtree. For some a memorial allows a safe return to trauma and the opportunity to make some sense of it; for others it preserves the trauma intact, undiminished by time or reparation. There are those who find the reversal of the Great Orton burial place to Watchtree Nature Reserve hard to credit, and on websites we find comments such as: 'MAFF still continue to speak with great pride at [sic] the marvellous work conducted at the recently renamed Watchtree. Plant a couple of trees and call it a nature reserve!' (Green 2002); 'The main dumping ground in Britain for slaughtered animals' (Airfields of Britain Conservation Trust n.d.); 'DEFRA are now trying to promote the site as a tourist attraction – called Watchtree' (Visit Cumbria n.d.); and 'Before proceeding any further with this page, I should point out that Wiggonby airfield (Great Orton) is now the site where upwards of 500,000 animal carcasses are buried; victims of the 2001 foot and mouth plague. This grim fact must be considered before a visit to this site is contemplated' (Barnes n.d.).

In George MacDonald Fraser's history of the reivers, he commends the Borderers' 'ability to endure, unchanging' despite 'the kind of continuous ordeal that has passed most of Britain by' (MacDonald Fraser 1986). Perhaps their history accounts for the quiet tolerance of the people who surround Watchtree, and the grace with which they have worked for its transformation from an icon of horror to a haven for wildlife, a place of literal and metaphorical recreation.

True, there is perhaps more hardware than you might expect to see at a nature reserve: the windmills swoop and hiss along the runway; the spiral cowls of ventilation pipes glitter in the sunlight; the dark bulk of the water treatment plant lurks among the pines. A solid wire perimeter fence echoes the wartime past, and there are notices forbidding entry to the areas over the burial cells. But on most days you will catch a glimpse of exhilarated cyclists whirring along the tracks. You may also hear the skylarks sing and see grazing cattle, flowering hay meadows, wetland birds, an exquisite barn owl, hares, bats and rare butterflies. This is, above all, a place where you stand under a vast sky, the earth tilting around you, aware of your own insignificance – your fleeting moment in Watchtree's centuries of endurance and recovery.

Acknowledgments

In researching this chapter I have journeyed to the outermost edges and darkest days of Cumbria, and I have seen Watchtree and the Solway Plain at their bleakest and most beautiful. I am indebted to Tim Lawrence, warden of Watchtree Nature Reserve, William Little, Chair of Great Orton parish council, and Patricia MacDonald, parish clerk, both of whom are trustees of the reserve and have been supporters of Watchtree from its inception, who all gave generously of their time. I thank the patient and helpful staff of Carlisle's Cumbria Archive Office, Tim Padley of Tullie House, and John Nixon, chronicler of Cumbria's airfields.

I interviewed and accessed materials from people who worked at Great Orton during and immediately after the period of burials, and I am grateful to Brian Jolly in particular for his willingness to recall harsh times.

Lastly I thank my erstwhile colleagues Ian Convery, Maggie Mort and Cathy Bailey, with whom I studied the 2001 Foot and Mouth Disease epidemic, for their advice and support.

BIBLIOGRAPHY AND REFERENCES

Airfields of Britain Conservation Trust, n.d. *Great Orton gets ABCT Memorial*, available from: http://www.abct.org.uk/news/great-orton-gets-abct-memorial [12 February 2014]

Anderson, I, 2002 *Foot and Mouth Disease 2001: Lessons to be Learned Inquiry Report*, The Stationery Office, London, available from: http://webarchive.nationalarchives.gov.uk/20100807034701/http://archive.cabinetoffice.gov.uk/fmd/fmd_report/report [12 February 2014]

Barnes, R, n.d. *Wiggonby Airfield Watch-office*, available from: http://www.users.globalnet.co.uk/~rwbarnes/defence/wiggonby.htm [12 February 2014]

Collinge, J M, 1984 List of commissions and officials: 1860–1870 (nos 95–136), *Office-Holders in Modern Britain: IX: Officials of Royal Commissions of Inquiry 1815–1870*, Institute of Historical Research, University of London, 62–91

Ferguson, R S, 1890 *A History of Cumberland*, Elliot Stock, London

Gavin (Dunbar) Archbishop of Glasgow, 1525 Letters of censure against insurgents [28 Oct 1525] [Enclosed in No 29] 289, *Henry VIII, Letters and Papers, Foreign and Domestic, 1509–46*, Cotton MS Caligula B II: 1st half of the 16th century, British Library Archives and Manuscripts Catalogue

Green, N, 2002 New Year comment by Nick Green & an article by Mike Sandersen, Appleby NFU Group Secretary, available from: http://www.warmwell.com/jan4nick.html [12 February 2014]

Hughes, J, 1971 The Plague at Carlisle 1597–98, *Transactions of the Cumberland and Westmorland Antiquarian and Archaeological Society* 71, 53

The London Gazette, 1866 [online], 2 February, available from: http://www.london-gazette.co.uk/issues/23065/pages/581 [12 February 2014]

MacDonald Fraser, G, 1986 *The Steel Bonnets*, Harvill, London

Ostle, J, 1866 *John Ostle's Journal, 1865–1866*, 24 March, available from: http://users.tinyworld.co.uk/peter-ostle/journal7.html [12 February 2014]

Scott, D, 1899 *Bygone Cumberland and Westmorland*, William Andrews & Co, London, available from: http://archive.org/stream/bygonecumberlan00scotgoog#page/n5/mode/2up [12 February 2014]

Visit Cumbria, n.d. *Foot and Mouth Disease in Cumbria – 2001*, available from: http://www.visitcumbria.com/foot-and-mouth-disease-in-cumbria/ [12 February 2014]

Ward, D, 2002 Nature casts a spell over site of slaughter, *The Guardian* [online], 8 November, available from: http://www.guardian.co.uk/uk/2002/nov/08/footandmouth.davidward [12 February 2014]

www.stevebulman.f9.co.uk, 2008 *Aikton (Wiggonby) School: Mr C I Elton's Report*, available from: http://www.stevebulman.f9.co.uk/cumbria/aikton_school.html [12 February 2014]

21

Earthquakes: People, Landscape and Heritage in Japan

Takashi Harada

In 1995 a major earthquake occurred in a densely populated area in the west of the main island of Japan. A year and a half later I met Ms Sakae, a 68-year-old woman who had become a refugee in the public shelter provided after the earthquake; I had worked there for six months as a volunteer, just after the earthquake hit (see Harada 2000). She related an episode that had taken place during a meeting of members at a house into which she had moved after spending three months at the shelter:

> At a meeting of the refugees living in this temporary apartment, I said 'It is not worth claiming against the municipal office about our condition here. It was our fault that we had our houses broken down by the quake. We ourselves had decided to live in such fragile houses.' There was no one who opposed me.

Her remark was so unexpected that the other members were too surprised to reply. I re-examine this episode after discussing earthquakes, which themselves are a form of natural heritage in Japan, and suggest that her remark gets to the heart of the matter.

Earthquakes and Personal History

After telling me about the meeting in 1996, Ms Sakae went on to describe her experiences in 1945, at the end of World War II. She talked vividly about her experience during the war, describing a sight she had seen over 50 years earlier and then comparing the two experiences:

> The next day, after the air raid, I walked around Osaka city searching for my friends. I saw a horse standing still, its skin was red.

> In case of an air raid, we had an alarm, but we had no alarm in case of earthquake. My apartment was twisted and broken within seconds.

She had been born in 1927 in Osaka, where her father started a business in 1947. Sakae became ill aged 27 and was hospitalised for two years, and then in 1956 her father died. After leaving hospital she joined the family business with her two brothers; their business was very successful in the 1960s and 1970s, but in 1982 the company became bankrupt. She recalled the loss of face and compared it to the impact of the earthquake:

> In the event of an earthquake I could still appear in public without fear [as everyone had suffered in the same way], but when bankrupt I hesitated; the newspapers played up the news... bankrupt [businesses] deserve the contempt of people.

She eventually recovered from this financial and social tragedy and rented an apartment, found a part-time job and enjoyed the company of her new colleagues. Her new network served her well following the earthquake:

> The day following the earthquake in 1995, one of my colleagues brought me some clothes and my boss drove his motorbike to bring me an eel bowl, kept warm with portable body warmers. I was delighted, but I don't like eel...

We laughed together. I'm certain she was satisfied with her relations with such people and again drew a comparison with her wartime experiences:

> In wartime we were patient [waiting] for food because it was the same for all the people throughout Japan. But in the case of the earthquake there was enough food in Osaka.

Osaka is located just 20 kilometres from the damaged area. People outside the earthquake zone area lived ordinary lives because nothing had happened to them. People living in the affected area received food, water and clothes from all over Japan and from foreign countries; hence the earthquake can be seen as a very local disaster. Sakae once again compared the impact of the earthquake to the problems with her business:

> After the bankruptcy, all of those [people] who had connections with our company stopped contacting me. [However] after the earthquake I met many good people, I have lived a full life.

She had survived not only the earthquake, but also the war and the bankruptcy of her business. The earthquake was major, with over 6000 victims, yet Ms Sakae's experience demonstrates that it evidently has a different history and meaning for each survivor. An earthquake is a personal disaster, affecting both physical and intangible aspects of an individual's heritage, and causing links to be made with other painful events experienced in life.

In 2008 I met another two former refugees of the public shelter, Akiko and her daughter Natsumi, and interviewed them about their experiences of the earthquake. In 1995 Akiko was in her 40s and Natsumi was a pupil at elementary school. Upon closure of the public shelter, they moved to temporary accommodation. They related this episode:

Akiko:　Two days before the quake, we stayed at my husband's parents' house and the following day, we stayed at my mother's house.

Natsumi:　I remember the red shoes my grandmother bought me on that day. Yes, I remember them. After we were evacuated to the public shelter I could not understand what was happening, and so I was worrying about the new shoes I had left at the entrance of our apartment.

Three weeks after the earthquake, the apartment was completely demolished, leaving them only a short time to collect their belongings from the debris. The earthquake had inverted the materials found in a house: those normally used every day were below those in store. All of their belongings, broken or twisted, are perhaps still there, but it is impossible to find and recover them. Akiko said:

> I went into the debris through a small hole which the owner had made. I found a cheap spare glass I had put in the chest, while the expensive one that I used every day and had put on the table was broken. As the quake completely destroyed our apartment, ordinary things that we used everyday were underneath the debris and those which we stocked on shelves were found easily. On the day of clearing the debris, I saw my daughter's clothes hanging in the air from a crane. We could do nothing to get them back.

The two-hour conversation between me, Akiko and Natsumi helped us remember the moment the earthquake struck – a moment that we had lost for 15 years. Then suddenly Akiko changed the topic:

Harada:	If we meet once again in the near future, would we talk about different topics from today's?
Akiko:	Well, we glorify things from the past and we always forget our bitter experiences. [She pauses] No! I remember that day. As we talked about the earthquake with you today, I remember it, just like Natsumi's experience about the smell.
Natsumi:	Debris!
Akiko:	We were buried under the debris... the debris of the wall of the apartment.
Natsumi:	Yes, we were under the debris. And on the day when they removed the debris, we were there then too.
Akiko:	I hate the smell of the clay, plaster and dust. Even now whenever I smell it, I say to my daughter that this was the smell of that moment. Even just a faint smell of the clay, perhaps from a person I pass in the street.
Harada:	Just a faint smell?
Akiko:	Only a faint smell.
Harada:	At a building demolition site?
Akiko:	No. When I walk in a crowded area such as Osaka metropolitan area, if I detect a smell of clay, for example on the clothes of strangers whom I pass, it is enough for me to remember the smell of the moment of the earthquake. It makes me depressed. We cannot escape nor overcome it. Those who experienced it would agree with me. Only those who experienced it would agree with me.

A few days after the interview, Akiko sent me a letter:

> Our two-storey apartment building was destroyed completely by the earthquake. We were buried beneath its broken roofs and walls for several hours, smelling the gas hissing from a cracked pipe. One neighbour, with a tiny saw, had cut through a thick wooden beam that lay across my back and we were able to escape from there. When we moved out through a tiny hole in the debris, I saw blood pouring down my husband's face and legs. As we were wearing

only nightwear, someone gave us coats in which we walked to the junior high school [a public shelter]. We walked in bare feet; another person we did not know offered us their shoes.

Fifteen years later they could laugh at some of the episodes of the aftermath, such as their experiences in the public shelter or in temporary housing. But even a faint smell of debris was enough to take them back to the exact moment of the earthquake, years earlier. Their expressions changed when they remembered their time under the rubble. They will doubtless continue to be sensitive to the smell of clay in the future, even 30 or 50 years after the event. Smell is a symbolic and material sense, which takes them back to the moment and place of the earthquake. In general, memories fade with time, but in the case of an earthquake, for those who suffered severely, tiny details can take them back to the moment; time does not necessarily ease the suffering and loss of their former lives, possessions and small but important aspects of their own personal heritage.

EARTHQUAKES AND NATURAL AND CULTURAL LANDSCAPES

The Japanese, just as foreign visitors, appreciate the country's beautiful landscapes and traditional cultures. For example, the '30 top experiences' recommended by *Lonely Planet Japan* (2011) include Onsen (hot springs), the Japanese Alps (high peaks), Mount Fuji, the Ogasawara archipelago, the Island of Yakushima, Daisetsuzan National Park (mountain chains and forests) and Kamikochi (a highland valley). All of these natural sites were created by seismic and volcanic activities over several million years on what is now the east coast of the Asian Continent. The guidebook also includes temples and gardens in the ancient capitals of Kyoto and Nara, and castles of samurai warriors. Modern architects inform us that traditional temples, shrines and castles were built on relatively solid ground using special techniques for wooden buildings, which has seen them survive many earthquakes, heavy rain and typhoons. Even the stone walls of castles feature anti-earthquake structures. The '30 top experiences' also include the architecture of modern Tokyo and the Shinkansen (bullet trains). Modern construction also employs earthquake-resistant structures, mainly developed from the time of the devastating Nobi earthquake in 1891 and the Kanto earthquake in 1923.

We think mountains are immobile and permanent. But earthquakes, volcanic eruptions, heavy rains and other 'natural phenomena' have continually twisted, buried, exposed and transformed the land; only now do we enjoy those landscapes as 'heritages'. Natural heritages are themselves always in the process of transformation. An earth scientist (Fujita 1985) has suggested that Mt Rokko, a popular place for outings for people in Kobe and the Osaka metropolitan area, has been created by about 1000 major earthquakes, which have lifted the land from sea level to 900 metres over the course of one million years. Many beautiful landscapes are the product of volcanic and seismological activity over long periods of time, even before the arrival of our ancestors on these islands. This fact indicates to us that the earthquake had been a natural phenomenon long before it became a social phenomenon. From the time that people settled in Japan earthquakes must be regarded as a social force.

For a long time our ancestors lived in the mountainous areas of Japan, on relatively solid ground. However, many of the modern Japanese metropolitan areas are built on the sedimentary rock that lies along the rivers and coastline – places where the geology is very young and fragile. Machida and Kojima (1996) suggest that the cost of strengthening these areas by artificial means is too great; natural consolidation would also take too long. Owing to the improvement of river courses and improvements in engineering the foundations of buildings, it is possible to build skyscrapers on fragile sedimentary rocks. However, this is not without danger: major earthquakes in such places would have a destructive impact.

EARTHQUAKES IN HISTORY

Let us return to our viewpoint at ground level. A tremor caused by an earthquake has a short duration, just several seconds or several minutes at most. In Japan we feel minor shakes caused by earthquakes in our daily lives without paying them much attention. Moreover, major earthquakes hit once every several hundred years or more only within a specific area, while outside that earthquake zone people feel only small tremors as routine occurrences and continue to live ordinary lives; in Japan it is rare for an individual to directly experience a great earthquake in her/his lifetime.

Therefore, and in a sense paradoxically, we have numerous descriptions of major earthquakes. Japanese scientists started collecting such descriptions from historical documents after a major earthquake in 1891, to investigate the space/time distribution of previous earthquakes in the Japanese islands. The first collection was published in 1904, and in that catalogue it is possible to read around 2000 descriptions of earthquakes. A revised edition was published in 1950 and included around 6400 descriptions. Earthquakes themselves – despite their destructive force – are now regarded as heritage in their own right, a historical phenomenon worth documenting.

The first description of an earthquake in Japanese historical documents appeared as a Chinese ideograph, kanji, which states 'there was an earthquake' (Aston 1972, vol 1, 317) in AD 664. There are more detailed descriptions, including the following, which describes an earthquake said to have occurred in AD 687:

14th day. At the hour of the boar (10 pm) there was a great earthquake. Throughout the country men and women shrieked aloud, and knew not East from West. Mountains fell down and rivers gushed forth; the official buildings of the provinces and districts, the barns and houses of the common people, the temples, pagodas and shrines were destroyed in numbers which surpass all estimate. In consequence many of the people and of domestic animals were killed or injured. The hot springs of Iyo were dried up at this time and ceased to flow. In the providence of Tosa more than 500,000 shiro of cultivated land were swallowed up and became sea. Old men said that never before had there been such an earthquake. (Aston 1972, vol 2, 365)

In Japan there are several folktales about earthquakes, regarded today as an important form of intangible cultural heritage. In general, earthquakes are something beyond human imagination and each ethnic group has their own folktale. It is said in India and East Asian areas that the land is supported by a giant serpent; in Southeast Asia and Japan, giant fish, serpents or dragons take the supporting role; when these beasts move underground, earthquakes occur. Usually, their power is suppressed but once in a while their movement shakes the land and causes earthquakes or volcanic eruptions. One famous episode refers to the major earthquake in 1855, when folk beliefs came to the fore; namazu-e (catfish pictures) were printed and sold until the government outlawed their production, for those pictures implied that the earthquake was a sign of heaven's displeasure with the politics of the government. The earthquake was regarded as 'yonaoshi' or 'world rectification' (Ouwehand 1964; Kuroda 2003; Ludwin and Smits 2007).

Such folktales are interesting; they remind us that earthquakes happen all of a sudden and, when they do, we like to discuss the cause. In this sense, our modern scientific findings are the same as those folktales. Twenty-first-century scientific approaches enable us to explain where and how strong an earthquake was immediately after it happened, and gradually scientists have begun to realise that major earthquakes only happen repeatedly in limited areas of the world. Torahiko Terada (1878–1935) was a Japanese scientist who experienced the Kanto Earthquake in 1923. He wrote in 1926:

> Things may happen which could never be experienced in other parts of the world where the simple and uniform structure of a large crustal mass is enjoying the golden era of peaceful stability. From time immemorial, the islands have, indeed, been the site of incessant displays of volcanic and seismic activities. (Terada 1926)

Modern earth science in the 1960s offered us plate tectonics theory:

> Earthquakes occur along well-defined belts of seismicity; these are particularly prominent around the Pacific Rim and along mid-oceanic ridges. We now know that these belts define the edges of the tectonic plates within Earth's rigid outermost layer. (Shearer 2009, 6)

Under the Japanese archipelago, four lithospheric plates converge on one another. More precisely, the Japanese archipelago is a product of the activity of these plates. Modern seismographs record 15 earthquakes a day in and around Japan. According to Usami (2003, 4), a seismologist who edited a list of 756 'destructive earthquakes' in and around the Japanese archipelago from AD 416 to 2001, we are 'fortunate'. Around the Japanese archipelago, earthquakes exceeding magnitude 6 occur 17 or 18 times a year; above magnitude 7 occur once or twice a year; and over magnitude 8 once a decade. Inland earthquakes over magnitude 6 may cause some destruction, as under certain conditions may those over magnitude 5. According to Usami, however, we are fortunate as the majority of major earthquakes have occurred below sea level.

In 2011 the East Japan Earthquake occurred. It happened under the sea but was a devastating earthquake and was followed by major tsunamis. One year later the president of the Japan Seismological Association recorded that 'it was our fault not to announce such a major earthquake in advance'. He also commented:

> Sometimes I fantasise, like in a comic story, that we could invent an epoch-making device to float our towns in the air, so escaping from the clutches of the occurrence of an earthquake. Then we would no longer be affected by any earthquakes. (Hirahara 2011)

How often and deeply we have been shaken, moved and our lives destroyed by earthquakes that last only several minutes and occur once every 200 or 1000 years in the same place without any advance warning. Modern scientific explanations are no better than traditional folktales in forecasting where and when earthquakes will strike.

EARTHQUAKES AS SOCIALLY DESTRUCTIVE FORCES

As noted above, statistically, few of us are injured by a major earthquake. These usually affect a limited geographical area and, even in damaged areas, individual buildings are affected to different degrees. Catastrophic earthquakes occur at very infrequent intervals in comparison with average human lifespan. Following an earthquake we therefore think, just like the ancient old man in the historical document, that 'never before had there been such an earthquake'. And, as time passes, we again become accustomed to thinking that we live on 'immobile' land and in nature-resistant buildings.

However, for those who have suffered, what is the meaning of an earthquake? For Akiko and Natsumi, the memory of their time trapped under debris remains. In our daily lives we do not notice the smell of the plaster and the bricks that make up the walls of our houses. Every day we see, touch and smell the doors, walls, floors and windows of our own houses or rooms, but we have no opportunity to directly see, touch and smell the materials that constitute the doors, walls, floors, or windows – the wood, glass, cement, clay, plastic board, paper and cloth. For Sakae, the earthquake was an event in her life that she compared to the painful experiences of an air raid and bankruptcy. Her remark – 'we ourselves had decided to live in such fragile houses' – demonstrates the essential feature of earthquakes. They are destructive insofar as they damage the building where we were at that moment in time. Although heavy rains, typhoons, hurricanes, or air raids have the same destructive power, they are predictable and in many cases we can escape them in advance. In the case of an earthquake, as Sakae explained, we have no warning and cannot escape.

Just as our ancestors did, we live our daily lives on the Japanese archipelago with its many elements of natural and cultural heritage. But we are different from our predecessors in that we have created buildings that may or may not be 'fragile'. Much depends on the strength of the buildings, the land and, of course, the magnitude of the earthquake. An architect suggests that we can construct perfectly earthquake-resistant buildings if we are prepared to spend a lot of money and not mind their external appearance (Kanda 1997). But, so far, that is as unreal a prospect as lifting towns up above the land to escape the earthquake's effects.

Thus we have socialised the earthquake. We have changed it from being a natural phenomenon into a destructive disaster because in modern times we live 'fragile' lives on 'fragile' land that is host to incessant displays of volcanic and seismic activity. Earthquakes have transformed the land and provided a wealth of natural heritages that we appreciate; the earthquake is both a natural heritage and a social disaster of our own making.

BIBLIOGRAPHY AND REFERENCES

Aston, W G, 1972 *Nihongi: Chronicles of Japan from the earliest Times to AD 697*, Charles E Tuttle, Tokyo

Fujita, K, 1985 *Hendo Suru Nippon Retto [The Japanese Archipelago in the Midst of Transformation]*, Iwanami Shoten Publishers, Tokyo

Harada, T, 2000 Space, materials, and the 'social': in the aftermath of a disaster, *Environment and Planning D: Society and Space* 10, London

Hirahara, K, 2011 Jishin Gakusha no Haiboku [Defeat of the seismologists], *Asahi Shimbun* (newspaper), 17 August, Tokyo

Kanda, J, 1997 *Taishin Kenchiku no Kangaekata [The Thought of Earthquake Resistant Construction]*, Iwanami Shoten Publishers, Tokyo

Kuroda, H, 2003 *Ryu no Sumu Nihon [Japan: a place where dragons live]*, Iwanami Shoten Publishers, Tokyo

Lonely Planet Japan, 2011, 12 edn, Lonely Planet Publications, Singapore

Ludwin, R S, and Smits, G J, 2007 Folklore and Earthquake: Native American Oral Traditions from Cascadia Compared with Written Traditions from Japan, in *Myth and Geology* (eds L Piccardi and W B Masse), Geological Society, Special Publications 273, London

Machida, H, and Kojima, K (eds), 1996 *Shizen no Moi [The Rage of Nature]*, Iwanami Shoten Publishers, Tokyo

Ouwehand, C, 1964, *Namazu-e and their Themes: an Interpretative Approach to Some Aspects of Japanese Folk Religion*, E J Brill, Leiden

Shearer, P M, 2009 (1999), *Introduction to Seismology*, 2 edn, Cambridge University Press, Cambridge

Terada, T, 1926 A Historical Sketch of the Development of Seismology in Japan, *Scientific Japan*, Tokyo

Usami, T, 2003 *Higai Jishin Soran [Materials for Comprehensive List of Destructive Earthquakes in Japan, 416-2001]*, University of Tokyo Press, Tokyo

22

Industrial Heritage and the Oral Legacy of Disaster: Narratives of Asbestos Disease Victims from Clydeside, Scotland

Arthur McIvor

Industrial heritage up and down the country is carefully preserved in museums, such as at Beamish in North East England, Styal Mills in the North West and the remarkable UNESCO New Lanark industrial village in west-central Scotland. These are nostalgia spaces where we are reminded of the hard graft, craftsmanship, community and camaraderie of work and life as it was. However, relatively little physical heritage survives of the traditional 'heavy industries' – the shipyards, iron and steel works and coal mines that once dominated the landscape and economy of whole industrial conurbations. These were places where the simple dignity of human toil was played out every day. But they were also sites of disasters big and small; of dangerous work, of trauma, pain and suffering; of fatal and disabling injuries, poisonous and toxic materials that damaged workers' bodies; maiming and killing – sometimes quite indiscriminately and obscenely in the pursuit of profit. This has been described aptly as 'economic violence' (Bourdieu 1991; 1993). In this context we are most aware, perhaps, of the coal mine disasters (the UK's largest occurred in Senghennydd in October 1913 with 440 deaths) – usually memorialised by a statue or plaque at the pithead and sometimes the only remaining indication that a thriving pit existed on that spot. These disaster memorials provide a sombre reminder that work could, and did, kill.

Recently scholars of occupational health have argued that definitions of disaster, which have tended to traditionally focus on natural disasters, need to be widened to embrace man-made, technological and socially constructed events, including longer-term insidious physical harm such as that associated with disease as well as 'accidents' such as mine explosions (see Mitchell 1996; Tucker 2006; McIvor and Johnston 2007). Such studies rarely, however, utilise an oral history approach based on the systematic interviewing of eye-witness participants (victims, regulators and perpetrators). Nor are the voices of disaster victims integrated substantively in the public history and heritage sector. With some notable exceptions, museums, heritage centres and commemorations do little to evoke the emotional history of disasters; the everyday lives of these workers, their feelings, emotions and cultures; how they navigated their way through hazardous labour processes, calculated risks, weighed options and how their lives were affected by disabling injuries, pain, suffering, trauma and insidious occupation-related industrial diseases and how their families and loved ones coped with their loss. Some mining museums and heritage centres constitute exceptions to the norm, such as the Senghennydd Heritage Centre, the State Museum of Pennsylvania and the Cape Breton Mining Museum – and increasing use of oral testimony is evident in such places and in the proliferation of museum websites, online museums and virtual archives.

Oral history approaches – talking to survivors and their families – can provide an intimate, more personal perspective that gets beyond the statistical 'body counts' to the meaning of disasters at an individual, family and community level. This has been done very effectively for a series of recent man-made disasters, such as Dunblane, Hillsborough, Piper Alpha and Bhopal, and an increasing amount of this recorded material is archived, constituting a significant body of oral heritage of disaster. Examples include the *Oil Lives* archive in the British Library Sound Archive and the occupational health archive in the Scottish Oral History Centre.

The methodology is applied less frequently to longer-term, more insidious disasters associated with chronic work-related disease and disability. This chapter explores the utility of oral history approaches in this context and the value of memories as public heritage, drawing upon a set of fieldwork interviews and using the case study of the asbestos disaster and the industrial conurbation of Clydeside (centred on Glasgow) as the focal point. Commemorating industrial disaster in museums and heritage centres can be of direct benefit to society, helping the public to understand the risks associated with work, thus placing the current political attack on an 'excessive' health and safety culture in perspective and enabling more effective policy responses to be shaped. Given the ongoing occupational health disasters in industrialising countries across the globe – such as India and China – and the 'new epidemic' of workplace stress in developed Western economies, there are clearly lessons here to be learnt from the past.

The Asbestos Disaster

The asbestos disaster is played out across the globe. Asbestos was extensively mined and its mineral fibres processed and manufactured from the late 19th century, with a wide range of uses as an insulating product and fire retardant. But breathing asbestos disables as well as kills – and in vast numbers. Inhaling asbestos fibres is associated with several deadly incurable diseases. Asbestosis usually results from prolonged exposure in the workplace and consists of a clogging up and resultant scarring and distortion of the lungs, leading to severe and disabling breathlessness and strain on the heart. Malignant mesothelioma is an aggressive cancer that usually develops on the outer lining of the lung (pleura), but also appears on the surface of the abdominal cavity (peritoneum). Those diagnosed with mesothelioma suffer higher levels of pain than most cancer victims and rarely live beyond 18 months after diagnosis. Asbestos-related lung cancer is the most common malignancy among individuals exposed to asbestos. Both lung cancer and mesothelioma can be caused by occupational or environmental exposures to asbestos fibres.

The World Health Organization (WHO) has recently estimated that the asbestos disaster accounts for more than 107,000 deaths every year and that an estimated 125 million people across the globe are currently being exposed to the deadly mineral through their occupations (World Health Organization 2010). According to one medical expert the final death toll from this disaster will top 5 million and may be as high as 10 million (Kazan-Allen 2007, 6).

Government regulations passed to control asbestos use (such as the first Asbestos Regulations in the UK in 1931) were limited, widely flouted and ineffective, and workers' bodies continued to be exposed, despite knowledge of the risks, for decades thereafter (see Brodeur 1985; Tweedale 2000). This was the pattern across Western industrial nations in the 20th century and today prevails in many countries across the globe (McCulloch and Tweedale 2008). Exposure affected workers but asbestos also constituted a wider public risk, contaminating the environment, not least in the vicinity of asbestos plants and when it was unsettled as in a natural disaster such as an earthquake, or an event such as a fire, explosion, war or a terrorist attack – as with 9/11 in New York.

THE ORAL HISTORY OF OCCUPATIONAL HEALTH DISASTERS

Studies of occupational health disasters, including asbestos, have tended to focus on forensic dissection of the extant documentary evidence – including company records – with interpretations ranging from pro-company to vitriolic critiques of capitalist exploitation (see for example Bartrip 2001 and Tweedale 2000). In such accounts, the voices of those affected are rarely directly heard. A number of studies have turned to oral evidence to elucidate occupational health disasters worldwide, including Clayson (2008), McIvor and Johnston (2007), Walker and LaMontagne (2004), Storey (2009) and Portelli (2012). A theme in Portelli's study of Harlan County, USA, is the 'black lung' disaster, which has also been the focus of a recent oral history-based study in the UK (McIvor and Johnston 2007). Another exceptional, stand out oral history-based investigation of disaster is Mukherjee's *Surviving Bhopal* (2010), which focuses on the stories of female survivors of the infamous Union Carbide gas leak in India in December 1984 (and see Sharma, Chapter 10, this volume).

By providing a view from the workplace through eye-witness testimony we gain valuable insights into the causes of such disasters – including the limited effectiveness of regulatory frameworks. We also invariably get a sense of the complexity of work-health and body cultures, the interplay of identities (such as gender and class), the agency of workers negotiating paths through the prevailing exploitative social relations and managerial, productionist work cultures and the impacts of such disasters on individuals and families (see also Walker 2011). Let me develop this argument with some brief comments on my own fieldwork (with my now retired colleague Ronnie Johnston) on asbestos in the Clydeside industrial conurbation.

RECONSTRUCTING THE IMPACT OF THE ASBESTOS DISASTER THROUGH MEMORIES

The port and city of Glasgow and the industrial conurbation surrounding it (Clydeside) was and remains an asbestos 'blackspot'. This was an area built on the heavy engineering and shipbuilding industries that were major consumers of asbestos and where asbestos was imported and manufactured, as well as used extensively in high-rise flat construction – including the highest in Europe in the early 1960s (at Red Road).

Oral testimonies from victims of the asbestos disaster and their families elucidate the economic, social, psychological and cultural impact asbestos had upon their lives – the physical pain and suffering; the shock of a terminal cancer diagnosis; the encroaching disability; the transition from independence to dependant; the emasculation. Narrators described the anxiety and stress associated with having to make adjustments to household budgets following the loss of a breadwinner and their struggles in claiming compensation and benefits. The trauma of coping with disability, the mutation of identities that disability invariably entails and coping with losing a loved one are all related in emotionally charged personal testimonies that put us virtually in the shoes of those experiencing the tragic consequences of the unfolding asbestos disaster. Some were fatalistic and stoic; others bitter and angry at the system that was responsible. As one Clydeside shipyard worker put it: 'It was like fighting an atomic war with a bow and arrow, you know. You hadnae a chance' (Interview by Ronnie Johnston, 3 February 1999, Scottish Oral History Centre Archive (hereafter SOHC),[1] SOHC/016/A18).

Eye-witness testimonies lay bare the realities of irresponsible and abusive power relationships at the point of production and the limited resources that workers could bring to bear upon their situation. The space in which workers toiled and the environment in which bodies were located was frequently vividly recalled, with dust, death and disability the recurring motifs in asbestos disaster work-life narratives. Such workers in the UK (and elsewhere) recalled asbestos dust suspended like a 'fog', or falling like 'snow' in their workplaces in the 1950s, 1960s and 1970s and of playing with the material (Johnston and McIvor 2000, 63–111). Information was withheld from workers by management, or only selective and sometimes misleading information was leaked out, such as the supposedly benign nature of white asbestos compared to blue and brown asbestos. While workers had some intuitive and lay knowledge, they were mostly ignorant of and not told of the extent of the dangers. Workers recalled feeling pressurised to work with the toxic material, to 'cut corners', ignore safety regulations and maximise productivity. In this sense these workers were victims of a Fordist, productionist culture that exalted hard graft and the maximisation of earnings at all costs.

This has to be understood, however, within a cultural framework – a milieu that facilitated the tolerance and persistence of abusive economic violence. This needs to be contextualised in the sustained and deep acculturation to undertaking dangerous work, a high risk threshold and a fiercely independent working-class culture where it was frowned upon for men to complain, grumble and 'make a fuss'. A dominant (or hegemonic) mode of 'hard man' masculinity was forged in such heavy industry workplaces where the 'top producers' and highest earners were lauded and praised. Those who sought to protect themselves could be pilloried as lesser men, subjected to peer pressure to take risks, to compete, to conform and to maximise earnings. Workplace disasters like asbestos destroyed lives – leaving in their wake a legacy of disability, premature death and deep psychological distress somewhat akin to other post-traumatic stress disorders. As a 64-year-old electrician with mesothelioma reflected: 'Until now I thought trauma was a fad imported from America and reserved for the middle classes. I am now wiser' (Interview by Ronnie Johnston, 10 January 1999, SOHC/016/A13). Oral interviewing methodologies enable this experience to be explored and elucidated – to get behind the sterile body counts to the human dimension, the lived reality. Oral testimonies of those suffering from ARDs (asbestos-related diseases) illuminate a hidden world of private grief, sadness, anger, frustration, disappointment, pain and suffering. In *Lethal Work* we reported on the 'blighted lives' of ARD victims in Scotland and argued that people were marginalised and stigmatised by their illness (Johnston and McIvor 2000, 177–208). Male narrators recalled the encroaching social isolation, within the feminised space of the home, and restricted social and physical activities (such as walking, sports and dancing). They spoke of the relative economic deprivation associated with income reduction, of the trauma associated with diagnosis by a GP or at a hospital, and of living and coping strategies as people struggled to adapt and survive with the news that they were going to die from an incurable cancer.

Speaking directly to participants enables a refocused history revealing much about the emotional journey involved in the transition from fit and able worker to disabled and dependent, with all

---

[1]   References in the text cited as SOHC/016/A** refer to interviews archived in the Scottish Oral History Centre, University of Strathclyde.

that that represented for gendered identities. What is being recalled is frequently an intimate, personal story of damage, loss, pain, adjustment – and of mutating identities. Lives invariably became narrowed as a consequence, with individuals, partners and families having to readjust their lives as injuries or chronic disease disabled workers. Dented pride, loss of self-esteem and embarrassment were among the emotions expressed. Again, however, there was agency here, albeit operating within the constraints of having to earn a living. For example, some workers with severe asbestosis chose to hide their disability and continue to work for as long as they could despite knowing that this could further damage respiratory function. Deindustrialisation, a lack of alternative job opportunities and the economic and cultural imperatives to act as men and provide for families influenced such decisions.

The physical and the psychological trauma of the asbestos disaster were inextricably intertwined. In a recent interview Phyllis Craig (Welfare Rights Officer for Clydeside Action on Asbestos since 1995) was asked about the impact of the mesothelioma disaster on the lives of her ARD clients:

> There's fear; there's pain; there's suffering; there's all sorts of anxieties; there's coping, there's worrying and if you add to that that someone else did this, such an anger because they are taking that person away from their partner; their children. … Their careers are ended; everything ends because they know they are going to die. … They are angry. It's devastating. … It's horrendous for them.
>
> (Interview by Arthur McIvor with Phyllis Craig, 28 January 2013, SOHC/016/A35)

The extent of the repetition of the point that an outside agent was responsible (eight times in the interview) serves to emphasise the significance of the issue to this particular professional eye-witness who has worked for almost 20 years with ARD victims. There was also a stigma attached to having a malignant disease and much 'anticipatory anxiety' about the cancer risk among those without any symptoms (wives, carers, family members and friends).

The asbestos disaster impacted upon what were distinctly gendered identities. For women with mesothelioma this challenged their femininity, corroding their capacity to act as nurturers, carers, mothers and wives. Phyllis Craig reflected that what her female ARD clients felt acutely was an eroding capacity to act as the 'homemaker' and family lynchpin. Women, Craig noted, 'fear for their children' (Interview SOHC/016/A35; see also Gorman 2000, 127–37). For men, ARDs could be deeply emasculating. If the traditional heavy industries provided an environment where working-class masculinities were forged, they also had the potential to emasculate as encroaching disability curtailed men's capacity to perform as men – as providers and breadwinners, as sexually active partners, as supportive parents and grandparents. The oral evidence brought to light the existence of a macho, individualist element in workers' culture that co-existed, sometimes uneasily, with the collective, mutual, class-conscious character of traditional working-class communities. Peer pressure determined that men should act in certain ways – including taking risks or taking work to fulfil the breadwinner role, even where this work was known to be dangerous to health. This was tied to a powerful, pervasive and enduring work ethic.

Men responded less directly to health education and hazards awareness campaigns than did women, and were generally more reluctant to admit they had a health problem and seek medical intervention, while when ill they could refuse to allow help or admit they needed help (Clayson 2008, 253). The wife of a quantity surveyor with mesothelioma reflected after his death that 'he never made a fuss. ... I was the one that used to see him sitting on the edge of the bed with his arms around himself rocking back and forward in pain' (Interview by Ronnie Johnston, 18 February 1999, SOHC/016/A20). A 61-year-old shipyard engineering worker with mesothelioma commented: 'A lot of it's my own problem. Too macho to be shouting out when I should be, you know, when I'm in pain. ... Personality change is quite enormous when I'm in pain. It's terrible... it also brings on the anxiety side of it' (Mr I, cited in Clayson 2008, 140). Emotions might be controlled by many men, except in private moments, as the wife of the quantity surveyor cited earlier recalled:

> You do your best to bolster them and keep going for them and make light of things. And he took my hand and said: 'I'm not going to see xxx as a bride'. Then we went up to bed together and we just cuddled and we both cried. And it's the one and only time that I saw my husband crying. (Interview SOHC/016/A20)

Of course, coping capacities and strategies ranged widely, but the oral testimonies consistently referred to the psycho-social distress and disruption to lives, commensurate to trauma, experienced by ARD victims and survivors. As with other major industrial disasters, there were deep, varied and unpredictable long-term consequences to be faced by individuals, families and communities (Mitchell 1996).

Some asbestos disaster victims withdrew into their private world of the family while others were radicalised and politicised by the experience – a notable response linked to the leftist, communitarian, socialist tradition embedded within working-class communities of the region – epitomised in the contested concept of 'Red Clydeside'.

Being 'killed' and the phrase 'murder', or 'mass murder', features in these ARD narratives, especially those of bereaved relatives (Clayson 2008, 4). The experience led some victims and family members directly into asbestos advocacy and activism. This could be expressed through forming and participating in pressure groups such as Clydeside Action on Asbestos, the Clydebank Asbestos Group and the Society for the Prevention of Asbestos and Industrial Diseases (Nancy Tait).[2] Others campaigned through their trade unions and participated in direct strike action, as with the 1966 Glasgow TGWU laggers' branch strike. However, there existed a wide range of union strategies on asbestos, with some processing unions opposing its ban (job losses were a significant consideration in a period of deindustrialisation and plant closures) and others, such as the GMB, who were aggressively anti-asbestos. On Clydeside, rank and file activists such as John Todd (who had several family members diagnosed with ARDs) campaigned to raise awareness in the 1960s and 1970s. Concurrently, the laggers' branch of the TGWU persistently stepped up pressure for reform, from pressing the wartime Minister of Labour during World War II (Ernest Bevin) for an inquiry into the use of asbestos (which Bevin rejected; see Dalton 1979, 98), to lobbying the Scottish Trades Union Congress at Perth in 1976 to adopt a more aggressive campaigning position on the growing asbestos disaster.

---

2   Nancy Tait (1920–2009) became an asbestos activist following the death of her husband (a Post Office engineer) from mesothelioma in 1968. She wrote a booklet, *Asbestos Kills*, in 1976 and formed SPAID in 1978 (which later changed its name to Occupational and Environmental Diseases Association).

Personal experience and reactions to disaster were intensely shaped by sense of place, with the radical tradition of Clydeside's proletarian neighbourhoods generating an embittered collective response and concerted direct action and organisation. This contributed to Scotland developing one of the most progressive policies in the world regarding welfare rights, civil law damages and statutory compensation for victims of the asbestos disaster in the early 21st century. This did nothing to prevent the asbestos catastrophe, but did assign culpability directly to the industry and did assuage, to some extent, the economic effects of the disaster.

CONCLUSION: INTERPRETING MEMORY NARRATIVES

An oral history methodology enables reconstruction of lost workplaces where occupational health disasters were incubated; elucidates the economic, social, psychological and cultural impacts of disaster; and evokes the personal meaning of disability and loss. Nonetheless there are issues with oral history that merit brief comment. One relates to coverage. The prevailing focus of such work is upon victims, whereas perpetrators also deserve to be interviewed for their perspective – as Jessee (2011) has persuasively argued in the context of genocide studies. Rigour needs to be applied at all stages of oral history projects, whether using existing archived oral material or generating new material through fieldwork interviews. Intersubjectivity comes into play in the relationship between interviewer and narrator, while leading and loaded questions need to be avoided. The material requires careful and sensitive handling and, like almost all document-based primary source material, there are biases and subjectivities to navigate.

In recalling their past in an interview context, narrators are filtering and sieving memories, constructing and composing their stories and mixing factual evidence with their own inter-pretations as they try to make sense of their lives in an active, dialogic and reflexive process of remembering (see Abrams 2010; Thompson 2000; Perks and Thomson 2006). Silences can be as significant as what is being said. Subsequent events can influence narrative construction and memory 'framing', including, in this context, the media coverage of the disaster and the rise in litigation and what some refer to disparagingly as the 'compensation culture'.

Nonetheless, with careful use, an oral history methodology – especially a 'life story' approach – has the capacity to provide a refocused history based on emotions, feelings and lived experience which can facilitate our understanding of disasters and their impact – including those disasters linked to work, such as asbestos, that have left such a grim legacy of disability, ill-health and death. These participant voices relating industrial disasters deserve to be heard, recorded and preserved as oral heritage, critically engaged with in disaster research and they merit a more prominent place in our public history in museums and heritage centres.

BIBLIOGRAPHY AND REFERENCES

Abrams, L, 2010 *Oral History Theory*, Routledge, London

Bartrip, P, 2001 *The Way from Dusty Death*, Athlone Press, London

Bourdieu, P, 1991 *Language and Symbolic Power*, Polity Press, Cambridge

Bourdieu, P, *et al*, 1993 *The Weight of the World: Social Suffering in Contemporary Society*, Polity Press, Cambridge

Brodeur, P, 1985 *Outrageous Misconduct*, Pantheon Books, New York

Clayson, H, 2008 The Experience of Mesothelioma in Northern England, unpublished MD thesis, University of Sheffield, 2008

Dalton, A, 1979 *Asbestos Killer Dust*, British Society for Social Responsibility in Science, London

Gorman, T, 2000 Women and Asbestos, in *Clydebank: Asbestos, the Unwanted Legacy* (ed T Gorman), Glasgow Clydeside Press, Glasgow, 127–37

Jessee, E, 2011 The Limits of Oral History: Ethics and Methodology amid Highly Politicized Research Settings, *Oral History Review* 38 (2), Summer–Fall 2011, 287–307

Johnston, R, and McIvor, A, 2000 *Lethal Work: A History of the Asbestos Tragedy in Scotland*, Tuckwell Press, East Linton

Kazan-Allen, L, 2007 *Killing the Future: Asbestos Use in Asia* [online], available from: http://ibasecretariat.org/ktf_web_fin.pdf [26 March 2014]

McCulloch, J, and Tweedale, G, 2008 *Defending the Indefensible: The Global Asbestos Industry and its Fight for Survival*, Oxford University Press, Oxford

McIvor, A, and Johnston, R, 2007 *Miners' Lung*, Ashgate, Farnham

Mitchell, J K, 1996 *Long Road to Recovery: Community Responses to Industrial Disaster*, United Nations University Press, Tokyo

Mukherjee, S, 2010 *Surviving Bhopal*, Palgrave, New York

Perks, R, and Thomson, A (eds), 2006 *The Oral History Reader*, 2 edn, Routledge, London

Portelli, A, 2012 *They Say in Harlan County*, Oxford University Press, New York

Storey, R, 2009 'They Have all been Faithful Workers': Injured Workers, Truth and Workers' Compensation in Ontario, 1970–2008, *Journal of Canadian Studies* 43 (1), Winter, 154–85

Thompson, P, 2000 *The Voice of the Past*, 3 edn, Oxford University Press, Oxford

Tucker, E (ed), 2006 *Working Disasters: The Politics of Recognition and Response*, Baywood Publishing, Amityville NY

Tweedale, G, 2000 *Magic Mineral to Killer Dust*, Oxford University Press, Oxford

Walker, D, 2011 'Danger was Something you were Brought up wi': Workers' Narratives on Occupational Health and Safety in the Workplace, *Scottish Labour History* 46, 54–70

Walker, H H, and LaMontagne, A D, 2004 *Work and Health in the Latrobe Valley: Community Perspectives on Asbestos Issues: Final Report*, Centre for the Study of Health and Society, University of Melbourne, Melbourne

World Health Organization, 2010 *Asbestos: Elimination of Asbestos-Related Diseases*, WHO, Geneva, available from: http://www.who.int/mediacentre/factsheets/fs343/en/index.html [26 March 2014]

23

Translating Foot and Mouth: Conveying Trauma in Landscape Photography

RUPERT ASHMORE

The Foot and Mouth Disease (FMD) epidemic that swept through rural Britain in 2001 had a catastrophic economic effect on many communities.[1] It also impacted upon social and cultural life, and it has since been suggested that the prolonged, negative experience of Foot and Mouth was potentially traumatic, both individually and communally (Convery *et al* 2008). Fundamental to this traumatic experience were the significant physical changes to the landscape. The disease, and the policies employed to control it, profoundly disrupted the relationship between communities and the spaces in which they live. It is fitting then, that the enduring cultural document of the outbreak is a landscape image: the photograph of burning livestock in an otherwise idyllic pastoral scene.

This chapter investigates how those traumatic changes were represented in landscape photographs produced in one of the worst-affected areas: Cumbria in North West England. It examines the dramatic pictures that flooded the media from the start of the outbreak, but also the work of photographers outside the press, such as John Darwell, Nick May and Ian Geering, who attempted to reveal a more detailed picture of the crisis, and who have provided some of the most widely exhibited images of the outbreak. It suggests that this imagery is essential to understanding a crisis that happened *within*, and *to* the landscape, and that it has played a significant role in contextualising and communicating FMD not only as a 'disaster', but as 'communal trauma'.

This trauma stemmed from the official policy to combat the disease as much as the disease itself. FMD outbreaks in Britain are contained by establishing a biosecurity cordon around infected premises and destroying all livestock within it, often *in situ*, and *whether infected or not*. Ultimately this 'contiguous cull' policy was successful, but it also led to the deaths of up to ten million animals (Mort *et al* 2005, 1234), and by the end of the crisis there were large areas of Cumbria totally devoid of livestock. The biosecurity policy also meant nationwide restrictions upon all animal movement, livestock and meat exports bans, and the cancellation of rural markets and events. This had clear financial implications for agriculture and its support economies, but strict access restrictions to the countryside also spelled disaster for tourism. FMD led to financial hardship across the rural economy, but also intense emotional and social upheaval.

Maggie Mort, Ian Convery, Cathy Bailey and Josephine Baxter have suggested that the epidemic caused a negative alteration of 'lifescape': the practical, social and cultural ways in

---

[1] FMD is a virus affecting cloven-hoofed animals, resulting in painful lesions in the mouth and on the extremities. Though extremely virulent, it is not fatal and most animals recover within a matter of weeks. However, long-term health can be affected, and bovine milk yields often do not recover to pre-infection levels.

which people normally experience their environment (Mort *et al* 2005). The stress of dealing with the disease was intensified by quarantine measures that curtailed much everyday social interaction and the cultural events that reinforce community.

On an individual level, FMD radically altered established patterns of life. For farmers, witnessing the destruction of their animals was particularly distressing and it disrupted the accepted 'emotional geography' of husbandry: slaughter took place not in the abattoir, but on their own premises, and the carcasses were often burned or buried *in situ*. Whole flocks or herds were slaughtered at once, destroying generations of animal lineage, and a connection to tradition spanning generations. This temporal and spatial disruption was compounded by the sheer scale and duration of the slaughter. The cull signified three potentially traumatic ruptures to normality: death in the wrong place; death at the wrong time; and death on the wrong scale (Convery *et al* 2005).

Furthermore, the government's initial inability to handle the logistics of the cull meant that slaughtered stock remained in fields before disposal, sometimes for weeks. So for months communities were continually enveloped by the sights and smells of death: slaughter, decomposing carcasses and the smoke and ash from pyres. The perception of the landscape was fundamentally altered, but also its use, as human movement was also curtailed through the quarantine of infected premises, or voluntary isolation through fear of spreading the virus. As Mort *et al* noted, those living in Cumbria suffered a 'disturbed relation between health and place; the changed significance of everyday places and spaces previously taken for granted', which extended beyond the cull to the eerie silence of a countryside without livestock, and left lasting changes in outlook (2005, 1237). This disorientation was intensified by feelings of impotence, the bureaucratisation of the cull process and the perception that government advice was often inadequate or inaccurate (2005, 1235–9).

Individuals experienced shock, stress and depression (Convery *et al* 2008, 89) and the cultural and social markers that reinforce communal identity were left with a palpable wound, which fundamentally affected social interaction. As Kai Erikson suggests, traumatic events deliver 'a blow to the basic tissues of social life that damages the bonds attaching people together and impairs the prevailing sense of communality' (Erikson 1995, 187), especially if the event carries notions of toxicity. Here, toxicity was physically manifested in biosecurity controls, but also internalised as a constant perception of contagion was projected on to both people and landscape. The epidemic was also economically divisive, raising a conflict between the interests of agriculture, best served by quarantine, and tourism that relies on access, and an iniquitous system of compensation that left permanent divisions within some communities (Hillyard 2009).

Mort *et al*'s analysis of these effects constitutes Jeffrey Alexander's 'cultural construction of trauma': the necessary acts of investigation and broadcast whereby 'disasters' become recontextualised as 'traumatic' events (Alexander 2004). This process entails social agents (the press, government, academics, or indeed, photographers) lodging a claim that trauma has taken place and has affected a certain group. An effective claim must address four integral points:

a) the nature of the trauma;
b) who was affected;
c) the relation of the affected group to the wider audience; and
d) who or what caused the phenomenon.

If effectively addressed, each point adds value to the claim, and if the proposition is communicated convincingly then an event will be accepted as communally traumatic (Alexander 2004, 12–15).

These points began to emerge during the height of the crisis through the forum of local radio. Traumatic events also engender community, and BBC Radio Cumbria devoted extensive programming and web facilities to FMD, becoming a site for *therapeutic gathering* as the affected used the platform to share their experiences, frustration and pain.[2] These testimonies were published as a compendium, *Foot and Mouth, Heart and Soul*, in September 2001.

The idea that FMD was traumatic also began to be constructed in the extensive press coverage of the crisis, propelled by the rapidly increasing number of infected herds and geographical spread, and rising sense of powerlessness. Its devastating consequences were symbolised by the dramatic image of the disposal pyre, and as one commentator suggested, the picture of 'Cumbria with four legs sticking up in the air with a bit of fire in the background' became the ubiquitous image of Foot and Mouth (Breakwell cited in Döring and Nerlich 2009, 7). These images of pyres, but also road blocks, warning signs and biosecurity equipment, filled local and national press reports. They provided documentary evidence that FMD resulted in animal deaths and human suffering, but also wrought changes to the landscape.

They also provided a stark *symbolic* contrast to the idealised image of Cumbria, perpetuated for centuries through landscape images of the Lake District. The images of FMD fundamentally undermined that vision, but this ideal also had practical effects upon the management of the epidemic. As restrictions and negative imagery affected tourism, efforts were made to protect the National Park from the disease, while simultaneously ensuring as much tourist access as possible. William Mitchell suggests that we approach any locality through three interlocking concepts: 'place' (the specific place, including the effects that order and planning have upon it), 'space' (the site in terms of the way that it is used through everyday human activities) and 'landscape' (the site encountered as image) (Mitchell 2002, vii–xii). In Cumbria, one might suggest that the lived 'space' correlates to *lifescape*, which was radically disrupted by biosecurity controls: the organising principles of 'place'. Furthermore, these organising policies were not only determined by the practical goals of biosecurity, but also the tourist idea of Cumbria, itself highly informed by cultural representations: 'landscape'.

However, we cannot say that the surrounding press reports conveyed a concrete message of trauma. They outlined some of Alexander's points, such as which communities were affected, and some of the hardships being faced. They also related the crisis to the wider economy. Yet reports tended to equate emotional distress to financial hardship or the waste of a life's work. They identified the destructive control policy as the cause of this distress, but this was uniformly framed as necessary and specific blame for the outbreak was not pinpointed but aimed variously at government, farmers, the food production infrastructure, even tourists. The result was not a clear message of trauma, but an ambiguous mix of messages, all symbolised by images of the pyre.

However, trauma itself defies logical processing, and provokes ambiguity and incomprehension. So the incongruous and disorientating image of the pyre may be an apt vehicle, not for explaining trauma, but for reflecting the *inexplicable* nature of those distressing events. In fact, both the epidemic and its imagery provoke the disorientating feelings of the sublime, prompting constant mental oscillation between the scientifically mapped and the incomprehensible, and the

---

2   This has been noted by a number of observers. Charles Fritz called these new networks 'therapeutic communities' (Fritz 1961 cited in Erikson 1995, 189). Erikson calls this action a 'gathering of the wounded' (ibid, 187). I shall combine these descriptions into the term *therapeutic gathering*: both a noun and a verb.

fear of disease, food safety and an increasingly technologically opaque environment only partially held in check by the distancing effect of the image. The methods used to contain the disease were just as inconceivable, with 18,000 animals being destroyed daily at one site near Carlisle at the height of the epidemic (Convery *et al* 2005, 104). So while some pyre imagery conflates incongruous symbolic associations by punctuating the pastoral scene with burning livestock, other photographs suggest the incomprehensible power of both FMD and the methods used to control it, by tightening the frame to a landscape engulfed by fire.

For Jean-François Lyotard it is the incompatibility between imagination and reason that engenders the sublime (Lyotard 1994, 123). This incompatibility means that the sublime phenomenon evades being subsumed into language, and instead is an unsettling, physically experienced *event*: it is experienced viscerally, through affect. Jill Bennett suggests that trauma is also recalled in feelings, 'sense memory', rather than 'common memory': the logical narrative way in which memories are normally used to make sense of the world (Bennett 2005, 24–7). In trauma these two types of memory conflict; there is incongruity between feeling and understanding. It is this conflict that should be the focus of any attempt to convey trauma. This conflict is paralleled in the image of the pyre. Our (culturally established) common memory dictates certain associations when we encounter the pastoral landscape. The pyre ruptures this with an incongruous visceral experience. Affect proves incompatible with rational understanding. Referring to this conflict at the heart of traumatic experience, Bennett proposes that images of trauma can only attempt an act of *translation*, rather than representation, acknowledging that traumatic experience can never be fully understood (2005, 121). Perhaps the image of the pyre is fitting: it may not help us to understand the details of FMD as trauma, but the way that it ruptures logical comprehension through affect certainly contributes to the message that it was trauma.

The pyre also features in the longer and more detailed documentary projects of John Darwell, Nick May and Ian Geering, but more often the focus is on smaller but equally intrusive changes in the landscape: the barriers, mats and buckets that betray infection, distant smoke, or merely emptiness. These landscapes communicate the effects of FMD in three interlinked ways: supplying pictorial evidence of the disrupted environment; symbolically alluding to bucolic ideals or the associations of scars and stains; and working through affect by conjuring the disorientating feelings of being in the changed landscape.

Unwilling to trespass on the suffering of others, and feeling that press coverage overlooked the realities of the epidemic, Darwell's initial response to the crisis was to investigate the altered landscape around his Cumbrian home (Darwell 2001, 125–6; Darwell 2009). In *Closed Footpath, Kirkstone Pass looking towards Ullswater* (Fig 23.1), the communion with nature promised by Lakeland scenes is subtly undermined by an anomalous strip of red tape barring access to the fells, and the notable absence of animal life. There is also something less tangible: a feeling of isolation and oppression. 'Common memory', the positive associations of the pastoral, are agitated by a contradictory emotional 'sense memory'.

In *Disinfectant Mat, Patterdale* (Fig 23.2), a tranquil scene is destabilised by the subtle, improvised biosecurity mat, which reveals a hidden human population, but also constitutes the physical, and public, manifestation of a community's sense of contagion. The scene also encapsulates a visual disruption. The eye is alternately pulled between following the lane to the (possibly infected) village ahead or to flee along the road veering to the right, provoking an increasing sense of disturbance. There is also a temporal conflict between the stasis of quarantine and the continued motion of the world outside the biosecurity barrier. So alongside visual evidence of a

FIG 23.1. JOHN DARWELL, *CLOSED FOOTPATH, KIRKSTONE PASS LOOKING TOWARDS ULLSWATER*, 2001.

FIG 23.2. JOHN DARWELL, *DISINFECTANT MAT, PATTERDALE*, 2001.

changed landscape, these photographs offer Deleuze's 'encountered signs': signifiers that cannot be readily deconstructed, but still stimulate feelings or affect. For Deleuze it is these signifiers that provoke further thought (Deleuze 1972, 160–3, cited in Bennett 2005, 7), and consequently it may be here that these photographs prompt a deeper enquiry into the experience of FMD.

Darwell's photograph of Ivegill (Fig 23.3) provides a stark record of the initial inefficiencies in carcass disposal. Here contagion is highly visible, but so is 'death in the wrong place': the rupture in the accepted emotional geography of husbandry and slaughter. The presence of this decomposing material suggests abjection: a continued unwanted contact with material that should be disposed of. The ultimate abject material is the cadaver, as its presence provides an affront to the boundary between the self and the unacceptable other (Kristeva 1982, 2–4). This undermines the sense of identity, and the trauma of FMD is similarly marked by the way that individual and communal narratives of identity were constantly and irreconcilably undermined by undesirable materials or events.

Significantly, this is not a sublime Lakeland landscape, but an unassuming scene. The tainted is not contrasted with the idyllic, but the *banal*. It reflects what Eric Rosenberg suggests is 'trauma's achievement – to forever shuttle signification between banality and upheaval: the quotidian and rupture' (Rosenberg 2006, 45). So these photographs undermined the idyllic view of Cumbria in another important way: they revealed parts of Cumbria eclipsed by the flood of Lakeland imagery, but also overlooked because they are agricultural, not tourist spaces. Consequently, these were the spaces decimated by the epidemic. The photographs revealed a Cumbria unknown to many; not idyllic and spacious, but inhabited, messy and ordinary.

FIG 23.3.   JOHN DARWELL, *AWAITING DISPOSAL, GAITSGILL ROAD, NEAR IVEGILL*, 2001.

Ian Geering attempted to convey the silence of this landscape, in which 'one could travel from Penrith to Carlisle and not see a single animal' (Geering 2002, 5). His photographs visualise the experience of 'the whole of the countryside… held in a silent, invisible grip' (ibid), and do not merely record absence but conjure a palpable presence of absence. They render this *absence as presence* by capturing the scars left on the landscape by the disposal pits and pyres. However, many emphasise emptiness, without any significant point of focus or action. These unremarkable scenes have been sites of destruction, and shadows and tyre imprints suggest the lingering stain of a violent event which, while no longer obvious, is preventing the landscape from returning to normality.

Taken together these photographs provide a panorama of Cumbria repeatedly punctuated by sites of destruction. Like traumatic 'flashbulb' memories, similar landscapes recur without resolution, and intensify feelings of silence and stasis, reflecting that a countryside suddenly devoid of all animals itself constituted a fundamental alteration of lifescape. As one Castle Carrock resident recalled, 'Suddenly there was nothing, absolutely nothing there. And I looked at the fields and I felt sick. And I just thought, this is not something sentimental about one particular sheep, or one particular flock of sheep. It was the whole lot had gone. The whole lot' (BBC Radio Cumbria 2002).

The photographs also contravene the pictorial space of the traditional Lakeland landscape, which excludes any feature that threatens the timelessness of the scene (but for a few well-placed tourists), whether that is vehicles, everyday detritus, or local people (Crawshaw and Urry 1997). It does not reflect the world beyond the frame but provides an escape from it. Conversely the FMD images highlight the detritus of agricultural life and *always* refer to the world beyond the frame; they reveal how outside powers have physical effects on the landscape.

So there is an underlying political comment in these landscapes. Another Darwell photograph captures a picnic site by the M6 motorway, cordoned off by police tape and inferring that a crime has been committed. Others show the tightly controlled opening of Lakeland fells to tourists in June 2001, offering a sardonic reflection on the government insistence of a countryside 'open for business'. The anger felt towards government is undisguised in the photographs of protest signs on fences and farm equipment. In Nick May's photograph, *MAFF Guy at King's Meaburn*, the livestock pyres of the previous months are echoed by the Guy Fawkes Night bonfire (Fig 23.4).[3] If Alexander proposes that conveying blame is an integral part of the message of trauma, then the effigy in official biosecurity overalls on this pyre reflects where many felt the responsibility for the crisis lay.

May also recorded the exhaustive and rigorously enforced process of disinfecting affected farms, which intensified feelings of stress, impotence and contamination. Like Darwell's series, his *'Til the Cows Come Home* follows the crisis through this purging process to the end of the epidemic and beyond. FMD had a clear narrative: infection and isolation was followed by decontamination, and then restocking and a restoration of social interaction and (potential) recovery. Appreciating the duration and many setbacks of this process is essential to understanding the traumatic experience of 2001, and both Darwell and May end their records suggesting that recovery may only ever be partial.

---

3   MAFF is an acronym for the Ministry of Agriculture, Forestry and Fisheries, which was rebranded as DEFRA (Department for Environment, Food and Rural Affairs) midway through the epidemic.

FIG 23.4.   NICK MAY, *MAFF GUY AT KING'S MEABURN, IN THE PENRITH SPUR, NOVEMBER 3RD 2001*, 2001.

All these projects include images of the human figure, which inevitably reveal other aspects of the photographic representation of trauma. However, the landscapes alone contribute greatly towards the four points Alexander outlines for a convincing message of trauma. Many Cumbrians' view of who was responsible for the crisis is communicated, and they vividly suggest the nature of that trauma, by conjuring feelings of infectiousness, isolation, abjection, shame, a loss of continuity and a fundamentally altered lifescape. They outline the group suffering this trauma: inevitably focusing upon farmers and other rural interests, but also conveying the experience of all who lived in the defamiliarised landscape. They also outline a relationship between those affected and the wider audience. The epidemic, and its imagery, highlight that the rural landscape is part of wider social, economic and cultural networks, and that the political or cultural ideologies that we invest in have real physical effects upon lived space. Whether these messages are appreciated by an audience depends upon our subjective responses to images and other factors of display. Yet together these photographs do outline a convincing message that Foot and Mouth was a potentially traumatic experience.

That message began to be disseminated in September 2001 when Darwell's images appeared in Radio Cumbria's *Heart and Soul*. The project was subsequently exhibited at the Royal Agricultural Show in 2004 and published as *Dark Days* in 2007. Geering's book, *The Aftermath*, was published in February 2002, and his photographs were exhibited on the website of Cardiff University's Centre for Business Relationships, Accountability, Sustainability and Society from 2003. Meanwhile, May's project toured Cumbrian galleries in 2005–6, finishing at Tullie House Gallery, Carlisle. In March 2006 all three exhibited at Manchester's Holden Gallery, alongside FMD photographs from around the UK. This exhibition explicitly linked these images to the message of trauma being outlined by other commentators. It accompanied a conference, *The Cultural Documents of Foot and Mouth*, which brought together sociologists, scientists, artists and those directly affected in order to achieve 'a wider public understanding about the health, economic and environmental impact and social consequences of the outbreak' (Hunter and Larner 2006, 1).

Indeed, five years on and despite a number of intervening inquiries, the photographs evidently still had an educational function. The visitors' book from May's exhibition at Tullie House includes such comments as 'terrible to watch, but learnt lots I didn't even no [sic] about it', and messages emphasising 'learning' (Visitors' Book 2006, n.p.). However, press coverage in 2001 had highlighted some significant FMD compensation awards, which inevitably undermined public sympathy. Another comment acknowledges that the epidemic was 'understandably traumatic', but also suggests that some farmers might have infected their farms intentionally 'to be sure they'd not miss the big cash bonanza' (ibid). Ambivalence towards the trauma of 2001 extended from how it was reported, to how it was remembered.

Indeed, May's project had commemorative as well as educational and therapeutic functions (May 2012), and its continued ability to communicate the traumatic memories of FMD is indicated by a recurring phrase in visitors' comments: 'Lest we forget' (Visitors' Book 2006, n.p.). After all, trauma leaves an indelible and unwelcome mark on identity, and so is met with a conflicting urge to both remember and forget. Both these impulses are evident in *Love, Labour and Loss: 300 Years of British Farming*, a 2002 exhibition at Tullie House Gallery, which aimed to boost tourism, counter the negative images of Foot and Mouth and 'lift the spirits of the community' (Bower 2002). However, while this exhibition stemmed from FMD, it also eclipsed it with a narrative that re-emphasised the symbolic associations of the idealised landscape. The resolute British attachment to idealised visions of Cumbria is also evident in subsequent tourist and artistic representation of this landscape, which has returned to the predominant imagery of Lakeland.

However, I conclude by suggesting that this reassertion of the Cumbrian idyll does not necessarily undermine the message of trauma. The FMD photographs convey dramatic shock or building discomfort because they disrupt cultural assumptions predicated on idealised landscapes. Similarly, the return to such conventions indicates the insistence with which British culture will protect those ideals (both through representations and through the physical measures to protect Cumbria's prime tourist locations during the crisis). Bookending the photographs of 2001 with images of the idealised landscape actually ensures that they remain incongruous, disorientating and able to communicate disturbance.

The photographs themselves convey a traumatic rupture – emotional, physical and cultural – between human, animal and landscape. The lasting image of that rupture is the countryside in flames, which vividly demonstrates what happens when a virulent disease strikes, and an official plan to control it is laid over real space. Nonetheless, the image does not convey a precise meaning, but may reflect many subjective experiences, or evade logical interpretation altogether and affect us viscerally. The landscapes of Geering, May and Darwell are equally essential in *translating* the feelings prompted by the insidious, as well as dramatic, changes to the landscape: isolation, contamination, stasis and numbness.

Indeed, all of these landscapes are fundamental to the cultural construction of Foot and Mouth as communal trauma, and the communication of that trauma to the wider public. By shuttling between embedded ideas of the rural and the dramatic unsettling of those ideas, they extend some experience of the crisis to all who have a cultural investment in the countryside. Ultimately, when the photographs of 2001 have been exhibited, comments reveal that they have conveyed the pain of the outbreak to those affected and an outside audience alike.

ACKNOWLEDGMENT

A version of this chapter originally appeared as '"Suddenly There Was Nothing": Foot and Mouth, Communal Trauma and Landscape Photography', in *Photographies* 6 (2), 2013, 289–306.

BIBLIOGRAPHY AND REFERENCES

Alexander, J, 2004 Toward a Theory of Cultural Trauma, in *Cultural Trauma and Collective Identity* (eds J Alexander, R Eyerman, B Giesen, N Smelser, and P Sztompka), University of California Press, Berkeley, 1–30

BBC Radio Cumbria, 2002 *A Sense of Place, Series One, Programme Three: The Truth about Sheep* [online], available from: http://www.bbc.co.uk/cumbria/sense_of_place/prog_3.shtml [17 April 2008]

Bennett, J, 2005 *Empathic Vision: Affect, Trauma, and Contemporary Art*, Stanford University Press, Stanford

Bower, S, 2002 *Interview with Clive Adams, A Curator's Perspective* [online]; available from: http://www.greenmuseum.org/generic_content.php?ct_id=77 [29 August 2010]

Breakwell, G, 2002 *Report to DEFRA: Public Perceptions Concerning Animal Vaccination: A Case Study of Foot and Mouth 2001* [online], available from: http://webarchive.nationalarchives.gov.uk/20130123162956/http://www.defra.gov.uk/science/documents/publications/mp0140.pdf [29 April 2014]

Convery, I, Bailey, C, Mort, M, and Baxter, J, 2005 Death in the Wrong Place? Emotional Geographies of the UK 2001 Foot and Mouth Disease Epidemic, *Journal of Rural Studies* 21, 99–109

Convery, I, Mort, M, Baxter, J, and Bailey, C, 2008 *Animal Disease and Human Trauma: Emotional Geographies of Disaster*, Palgrave, London

Crawshaw, C, and Urry, J, 1997 Tourism and the Photographic Eye, in *Touring Cultures: Transformations of Travel and Theory* (eds C Rojek and J Urry), Routledge, London, 176–95

Darwell, J, 2001 *Foot and Mouth, Heart and Soul: A Collection of Personal Accounts of the Foot and Mouth Outbreak in Cumbria 2001* (ed C Graham), Small Sister/BBC Radio Cumbria, Carlisle, 125–30

— 2009 Personal communication (interview with the author), 8 July

Deleuze, G, 1972 *Proust and Signs* (trans R Howard), George Brazillier, New York

Döring, M, and Nerlich, B, 2009 From Mayhem to Meaning, in *The Social and Cultural Impact of Foot and Mouth Disease in the UK in 2001: Experiences and Analyses* (eds M Döring and B Nerlich), Manchester University Press, Manchester, 3–18

Erikson, K, 1995 Notes on Trauma and Community, in *Trauma: Explorations in Memory* (ed C Caruth), Johns Hopkins University Press, Baltimore, 183–99

Fritz, C, 1961 Disaster, in *Contemporary Social Problems* (eds R Merton and R Nisbet), Harcourt Brace, New York

Geering, I, 2002 *Foot and Mouth: The Aftermath*, Geerings of Ashford, Ashford

Hillyard, S, 2009 Farmers and Valuers: Divisions and Divisiveness and the Social Cost of FMD, in *The Social and Cultural Impact of Foot and Mouth Disease in the UK in 2001: Experiences and Analyses* (eds M Döring and B Nerlich), Manchester University Press, Manchester, 81–94

Hunter, I, and Larner, C, 2006 *The Cultural Documents of Foot and Mouth*, conference and exhibition poster [online], available from: http://www.littoral.org.uk/HTML01/conferences/Cultural_Documents_FMD.pdf [14 April 2006]

Kristeva, J, 1982 *Powers of Horror: An Essay on Abjection*, Columbia University Press, New York

Lyotard, J-F, 1994 *Lessons on the Analytic of the Sublime* (trans E Rottenberg), Stanford University Press, Stanford

May, N, 2012 Personal communication (email exchange with the author), 15 July

Mitchell, W J T, 2002 *Landscape and Power*, 2 edn, University of Chicago Press, Chicago

Mort, M, Convery, I, and Bailey, C, 2005 Psychosocial Effects of the 2001 UK Foot and Mouth Disease Epidemic in a Rural Population: Qualitative Diary Based Study, *British Medical Journal* 331, 1234–9, available from: http://www.bmj.com/cgi/content/full/331/7527/1234 [14 March 2007]

Rosenberg, E, 2006 Walker Evans' Depression and the Trauma of Photography, in *Trauma and Visuality in Modernity* (eds L Saltzman and E Rosenberg), Dartmouth College Press, Hanover NH, 28–50

Visitors' Book, 2006 Exhibition: Nick May, '*Til the Cows Come Home*, Tullie House Museum and Art Gallery, Carlisle, May–June

Displaced Natural Heritage

24

Changing 'Red to Grey': Alien Species Introductions to Britain and the Displacement and Loss of Native Wildlife from our Landscapes

Peter Lurz

'Squirrels, beware, beware; Red squirrel run!…' – Joseph Langland (1963)

Introduction

In our hearts and minds we associate and often inextricably link local features and places, or the passing of the seasons, with the reproductive status of plants or the appearance of familiar animals – whether these are the first flowering messengers of spring, such as winter aconites or snowdrops, winged migratory or regular visitors to our bird feeders in the garden, or the haunting and beautiful calls of curlews across the Cumbrian fells in spring. Our feelings for a place and our sense of home do not simply relate to a location – a geography and 'what people feel, know and do', to paraphrase Albert Camus – but also to the birds, mammals and plants that we have become familiar with and accustomed to. What I mean is perhaps captured by Michael Longley's link to Carrigskeewaun in Ireland in his poem 'Greenshank' where he says: '… If I had to choose a bird call for reminding you, the greenshank's esturial fluting will do' (Longley 2011).

In each place, with its local geology, climate and history, a combination of human influences, chance events, natural processes, individual species journeys and their interactions determine its unique character and our experience of it. Places are both physical and emotional (see for example Kahn 1996), and we cherish and cling onto these impressions and memories as we do to the plant and animal species that belong to them. These sentiments are also inherent in the choice and use of words that describe our efforts to 'conserve' wildlife. The very definition of 'conservation', to cite *The Collins English Dictionary*, is: 'the act of conserving or keeping from change...' (Hanks 1986). In many ways, the act of conservation is to apply or try to impose static concepts on processes that have always been in flux and changing, albeit over timescales that are usually beyond our normal frame of reference. The Greek philosopher Empedocles, who lived in the west of Sicily around 492–432 BC, acknowledged life's constant change. He imagined and spoke of cosmic cycles with an alternating pattern of unity and plurality (Osbourne 2004). Change is also reflected in the landscapes and rocks around us. A close look at the fossil record would show us that the mean life span of a species, based on studies of mammals and marine invertebrates, ranges between 1 and 10 million years (May *et al* 1995). The periods (and forces) of warming and cooling, the ice ages that helped shape many northern landscapes and the species that inhabit them (see for example Hourlay *et al* 2008), are measured in tens of thousands of years. Yet the combined recent impacts of human-induced land use or climate change and introductions of non-native species have the potential to cause much more rapid and sometimes **catastrophic, large-scale displacement and loss of native wildlife**. Next to habitat destruction, alien species are one of the biggest threats to biodiversity worldwide, and a major force of global change (see for example Vitousek *et al* 1997).

Given current trends, May *et al* (1995) estimated that the lifespan of current bird and mammal species previously measured in millions of years has been reduced to 200–400 years. But it is not just archaeology and bones hidden deep in the ground that illustrate a long sequence of loss and change. Our landscapes, our place-names and familiar, local points of reference are a history, a verbal record of change. They document and remind us of long-lost residents of the British Isles. Among these are elegant birds, such as cranes, ecosystem engineers, such as beavers, and species now more commonly linked to fairytales and legends, such as the wolf. Derek Yalden, in his studies on Celtic, Anglo-Saxon, Norse and Old English place-names that are connected with former birds and mammals of the British Isles, showed that there are, for example, more than 300 place-names linked to cranes. Evidence from medieval bestiaries, illustrated manuscripts and bones from 78 actual excavations all indicate that cranes were once a familiar presence and were breeding in Britain (Boisseau and Yalden 1998). Anglo-Saxon place-name evidence for wolves also encompasses over 200 locations in England alone. However, Aybes and Yalden (1995) could only find 20 places-names for beavers. The latter suggests that while wolves were still widespread in Anglo-Saxon times, beavers were already rare. Though not a new phenomenon, the displacement and loss of recognised and cherished species is sadly becoming an increasing feature of our times, and this is aptly illustrated by the case of the British red squirrel (*Sciurus vulgaris leucourus*).

THE EURASIAN RED SQUIRREL IN BRITAIN

Red squirrels colonised the British Isles from the Continent after the end of the last ice age over 10,000 years ago, when forest cover moved northwards again. This distinctive member of our mammal fauna would have been widespread and established in mature woodland habitats throughout the country, and approximately 4500-year-old fossil remains have been found on the Isle of Wight (Lurz *et al* 1995; Lurz 2010). While little is known about prehistoric times, red squirrels have been a recognised economic and cultural aspect of societies in Britain, stretching from Anglo-Saxon stone sculptures of squirrels on the 8th-century Bewcastle Cross, Cumbria (Bailey and Cramp 1988), to trading records dated 1243, which show that red squirrel fur was used to line coats and make warm clothing. The word 'scorel' and 'squerel' appeared in the first English–Latin dictionary in East Anglia around 1440 (Shorten 1954); more recently we have Beatrix Potter's 'Squirrel Nutkin' and the Royal Society for the Prevention of Accidents' 'Tufty Fluffytail', who introduces road safety messages for children.[1]

In the 18th century, wide-scale forest destruction brought about a decline and regional extinctions of the red squirrel in the British Isles. However, populations recovered with reforestation efforts and the advent of large plantings of non-native conifer trees at the beginning of the 19th century (Shorten 1954). In fact, red squirrels had become so abundant by the late 19th century that they were locally considered a forest pest in places such as the Highlands, the New Forest

[1]   See: http://www.rospa.com/about/history/tufty.aspx [25 February 2014].

FIG 24.1.   FEEDING SQUIRRELS IS ONE OF THE FEW AND POSITIVE WILDLIFE ENCOUNTERS OF
PEOPLE LIVING IN CITIES.

and Cornwall, and 'Squirrel Clubs' were formed to control their numbers (ibid). Their well-publicised and inexorable decline since then has chiefly been due to the arrival and interactions with the non-native North American gray squirrel (*Sciurus carolinensis*, henceforth referred to as 'grey squirrel') and what has been described as a form of 'disease-mediated competition'. Grey squirrels are native to the eastern part of North America and are superbly adapted to deciduous oak woodlands. They are considered by the World Conservation Union (IUCN) (Lowe *et al* 2000) to be among the world's 100 worst invasive alien species, and were introduced to Britain in the late 19th century. While the first recorded introduction to Cheshire in 1876 has received a lot of attention, the introduction to Woburn and successive subsequent translocations from that thriving population (minimum of eight known instances; Middleton 1930) contributed significantly to their spread. In total there were over 30 known separate introductions or translocations of grey squirrels across the British Isles, including three to Scotland (Loch Long, Edinburgh and Fife) and one translocation to Ireland (Castle Forbes).

Grey squirrels not only compete with red squirrels for resources but also act as a carrier and vector of a deadly squirrel poxvirus (SQPV). The virus does not cause any obvious signs of disease in grey squirrels,[2] and positive antibody tests of blood samples from greys in their native range (Wisconsin, USA) suggest that it arrived with them from North America (McInnes *et al* 2006). However, in red squirrels disease is invariably lethal within 2–3 weeks for most affected individuals,

---

2   For a detailed description and discussion of the minor impacts of the virus on grey squirrels, see Fiegna (2011).

often as a result of secondary infections and a severely compromised immune system. There has been a lot of speculation with regard to the competitive interactions between the two species over the years. What can be measured in research studies using live trapping, radio tracking and observational methods is that the main impact of competition locally is on red squirrel reproductive success and juveniles. In the presence of grey squirrel competition, red squirrel breeding is diminished and females often only have one (rather than the normal two) litters per year. The recruitment into the population of the fewer young that are born is also markedly reduced and the combined effects can lead to a declining population that dies out within a few years, especially in deciduous woodland habitats.

Competition between the two squirrel species is not a simple process or interaction; it is influenced by local forest composition, annual seed crop patterns, landscape structure and the presence of disease, which will speed up decline dramatically (Bosch and Lurz 2012). However, the combined onslaught of lethal disease and competition for resources means that red squirrels, bar some small remnants on islands, have now all but disappeared from southern and central England. They are scarce in Wales and declining and under threat in northern England and parts of Scotland, and fewer and fewer of us will know, be excited about and cherish them as visitors to gardens and bird tables, or be 'greeted by them' in local parks (see Fig 24.1). Kenward and Holm (1989) warned of potential extinction in mainland Britain within the near future, and more recent claims predict possible extinction within the next 20 years (Booth 2011).

NON-NATIVE OR ALIEN SPECIES

Grey squirrels in Britain and the loss of red squirrels are symptomatic of a much larger issue of human-induced alien species introductions and change to wildlife and ecosystems on a national and global scale. The Parliamentary Office of Science and Technology in the UK illustrates the enormity of the non-native species problem for England alone (see POST 2008), where 2271 non-native species were recorded in 2005. Of these, 122 species had a negative impact on the environment and 188 species had an economic impact, with an estimated cost to the economy of several billions (of pounds sterling) per year. If UK territories and crown dependencies worldwide are included, then the number increases to 2900 (Varnham 2006). Alien species' impacts on native wildlife tend to be varied and may encompass a range of different interactions (Manchester and Bullock 2000). They are known to hybridise (eg Japanese sika deer and red deer), compete with or eat native species outright (eg American mink and water vole), or act as carriers of disease (eg red and grey squirrel). Their presence can be localised (eg rose-ringed parakeets in London) or ubiquitous, with nationwide distributions (eg Canada goose or *Rhododendron ponticum*). Similarly, their behaviour and impacts can be neutral or highly damaging, with local or ecosystem-wide effects.

Most research on the interactions of red and grey squirrels has to date focused on competition, disease, bark-stripping behaviour of grey squirrels, which causes significant economic damage in forest plantations all across the UK, and some initial attempts to quantify impacts on other species such as woodland birds (see for example Newson *et al* 2010). Little attention has so far been given to wider, long-term ecosystem impacts or potential trophic-level interactions. We are not just shaped by the physical environment around us, but also very much by the other species (and competitors) there with us. Darwin, in his *Origin of Species*, notes that:

… though Nature grants long periods of time for the work of Natural selection, she does not grant an indefinite period; for as all organic beings are striving to seize on each place in the economy of nature, if any one species does not become modified and improved in a corresponding degree with its competitors, it will be exterminated… (Darwin 1972)

Fascinating studies on the Iberian Peninsula and Mediterranean Islands on 'crossbill–red squirrel– Aleppo pine tree' systems illustrate these interdependencies and how the presence, or in this case absence, of squirrels can influence how trees evolve measures to protect their seeds from preda- tion and how this can shape the morphology of competitors (eg beak morphology). Where both squirrels and crossbills were present on the mainland, trees responded to squirrels and cones were larger, forcing crossbills to feed on the seeds of other conifers. On Mallorca where only crossbills were present, cones were smaller but had thicker scales, and crossbills specialising on Aleppo pine tended to have thicker beaks. On Ibiza where neither seed predator was present, investment in physical cone defences was reduced and cones tended to be small with thin scales (Mezquida and Benkman 2005). With respect to the British Isles, we simply do not know and are unable to predict how the arrival and presence of grey squirrels in Britain and their seed preferences will influence and shape our forest ecosystems and the species within them, in the long term.

But the displacement and loss of red squirrels is not just a matter of scientific understanding of what shapes the species around us. Red squirrels, like no other endangered animal I have worked on, typify the complex attitudes and tensions inherent in conservation efforts. Their predicament crystallises that our motivations for saving species go far beyond numbers, indices of biodiversity or probabilities of extinction. They relate to an experienced local, regional and national 'Natural History and Heritage', and loss is always an individual and personal experi- ence. While red squirrels may have been a forest pest in the past, or largely unremarked upon and their presence taken for granted, it is their relentless disappearance and loss that has evoked and precipitated this sense of belonging and identity (see also Barkan and Bush 2002). Nick Mason, manager of a red squirrel conservation project in the North of England, epitomises this sentiment in an interview by CBC on top of an 'historic line in the sand' (Hadrian's Wall) near Kielder Forest, one of the largest remaining strongholds for red squirrels in England: 'People look at red squirrels as a heritage feature that they want to see retained, something that links clearly to several centuries of cultural identity. They do not want to see yet another thing change…' (*The Nature of Things* 2013).

Red squirrel decline and grey squirrel spread has not been a swift or instantaneous process, a sudden disappearance that can be mourned. Expansion has been measured in decades regionally and a century nationally, resulting in a geography of loss. My 20 years of experience in squirrel research and conservation have shown that people particularly in the north of the country often link squirrels to places and experiences. Seeing a red squirrel is 'news to be shared', whether the recipient is interested or not (!), whereas in places where grey squirrels have been present for a long time, attitudes tend to be more general with an abstract sense of concern over losing biodiversity and rare species. A sighting can even be met with bemusement by members of the younger generation in the south, who only see red squirrels while on holiday and cannot help but wonder why an animal that introduces road traffic safety to children is so frequently seen dead by the roadside in the North. While life is characterised by change, the scale and speed of non-native species introductions and human impacts on ecosystems mean that displacement and loss of familiar wildlife, and changes to our physical and remembered, emotional landscapes are likely to become a common feature of our future. It highlights a need for debate on current conservation approaches and our responses to the displacement of iconic species that belong, and perhaps a need and mechanism for sharing the sense of loss that this entails.

The scientist in me cannot help but notice that there is also a time dimension linked to what we consider as native and part of our culture and life, and that each generation makes their own connections. Few of us would consider the brown hare an alien species, yet it may have been introduced to Britain during the Iron Age (Yalden 1999); the rabbit, like the fallow deer, was probably brought to Britain by Normans and is now so established and integrated in our ecosystems that eradication could lead to serious consequences for rare native predators (see for example Lees and Bell 2008); and with respect to grey squirrels and other introduced mammals in Britain, Macdonald and Burnham (2010) have asked the question: 'When are non-natives part of the natural community?' Personally, I have lived in the North of England for close to three decades and have watched, been fascinated by and tried to contribute to red squirrel conservation, and I do mourn their disappearance from my old garden in Cumbria and would sorely miss them if they disappeared completely. There is a pub called the 'Red Squirrel' in my favourite Scottish city, and I wonder if, like the 'Deacon Brody', 'Ensign Ewart', or the 'Old Dukes' and 'Kings' dotted around the country, the pub and the red squirrel will one day also be a part of history.

ACKNOWLEDGMENTS

I am indebted to Catherine Sylvain and Oliver Lurz for support, ideas and inspiration and would like to thank Linda, Lucy and Thomas DaVolls for their really helpful comments and thoughts.

BIBLIOGRAPHY AND REFERENCES

Aybes, C, and Yalden, D W, 1995 Place-name evidence for the former distribution and status of wolves and beavers in Britain, *Mammal Review* 25, 201–27

Bailey, R N, and Cramp, R, 1988 *Corpus of Anglo-Saxon stone sculpture II: Cumberland, Westmorland and Lancashire North-of-the-Sands*, Oxford University Press, Oxford

Barkan, E, and Bush, R, 2002 *Claiming the Stones, Naming the Bones: Cultural Property and the Negotiation of National Identity and Ethnic Identity*, Getty Publications, Los Angeles

Boisseau, S, and Yalden, D W, 1998 The former status of the Crane *Grus grus* in Britain, *Ibis* 140, 482–500

Booth, R, 2011 Red squirrel 'could be extinct within next 20 years', *The Guardian*, 25 September, available from: http://www.theguardian.com/environment/2011/sep/25/british-mammals-red-squirrel-extinction [3 March 2014]

Bosch, S, and Lurz, P W W, 2012 *The Eurasian Red Squirrel*, Westarp Wissenschaften, Hohenwarsleben

Darwin, C, 1972 *The Origin of Species*, Dent, London

Fiegna, C, 2011 A study of squirrelpox virus in red and grey squirrels and an investigation of possible routes of transmission, unpublished PhD thesis, University of Edinburgh

Hanks, P, 1986 *Collins Dictionary of the English Language*, 2 edn, Collins, London

Hourlay, F, Libois, R, D'Amico, F D, Sarà, M, O'Halloran, J, and Michaux, J R, 2008 Evidence of a highly complex phylogeographic structure on a specialist river bird species, the dipper (*Cinclus cinclus*), *Molecular Phylogenetics and Evolution* 49, 435–44

Kahn, M, 1996 Your Place and Mine: Sharing Emotional Landscapes in Wamira, Papua New Guinea, in *Sense of Place* (eds S Feld and K H Basso), School of American Research Press, Santa Fe, 167–96

Kenward, R E, and Holm, J L, 1989 What future for the British red squirrel? *Biological Journal of the Linnean Society* 38, 83–9

Langland, J, 1963 Sacrifice of a Red Squirrel, *New Yorker* 38 (49), 26 January, 32

Lees, A C, and Bell, D J, 2008 A conservation paradox for the 21st century: the European wild rabbit *Oryctolagus cuniculus*, an invasive alien and an endangered native species, *Mammal Review* 38, 304–20

Longley, M, 2011 *A Hundred Doors*, Jonathan Cape, London

Lowe, S, Browne, M, Boudjelas, S, and De Poorter, M, 2000 *100 of the World's Worst Invasive Alien Species: A selection from the Global Invasive Species Database*, The Invasive Species Specialist Group (ISSG), a specialist group of the Species Survival Commission (SSC) of the World Conservation Union (IUCN), updated and reprinted version, November 2004

Lurz, P W W, 2010 *Red Squirrels, Naturally Scottish*, Scottish Natural Heritage, Battleby

Lurz, P W W, Gurnell, J, and Magris, L, 1995 *Sciurus vulgaris*, *Mammalian Species* 769, 1–10

Macdonald, D, and Burnham, D, 2010 *The State of Britain's Mammals: a focus on invasive species*, People's Trust for Endangered Species, London

Manchester, S J, and Bullock, J M, 2000 The impacts of non-native species on UK biodiversity and the effectiveness of control, *Journal of Applied Ecology* 37, 845–64

May, R M, Lawton, J H, and Stork, N E, 1995 Assessing extinction rates, in *Extinction Rates* (eds J H Lawton and R M May), Oxford University Press, Oxford, 1–24

McInnes, C J, Wood, A R, Thomas, K, Sainsbury, A W, Gurnell, J, Dein, J, and Nettleton, P F, 2006 Genomic characterisation of a novel poxvirus contributing to the decline of the red squirrel (*Sciurus vulgaris*) in the UK, *Journal of General Virology* 87, 2115–25

Mezquida, E T, and Benkman, C W, 2005 The geographic selection mosaic for squirrels, crossbills and Aleppo pine, *Journal of Evolutionary Biology* 18, 348–57

Middleton, A D, 1930 The ecology of the American Grey Squirrel (*Sciurus carolinensis*) in the British Isles, *Proceedings of the Zoological Society of London 1930*, 808–43

The Nature of Things, 2013 Nuts About Squirrels, broadcast on CBC (Canadian Broadcasting Corporation), 4 February

Newson, S E, Rexstad, E A, Baillie, S R, Buckland, S T, and Aebischer, J, 2010 Population change of avian predators and grey squirrels in England: is there evidence for an impact on avian prey populations?, *Journal of Applied Ecology* 47, 244–52

Osbourne, C, 2004 *Presocratic Philosophy: A very short introduction*, Oxford University Press, Oxford

POST (Parliamentary Office of Science and Technology), 2008 *Postnote* 303, April, available from: http://www.parliament.uk/documents/post/postpn303.pdf [3 March 2014]

Shorten, M, 1954 *Squirrels*, Collins, London

Varnham, K, 2006 Non-native species in UK Overseas Territories, *JNCC report* 372, Peterborough

Vitousek, P M, D'Antonio, C M, Loope, L L, Reimánek, M, and Westbrooks, R, 1997 Introduced species: a significant component of human-caused global change, *New Zealand Journal of Ecology* 21, 1–16

Yalden, D, 1999 *The History of British Mammals*, T and A D Poyser Ltd, London

25

Displacing Nature: Orang-utans in Borneo

Marc Ancrenaz and Isabelle Lackman

The forests of Borneo and Sumatra are a major hot-spot for biodiversity. They are home to the only two species of Asian great apes that survive today: the Sumatran orang-utan (*Pongo abelii*) and the Bornean orang-utan (*Pongo pygmaeus*). However, huge tracks of natural forests are being degraded by extractive industries (mining, timber harvesting, etc) or replaced with small-scale subsistence crops and large-scale industrialised monoculture (oil palm, acacias, eucalyptus, etc) throughout the region. As a result, entire wildlife populations are displaced or wiped out.

The destruction of these forests is a multi-faceted disaster for the orang-utan and for myriad other species. Indeed, the intense degradation and destruction of food resources lead to starvation, increased sensitivity to disease and stress, or to decreased reproduction. During waves of forest conversion, animals that are not killed are displaced and take refuge in remaining forest patches, resulting in significant alteration of the dynamics and structure of surviving wild populations. The newly created man-made landscape results in increased contact and conflict between people and wildlife. Crop-raiding orang-utans are often perceived as pests and are killed, although they are officially protected in both Indonesia and Malaysia. Surprisingly, our work in Kinabatangan (Sabah, Malaysian Borneo) indicates that orang-utans show an unexpected resilience to these disasters. To a certain extent, the animals are able to adapt their behaviour and ecology to these new environmental conditions. From a conservation point of view, it is essential to investigate this resilience in order to design and promote wiser land management practices that will reconcile economic growth with the long-term survival of orang-utans and other animal species.

It is also crucial to investigate how local communities' perceptions of wildlife are being influenced by recent landscape transformation. The communities living in the Kinabatangan floodplain used to consider orang-utans as non-detrimental and peaceful creatures living in the depths of the jungle. The recent increase in conflicts between crop-owners and orang-utans has greatly changed this perception: from being part of their natural heritage, orang-utans are now a competitor and a hindrance to economic growth. Reconciling people and wildlife and identifying ways for local communities and displaced animals to cohabit peacefully is the major conservation challenge in these fractured landscapes.

The Changing Environment of the Orang-utan

Great apes are our closest living relatives (Harrison 2010). Orang-utans (literally 'men of the forest') occur solely in Borneo (*Pongo pygmaeus*) and in Sumatra (*Pongo abelii*). The orang-utan is the largest arboreal mammal in the world. Animals spend most of their life travelling in the canopy, foraging primarily for fruits but also feeding on leaves, barks and insects (termites, ants). They are often considered as the 'gardeners of the forest', dispersing seeds over large distances and

FIG 25.1.	MOSAIC LANDSCAPE OF THE LOWER KINABATANGAN SHOWING LARGE INDUSTRIAL
OIL PALM PLANTATIONS, SMALL-SCALE AGRICULTURE FIELDS AND FRAGMENTED FORESTS.

creating gaps in the canopy when they break branches, allowing light penetration and favouring under-canopy tree growth (Rijksen and Meijaard 1999). They live a semi-solitary life and rarely aggregate in groups. Males are the dispersing sex: upon reaching sexual maturity (around 10–12 years old), they leave the area where they were born to establish large territories that cover several hundred hectares. Females' territories are smaller, depending on forest type and availability of food resources. Historically, orang-utans were most abundant in primary dry and semi-inundated lowland dipterocarp forests where mosaics of different habitat types could buffer food availability throughout the year.

Orang-utans are very slow breeders; they produce on average one offspring every six to eight years, explaining their extreme sensitivity to hunting pressure. Recent studies show that climatic changes and human pressure have significantly reduced the distribution range and number of orang-utans in recent centuries. Today it is estimated that only 60,000 orang-utans survive in Borneo and Sumatra, a mere 10% of the number that are thought to have existed in recent prehistoric times (Wich *et al* 2008). Both species are threatened with extinction: the Bornean orang-utan is 'Endangered' while the Sumatran species is 'Critically Endangered', according to the IUCN Red List.[1]

Major threats to the orang-utans include habitat loss, forest degradation and illegal killing (to mitigate conflicts, for pet trade, bushmeat, or traditional medicine). These threats are a direct

---

[1]	See: http://www.iucnredlist.org/ [28 January 2014].

consequence of (1) the world demand for oil palm and other commodities, and (2) the rapid economic growth experienced by the only two countries home to wild orang-utan populations: Indonesia and Malaysia. Predictably, natural resources fuel this growth. Forests are increasingly converted to other types of land uses. Agricultural ecosystems are becoming the dominant landscape in these two countries (as well as in many countries of the inter-tropical belt), just as they have dominated for centuries in temperate regions.

Compared to intact forests, secondary forests (or forests that have been exploited by people) and agriculture landscapes show an overly simplified structure and composition: tree density and diversity, as well as tree canopy layers, are reduced (canopy layers are completely suppressed in annual crops). Top soil is stripped by erosion or damaged by compaction and microclimate conditions often become drier and hotter. These new conditions bring changes in wildlife community structure and abundance, with endemic and specialised taxa most at risk of being replaced with invasive and generalist taxa (Meijaard *et al* 2005). Degradation of ecosystems is exacerbated by climate change and other large-scale man-made ecological disasters, such as overharvesting of forest products and unsustainable rates of hunting (Sodhi *et al* 2010). The long-term impacts of these changes on complex processes are still not fully understood: the relationship between wildlife and forest regeneration (seed dispersal and germination); changes of trophic cascades (prey–predator relationships); shifts in population distribution owing to climate change; emerging diseases. However, what is known is that changes in wildlife community composition affect the functionality of ecosystems and impoverish the value of the services they provide.

FUNCTIONS AND LIMITS OF PROTECTED AREAS FOR ORANG-UTAN AND BIODIVERSITY CONSERVATION

For a long time, it was believed that the orang-utan was extremely sensitive to forest degradation and that this species was strictly dependent on primary forests to survive. Because of this, setting aside protected forests became the cornerstone of orang-utan conservation. More generally, the dichotomy opposing human development and biodiversity conservation has resulted in land-sparing strategies, with the creation of protected areas free of human influence to halt the loss of biodiversity in the world (Chape *et al* 2008). A protected area is defined by IUCN as 'a clearly defined geographical space, recognised, dedicated and managed, through legal or other effective means, to achieve the long-term conservation of nature with associated ecosystem services and cultural values' (Dudley 2008, 8). Today more than 150,000 protected areas exist worldwide, totalling around 13% of the overall land mass. Signatories of the Conservation on Biological Diversity have set 17% as the target to be reached by 2020.

Without any doubt, protected areas are an effective tool to conserve biodiversity. However, the capacity of protected areas to preserve biodiversity in the long term is challenged by many factors. First, many taxa and ecosystems are poorly represented in the current network of protected areas. Most protected areas occur in highland areas with steep slopes, or in remote places offering a lower suitability for agriculture. However, protecting lowland areas would be much more effective in mitigating the current land-use transformations that threaten orang-utans and most species worldwide. Second, this network is too fragmented to provide enough habitat for wide-ranging populations, such as large mammals and migratory birds. Third, it is yet to be seen whether the network of protected areas will be able to maintain biodiversity in a changing environment impacted by dynamic processes such as global warming. Fourth, many protected areas are in crisis, threatened by direct encroachment and illegal activities, lack of support from local communities, poor management and under-funding. Last but not least, the integrity of protected areas depends on complex ecological processes that stretch well beyond their geographical boundaries. Their functionality depends not only on changes that occur within their boundaries (hunting, logging and other encroachment) but just as importantly on changes that occur outside.

Over the past 20 years, the current status quo opposing protected and non-protected areas has led to increasingly degraded natural ecosystems inside protected areas, and to the proliferation of man-made landscapes outside. There has been an increasing emphasis on human development and poverty alleviation at the expense of wildlife conservation. Importantly, the land-sparing strategy ignores valuable biodiversity and ecosystem services outside of protected areas. A paradigm shift about our comprehension of orang-utans and biodiversity conservation is sorely needed.

It is clear today that the vast majority of protected forests occupied by great apes are not free of human influence. Most of these areas were exploited in the past before being afforded protection status, and are currently encroached for various reasons (illegal logging, mining, poaching, etc), or suffer from fires and climate change. It is estimated that by 2030 only a few per cent of the current orang-utan habitat will be left undisturbed by infrastructure development (Gaveau *et al* 2013). These facts clearly show that the long-term survival of the species in Borneo will rely heavily on improved management of non-protected forests and minimising losses among orang-utan populations living in these areas. This requires a major strategic change in how orang-utan conservation is practised.

Orang-utan Resilience to Ecological Changes

More and more empirical evidence shows an unexpected resilience of many biodiversity elements (including orang-utans) to human disturbance. Although primary forests are irreplaceable for some wildlife and plant species, selectively logged forests are also able to maintain a relatively high level of biodiversity. For example, forests exploited for timber can support viable orang-utan populations when sustainable timber practices are implemented. Such practices include reduced impact logging and directional felling, low extraction rate, precise inventory of harvestable timber, careful road planning, etc. Coupled with a strict control of illegal killing, these sustainable logging practices can reconcile forest exploitation with orang-utan conservation (Ancrenaz *et al* 2010). Better forestry practices could significantly improve biodiversity in commercial forests.

The resilience of the orang-utan and its ability to cope with drastic habitat changes is further illustrated by the presence of the species in acacia plantations in East Kalimantan (Meijaard *et al* 2010) and in mature oil palm plantations in Sabah (Ancrenaz *et al* forthcoming). Other great apes are also found in agricultural landscapes: mosaic of subsistence agriculture (all species) or modern agro-industrial plantations (chimpanzees) (Hockings and Humle 2009).

Given the drastically altered structure of agricultural landscapes that orang-utans are surviving in, it is not surprising that their behavioural and feeding ecology differ markedly from what is known of orang-utans in primary forests. Animals found in altered landscapes often (although not always) have smaller home ranges and shorter daily distances travelled. They walk on the ground more often although terrestrial behaviour is also present in primary intact forests (Ancrenaz *et al*

Fig 25.2.   Nest built by an orang-utan in an oil palm plant.

2014). Generally the animals spend less time foraging and their diet is dominated by cultivated fruits (although they can survive on acacia bark and fronds of oil palm in periods of fruit scarcity). Their nests can also be found in acacias and in palms when no other tree is available for nesting (see Fig 25.2, an orang-utan nest in an oil palm plant). Importantly, the animals show an increased tolerance of close human proximity. Similar changes also occur with African great apes living in agriculture landscapes.

ARE ORANG-UTANS ABLE TO SURVIVE IN AN AGRICULTURAL LANDSCAPE?

The presence of patches of natural forests is necessary to maintain great apes in heavily altered landscapes in Africa and Asia. Research in Borneo shows that extensive industrial monoculture alone cannot sustain viable populations of orang-utans. Indeed, patches or corridors of natural forests need to maintain some sort of connectivity between sub-populations in this highly altered matrix (McShea *et al* 2009; Ancrenaz *et al* forthcoming). These patches of natural forest are used by the animals to feed (fruits, leaves and barks), to rest and to move across the landscape (they use the small patches as stepping stones during movement).

This knowledge clearly shows that conserving orang-utans in Asia and other great apes in Africa must include the preservation and/or restoration of small patches of forest within the newly developed agro-industrial landscape. Used as corridors or stepping stones, these small forest patches, even though degraded, play an important role in sustaining the animals.

Recent Borneo-wide analysis showed that approximately 22% of the current orang-utan distribution is protected while the rest (78%) falls outside of protected areas (Wich *et al* 2012). About a third of the entire orang-utan range lies within commercial forest reserves exploited for timber, and about 45% falls within forest areas earmarked for conversion to agriculture and to other land uses. It is highly unlikely that all forests containing orang-utans in Borneo are going to be protected in a foreseeable future. On the contrary, a business-as-usual scenario, where non-protected forests are converted according to current development plans, will result in the loss of more than half of the current orang-utan range on Borneo island. Since transformation of forest landscape into non-forest is primarily done without large scale land-use planning, such a development process would have a devastating impact on resident orang-utan populations since most animals are killed when natural forests are converted to other types of land uses (Meijaard *et al* 2011a).

DYNAMIC CHANGES BETWEEN PEOPLE AND ORANG-UTANS IN MULTIPLE-USE LANDSCAPES

For thousands of years, local communities and wild great apes have lived together and shared the same environment relatively peacefully. Forest reduction and loss result in a closer proximity between people and animals. With the decrease in natural orang-utan food availability comes an increase of crop-raiding activities, which inevitably lead to conflicts between people and apes. These conflicts lead to emotional distress of local inhabitants and, sometimes, significant economic losses within local communities (Campbell-Smith *et al* 2010). Orang-utans can consume entire fruit crops in orchards belonging to local villagers. Economic losses are also felt by agro-industrial plantations: apes kill acacias by stripping bark and cambium (Meijaard *et al* 2010); they pull out stems and destroy young palms to feed on the heart of the plant. The occurrence of conflicts results in orang-utan killings and creates a negative perception towards wildlife, which has been a major impediment to building local support for conservation.

The intense and rapid transformation of the environment impacts not only orang-utans and other wildlife populations, but also human communities who share the same habitat. The destruction of natural resources used by rural communities (game species and other food sources, timber, non-timber forest products, etc) as well as the increase in human–wildlife conflict exacerbates the hardships already faced by the poorest human groups. The disappearance of local ecological knowledge and cultural values linked to orang-utans results in a lower acceptance level for cohabitation with wildlife. It also increases the disconnection felt by many local communities between our modernisation age and people's identity, roots and cultural values and heritage. The symbolic importance of orang-utans and other wildlife species disappears, and the animals are more and more perceived as a commodity one can see in zoos or in the media. Environmental change and the resulting hardship culminates in mass migration from rural to urban areas. The social ramifications of this have yet to be seen, as this move means the loss of natural heritage, tradition and culture valued by local communities (Meijaard *et al* 2013).

Considering that the needs and aspirations of local communities are the ultimate drivers of conservation success or failure outside of protected forests, it becomes clear that we need to encourage these groups to become active participants in conservation and not simply beneficiaries of what it can offer. Communities need to be empowered to act.

In the Kinabatangan floodplain in north-eastern Borneo, 15 years of community-based conservation has proven that a negative perception of wildlife by local communities resulting from increased competition for the same natural resources and from increased human–wildlife conflicts created by the changing landscape can be overcome. A strong and genuine engagement between local NGOs, government agencies and local communities who share the land with the orang-utans led to the development of peaceful techniques that were efficient in mitigating conflicts opposing smallholders and displaced animals (Ancrenaz *et al* 2007). Empowering the villagers and seeking their active involvement in conflict mitigation and law enforcement activities through intense capacity building, awareness campaigns and the development of alternative economic opportunities using orang-utans and their forest habitat as a product (eco-tourism, research) changed the general perception of the value of wildlife in the area.

By acknowledging that the fate of most wildlife populations and local human communities are intrinsically linked, there is an urgent need to shift our focus from conserving sites and species to respecting landscapes and processes.

NEED FOR NEW CONSERVATION LANDSCAPES

The orang-utan's ability to cope with drastic changes in its environment offers hope in the struggle to balance the species' survival and the needs of human development. The first step in creating this balance requires industry to embrace better management practices that minimise the strong negative impacts of forest exploitation and agro-industrial development on orang-utan survival. The ultimate and long-term impact of human disturbance on the original biodiversity is strongly influenced by the general configuration of the landscape after habitat loss and alteration (Lindenmayer and Fischer 2006). As a result, remaining forests that sustain key orang-utan populations should be identified as 'High Conservation Value Forests' and maintained as forests. Retaining forests within an agro-industrial landscape is key in preserving ecosystem functionality, improving meta-populations of many wildlife species and facilitating dispersal and survival of many species such as the orang-utan (McShea *et al* 2009).

However, to ensure the long term co-existence of both human and animal populations, a truly holistic approach must be taken. Development planning must include not only social and economic parameters but must also take into account ecological connectivity. For the island of Borneo, this means developing a conservation/development blueprint or road map that goes beyond the political boundaries of Malaysia, Indonesia and Brunei (Meijaard *et al* 2011b). Taking such an approach requires combining and refining traditional views in conservation and development. Protected areas would be the core of the entire system but their delineation should also consider their surroundings, while ensuring that hydrologic, ecologic and socio-economic characteristics are taken into account. These core protected areas could be connected with non-protected forests exploited for timber adhering to sustainable and reduced-impact practices. In turn, these non-protected natural forests could be buffered by low-intensity and mosaic plantations such as acacia, pulp and paper and other mixed industrial tree plantations. This landscape would then be connected to high-intensity use areas such as agro-industrial schemes and areas where infrastructures, roads and small-scale agriculture are concentrated and where most people live. Designing and establishing such a hypothetical functional landscape requires the development of new alliances (between government, conservation, local community and industry) and new approaches that will influence and change how contemporary human societies perceive orang-utans and biodiversity conservation.

CONCLUSION

Protected areas reflect a binary view of the world. This predisposition results in all-or-nothing judgements and biases the vision of the natural world that we would like to preserve. As a result, society emphasises the value of pristine habitats for biodiversity conservation and subsequently neglects modified habitats.

Although the role of exploited forests for biodiversity conservation is still debated today, many scientists are convinced that the ecological value of secondary forests is worth protecting. Similarly, even though croplands retain less diverse and abundant biodiversity than natural forests, it is important to ensure that these agricultural landscapes can retain some functional ecological role to guarantee a certain level of ecosystem services. Well-managed plantations can provide foraging resources and dispersal opportunities for numerous species. They can also provide important ecosystem functions.

Securing viable orang-utan populations in Borneo must incorporate the need to manage wild populations that are found outside of protected areas. This means in particular moving away from the original vision of pristine environments to study the numerous new challenges created by modern development in newly human-transformed landscapes. Reconciling people and wildlife, and identifying ways for local communities and displaced animals to cohabit peacefully, is the major conservation challenge in these fractured landscapes. In these fractured landscapes, both orang-utans and people are facing huge challenges, which are in many respects remarkably similar. In order to resolve these challenges innovative and adaptive behaviours are required. New balances need to be identified and adjusted to the ever-changing environment of the 21st century. This paradigm shift requires innovative and alternative conservation strategies involving new partnerships and new thinking.

BIBLIOGRAPHY AND REFERENCES

Ancrenaz, M, Dabek, L, and O'Neil, S, 2007 The cost of exclusion: recognizing a role for local communities in biodiversity conservation, *PLoS Biology* 5 (11), e289

Ancrenaz, M, Ambu, L, Sunjoto, I, Ahmad, E, Manokaran, K, Meijaard, E, and Lackman, I, 2010 Recent Surveys in the Forests of Ulu Segama Malua, Sabah, Malaysia, Show That Orang-utans (*P. p. morio*) Can Be Maintained in Slightly Logged Forests, *PLoS ONE* 5 (7), e11510

Ancrenaz, M, Sollmann, R, Meijaard, E, Hearn, A J, Ross, J, Samejima, H, Loken, B, Cheyne, S, Stark, D J, Gardner, P C, Goossens, B, Mohamed, A, Bohm, T, Matsuda, I, Nakabayasi, M, Lee, S K, Bernard, H, Brodie, J, Wich, S A, Fredriksson, G, Hanya, G, Harrisson, M, Kanamori, T, Kretzschmar, P, Macdonald, D W, Riger, P, Spehar, S, Ambu, L, and Wilting, A, 2014 Coming down the trees: Is terrestrial activity in orangutans natural or disturbance-driven?, *Nature Scientific Reports* 4, 4024, 1–4, DOI: 10.1038/srep04024

Ancrenaz, M, Oram, F, Ambu, L, Lackman, I, Ahmad, E, Elahan, H, Kler, H, and Meijaard, E, forthcoming *Of pongo, palms, and perceptions – A multidisciplinary assessment of Bornean orangutans in an oil palm context, Kinabatangan, Sabah, Borneo*, Oryx

Campbell-Smith, G, Simanjorang, H, Leader-Williams, M, and Linkie, M, 2010 Local attitudes and perceptions towards crop-raiding by Sumatran orangutans (*Pongo abelii*) and other non-human primates in Northern Sumatra, *American Journal of Primatology* 72, 866–76

Chape, S, Spalding, M, and Jenkins, M, 2008 *The world's protected areas: status, values, and prospects in the twenty-first century*, University of California Press, Berkeley

Dudley, N, 2008 *Guidelines for Applying Protected Area Management Categories*, IUCN, Gland, Switzerland

Gaveau, D, Kshatriya, M, Sheil, D, Sloane, S, Molidena, E, Wijaya, A, Wich, S, Ancrenaz, M, Hansen, M, Broich, M, Guariguata, M, Pacheco, P, Potapov, P, Turubanova, S, and Meijaard, E, 2013 Reconciling Forest Conservation and Logging in Indonesian Borneo, *PLoS ONE* 8, e69887

Harrison, T, 2010 Apes among the tangled branches of human origins, *Science* 327, 532–4

Hockings, K, and Humle, T, 2009 *Best Practice Guidelines for the Prevention and Mitigation of Conflicts Between Humans and Great Apes*, Occasional Paper of the IUCN Species Survival Commission No 37, Primate Specialist Group, IUCN, Gland, Switzerland

Lindenmayer, D, and Fischer, J, 2006 *Habitat Fragmentation and Landscape Change: an ecological and conservation synthesis*, Island Press, Washington DC

McShea, W, Stewart, C, Peterson, L, Erb, P, Stuebing, R, and Giman, B, 2009 The importance of secondary forest blocks for terrestrial mammals within an Acacia/secondary forest matrix in Sarawak, Malaysia, *Biological Conservation* 142, 3108–19

Meijaard, E, Sheil, D, Nasi, R, Augeri, D, Rosenbaum, B, Iskandar, D, Setyawati, T, Lammertink, M, Rachmatika, I, Wong, A, Soehartono, T, Stanley, S, and O'Brien, T, 2005 *Life after logging: reconciling wildlife conservation and production forestry in Indonesian Borneo*, Center for International Forestry Research, Jakarta

Meijaard, E, Albar, G, Nardiyono, Rayadin, Y, Ancrenaz, M, and Spehar, S, 2010 Unexpected Ecological Resilience in Bornean Orangutan and Implications for Pulp and Paper Plantation Management, *PLoS ONE* 5 (9), e12813

Meijaard, E, Buchori, D, Hadiprakarsa, Y, Ancrenaz, M, *et al*, 2011a Quantifying Killing of Orangutans and Human–Orangutan Conflict in Kalimantan, Indonesia, *PLoS ONE* 6 (11), e27491

Meijaard, E, Wich, S, Ancrenaz, M, and Marshall, A, 2011b Not by science alone: why orangutan conservationists must think outside the box, *Annals of the New York Academy of Science*, 1–16

Meijaard, E, Mengersen, K, Abram, N, Wells, J, and Ancrenaz, M, 2013 Without forest we are not well: Local perceptions on the importance of forests for people's livelihoods and health in Kalimantan, Indonesia, *PLoS ONE* 8 (9), e73008

Rijksen, H, and Meijaard, E, 1999 *Our vanishing relative*, Kluwer Academic, Dordrecht

Sodhi, N, Posa, M, Lee, T, Bickford, D, Koh, L, and Brook, B, 2010 The state and conservation of Southeast Asian biodiversity, *Biodiversity Conservation* 19, 317–28

Wich, S, Meijaard, E, Marshall, A, Husson, S, Ancrenaz, M, *et al*, 2008 Distribution and conservation status of the orang-utan (*Pongo spp.*) on Borneo and Sumatra: how many remain? *Oryx* 42 (3), 1–11

Wich, S, Gaveau, D, Abram, N, Ancrenaz, M, Baccini, A, *et al*, 2012 Understanding the Impacts of Land-Use Policies on a Threatened Species: Is There a Future for the Bornean Orang-utan? *PLoS ONE* 7 (11), e49142

26

Better to be a Beast than Evil: Human–Wolf Interaction and Putting Central Asia on the Map

Özgün Emre Can

Blessed is the face of the wolf…
You, whose sun rises when dark night falls,
Who stands up like a man in the snow and the rain…

– Dede Korkut

(The Book of Dede Korkut, epic stories of Oğuz Turks
transcribed in 13th century)

Large predatory mammals, destructive to livestock and game,
no longer have a place on our advancing civilization.

– E A Goldman

(Meeting of the American Society of Mammalogists, 1924)

On 1 April 2013, I was in the Langtang National Park in the Nepalese Himalayas. Located about 30km south of Tibet, covering an area of 1710km² and between 1300 and 7245 metres above sea level, Langtang is the first national park established in the Himalayas. However it is probably the least explored. In the company of two wildlife officers from Nepal's national wildlife authority, I was exploring the park to assess the presence of the clouded leopard and a few other large carnivore species. On this particular day, my mission was to visit the local people living a nomadic lifestyle within the park. After hiking for half the day, we finally reached a site where a group of families were camping. One of the families welcomed us to their tent and the three of us entered. The mother, wife of the livestock owner, and her two children were boiling milk on the fire. We sat on the floor and conversations started; I joined in as much as I could, with my companions translating from Nepalese to English and from English to Nepalese.

There was a word that our host, the father of the family, used frequently. Was it my mind making this up or could such similarity exist between the Nepalese language and my own native tongue, Turkish? I could not resist interrupting the conversation to ask one of my companions to write down this particular word in my notebook. He wrote the word in Latin letters so that I could understand it and returned the notebook to me. I was simply astonished.

What was written on the notebook was 'canavar' – the same word used by rural people in Turkey to describe wolves and other large predatory species. The Turkish word 'canavar' was read and used in the same way by this Nepalese family. This discovery simply astonished me, but what led a Nepalese nomad to use a Turkish word to describe wolves and similar predators? The Nepalese nomad was of Tibetan origin, and then I guessed that the word might have travelled

Fig 26.1.   Photograph of a wolf, taken in 2006 in Yeñice Forest, Karabük, Turkey's largest intact forest habitat.

from Central Asia, where Turkic-speaking peoples live, to the Tibet region, and then from Tibet to Nepal.

The word canavar ('beast' in English) reflects the human perception of wolves in this part of the world, and it does not sound like a very positive term. However, in Western culture, people's perception of wolves is much more negative, with wolves historically considered as evil creatures (see Marvin 2012). Why does the wolf have such a bad reputation? This chapter aims to answer this question. First, I will explore wolf origins and behaviour before then providing a global snapshot of human–wolf interactions. In doing so I will try to explain why wolves have such a negative image and why humans have historically referred to wolves as 'beasts' or 'evil'. Finally, I will highlight the need for future studies on wolves and human–wolf interaction in Central Asia.

What is a Wolf?

The wolf (*Canis lupus*) might be considered as the original dog (see Fig 26.1). Although the size and weight of a wolf varies across its range, between countries or within a given country, an average healthy adult wolf is not necessarily larger than a German shepherd dog (Boitani 2000). An adult wolf may weigh between 13 and 78kg, travel 72km per day and live up to 13 years in the wild, in environments where temperatures range between −56° and +50°C (Mech and Boitani 2003). Wolves live in packs, and this provides particular advantages in relation to hunting behaviour and strategy, enabling them to hunt much bigger prey than themselves. Wolves' primary prey species are large ungulates such as elk, reindeer, deer, wild boar and others but they also consume livestock, carrion and refuse, and may fast for months if food sources are scarce (Mech and Boitani 2003). On average, a wolf requires 3–5kg of meat per day (Boitani 2000). Wolves are usually territorial and may have territories that are as small as 33km² or as large as 2600km² (Mech and Boitani 2003). Pack size can vary from a few individuals to over 40 animals, though wolves can also live alone (ibid).

Historically, wolf populations ranged from the entire Northern Hemisphere including central Mexico, the Arabian Peninsula and southern India (Mech and Boitani 2003). Throughout history, in places where wolves and humans lived together, the two species competed for resources such as space and wild prey (Treves *et al* 2013). As the human population increased, human presence and activities extended into natural habitats, often leading to an increase in the extent and intensity of conflict. Over time this conflict resulted in wolves becoming extirpated from many parts of their historic range. As a result, humans have largely shaped wolf ecology and behaviour and have determined where wolves can/cannot exist. Human attitudes towards wolves were (and still are) complex, ranging from reverence to hatred (Fritts *et al* 2003). The wolf was sometimes evil, sometimes the beast. Today, wolf distribution is mainly confined to forest, tundra, mountain, desert and swamp habitats of North America, parts of Europe and Asia (Mech and Boitani 2003). The Wolf Specialist Group of the International Union for the Conservation of Nature (IUCN) considers the global status of the wolf as a species of 'Least Concern', an official status given to species that are widespread and abundant (Mech and Boitani 2010).

The wolf has been the subject of scholarly research in biology but also in diverse fields such as folklore, literature, psychology, sociology and history (Smith 2007). A quick search in the scientific literature database Web of Knowledge produces 5756 papers that have been published about wolves since 1864. This number may seem high, but during the same period 6500 papers were published on dung beetles – so perhaps wolves have not received the scientific attention they deserve. The first detailed field studies on wolves in the English-speaking world were conducted in the 1930s, but wolf research became a main scientific endeavour only in the 1960s (Fritts *et al* 2003). To date, most published studies on wolves are from North America and Europe, regions that host about 37% (61,000) and 9% (14,410) of global wolf population respectively. The Central Asia region, with 31% (50,000) of global wolf population – the second biggest wolf population after North America – has attracted very little attention for wolf research. Only 1% (around 70 papers) of wolf literature published since 1864 originate from Central Asia. As a result much of what we know about wolves and their interaction with humans is based on studies conducted in North America and Europe, the so-called WEIRD (Western, educated, industrialised, rich and democratic) societies (Henrich *et al* 2010). We know very little about the 50,000 or so wolves of Central Asia. This represents a potentially important gap in our current understanding of wolves and how they interact with humans, and raises the question: what do we really know about human–wolf interaction?

HUMAN–WOLF INTERACTION: WOLF AS AN EVIL ANIMAL

In the New World, the wolf was a respected legendary animal in Native American culture (Stout 2001; Fritts *et al* 2003; Marvin 2012). Wolves were hunted but they were also respected. According to Fritts *et al* (2003), ethnographic studies show that Native Americans were not in conflict with wolves. In Europe, wolves were first viewed positively in the mythology of the

Celts, Greeks and also in the mythology of Sabines, who regarded wolves as a totem animal (Stout 2001; Fritts *et al* 2003). Romulus and Remus, characters of Rome's foundation myth, were saved from starvation by a she-wolf who fed them with her milk (Hunt 2008). Later on, socio-economic relationships between human society and the environment determined the perception of wolves in Europe. According to Stout (2001), as humans started to see themselves as the master of nature, people believed they had the right to destroy whatever did not serve them, wolves became a symbol of threat to the Roman Catholic Church and finally became the 'evil species' in medieval minds (Fritts *et al* 2003; Mech and Boitani 2003). Supernatural powers were attributed to lone wolves, and werewolves were believed to be 'hell-possessed', 'devil-driven' and equipped with the power of Satan (Fogleman 1989). In medieval Europe, it was believed that evil power could turn innocent people into wolves, which were then called werewolves (ibid). In brief, the wolf was perceived as an evil-intentioned animal (Linnell *et al* 2002; Marvin 2012). Human fatalities attributed to wolves also understandably contributed to the fear and shaped human perceptions of the wolf. In Europe, the earliest record of a wolf attacking a human is from 1557 (in Germany), and more than 1539 people were killed by wolves in the 18th and 19th centuries (Linnell *et al* 2002). Efforts to exterminate wolves from various parts of Europe lasted for more than a thousand years (see Table 26.1). All possible methods were used to exterminate wolves. They were tracked on foot and by horses; hunted and killed by guns, traps, pitfalls, poisons, set guns, snares, spears and steel traps (Fuller *et al* 2003). These extermination efforts sometimes went to extremes, such as the burning of Caledonian pine forest in Scotland to aid wolf eradication (Mech and Boitani 2003).

European colonists also took their myths, folklore and fears to America (Fritts *et al* 2003). The same conflicts over issues such as livestock depredation, threats to game species etc occurred in the New World, and it was not long before the colonists initiated campaigns to exterminate wolves (Fogleman 1989). Extermination efforts started in the 1630s and constituted 'the longest, most relentless, and most ruthless persecution one species has waged against another' in history (Hampton 1997 cited in Fritts *et al* 2003). In fact, according to Brenton (2004), part of American heritage was defined as the nation's war against the wolf. However, almost all of what was known about wolves was based on folklore rather than biology. It has taken centuries for historically negative attitudes towards wolves to change. The early 1960s saw a renewed interest in wolf ecology but occurred during a time when wolf numbers globally were greatly reduced (Fogleman 1989). The US *Endangered Species Act of 1973*[1] signalled the end of campaigns to exterminate wolves.

Today, wolves are making a comeback in many Western countries. In the US, the wolf is under a mosaic of management regimes of federal, state and tribal authorities (Treves and Bruskotter 2011). The wolf's existence requires active management, which includes the payment of compensation for damage to livestock caused by wolves (Muhly and Musiani 2009; Steele *et al* 2013). In the USA people's tolerance of wolves is declining, which threatens the future of wolf populations (Treves *et al* 2013). In Europe, the wolf has also been making a comeback, though population size varies greatly by country, ranging from a few individuals (Austria 2–8 wolves; Germany 4 wolves; Switzerland 8 wolves) to several hundred (Finland 166 wolves; France 250 wolves; Poland 723 wolves) to a thousand or more individuals (Bulgaria 1000 wolves; Romania 2300 wolves) (see

1   See: http://www.epw.senate.gov/esa73.pdf [28 April 2014].

Table 26.1: A brief history of war with wolves. The current wolf population size is given in parenthesis if wolves came back from extinction by recolonisation in that particular country

6th century BC	First reports of planned extermination efforts by Solon of Athens
3rd century BC	Ancestors of modern Irish wolfhound dogs were bred for hunting wolves
9th century	An elite corps established for wolf control, remaining operative until 1787 in France (current wolf population size is about 250 individuals)
1272	King Edward I ordered the total extermination of all wolves in his kingdom
12th and 13th century	Wolf bounties were paid in Italy (current wolf population size is 667–867 individuals)
Early 16th century	The wolf became extinct in Scotland
Early 16th century	Last wolf was killed in England
1630	First bounty on wolves in European settlements in US
1647	Wolf bounty was introduced in Sweden (current wolf population size in Scandinavia is 260–330 individuals)
1770 or 1786	The wolf became extinct in Ireland
1772	Last wolf was killed in Denmark.
1818	'War of Extermination' declared against wolves in Ohio
1818–1884	Bounty systems established in Michigan, Texas, Colorado, Minnesota, Iowa, Wisconsin, Wyoming and Montana
1870–1877	Professional and civilian wolf hunters killed an estimated 100,000 wolves per year during this period in the US
1876	2825 wolves were killed in response to fatal wolf attacks on humans in India
1900	The wolf became extinct in Poland (current wolf population size is about 723 individuals)
1930	The wolf had disappeared from almost all 48 contiguous states of the US, including Yellowstone National Park (current wolf population size is 9000 individuals)
1937	The last wolf in France was killed (current wolf population size is 68 individuals)
1955–1961	Up to 17,500 wolves were poisoned in Canada (current wolf population size is 52,000–60,000 individuals)
1995–1965	Up to 2800 wolves killed in Romania (current wolf population size is about 2300 individuals)
1966	Last wolf killed in Sweden
1976	Last wolf killed in Norway (current population size is 46–49 individuals)

Data sources: Madonna 1995; Hickey 2000; Coleman 2003; Marie-Claire 2005; International Wolf Center 2013; Mech and Boitani 2003; Stanmore 2011; Large Carnivore Initiative for Europe 2004; European Commission 2012; 2013.

European Commission (2012) for wolf population sizes for European countries). The human–wolf conflict is still the main reason for wolf mortality in Europe (Boitani 2000).

In the Western world, the perceived threat from wolves exceeded the real threat, largely because the wolf had become demonised and was regarded as an evil animal (Fogleman 1989; Linnell *et al* 2002; Mech and Boitani 2003; Marvin 2012). This led to large-scale extirpations across the wolf range. Paradoxically, as wolves have been exterminated in many parts of their historical distribution range in North America and Europe, they have become an important symbol of nature conservation (Boitani 2003). However, the human–wolf conflict, whether perceived or real, still threatens the future of wolves in North America and Europe (Boitani 2000; Treves *et al* 2013). In summary, the wolf still carries its folklore burden of being an evil animal, and this heritage hinders management and conservation efforts in the Western world.

Human–Wolf Interaction: Wolf as a Beast

Unlike the case with North America and Europe, historical information about human–wolf interactions in Asia is very limited. It has been suggested that wolves have been exterminated from 80% of their historical range in China and India (Mech and Boitani 2003). In Japan, farmers considered wolves beneficial during the period AD 710–1867 as they preyed upon deer and other crop-damaging wildlife. In the 1600s, at the same time as wolves were being exterminated in Scotland and England and wolf bounties were in place in the US, Sweden and Italy, Japanese farmers were praying for wolves to kill the crop-eating wildlife (Fritts *et al* 2003). However, in the late 18th century, Western advisors (who had been brought to Japan to modernise agriculture) advised the use of poison to control the wolf population (Fritts *et al* 2003), and so the wolf, a holy animal for more than a thousand years, became the hated wolf in less than a hundred years.

It is generally believed that our perception of wolves is largely based on the experiences of prehistoric human societies' relationship with the environment, and that communities that were vulnerable to wolf depredation grew to hate wolves (Fritts *et al* 2003). However, I would argue that this is an oversimplification and that human–wolf interactions in Central Asia are complex.

The wolf is probably the most important animal in ancient Turkish mythology (Hyo-Joung 2005; Roux 2011) and probably the most respected animal of the Turkic-speaking peoples in Central Asia and the Caucasus (see Hunt 2008 and Kalafat 2009 for an overview). Evidence of the wolf's uniqueness in Turkish culture is threefold: first, the various legends of a clan primogenitor being born from a wolf or nourished by a wolf (such as the Ergenekon Epic); second, legends and other evidence of respect shown to wolves, as is evident in folklore (such as The Book of Dede Korkut, epic stories of Oğuz Turks, transcribed in the 13th century); and third, the desire by people to imitate wolves (such as the Orhon-Yenisey Transcriptions) (Esin 2001; Armutak 2002; Bayram 2006; Kalafat 2006; Hunt 2008; Kalafat 2009; Juraev and Nurmuradova 2010; Özkartal 2012; Koca 2012).

Today, the relationship between Turkic-speaking people and the wolf is unique and different from that of the Middle East and North Africa. This results from Turkic-speaking peoples' beliefs and practices being inspired by a cross-fertilisation of Sufism, Animism, Shamanism, Buddhism and Islam (Zarcone 2005). For example, a study in 1976 by linguists Vereschagin and Kostomarov showed that when the fable of 'The Wolf and the Lamb' by Ivan Andreyevich Krylov was dramatised in Russian and Kyrgyz schools, none of the Russian children wanted to play the wolf, whereas almost all of the Kyrgyz children wanted to play the wolf (Hunt 2008).

The word 'wolf' aroused different associations in Russian and Kyrgyz children (see Hunt 2008 for a snapshot of diverging perceptions of the wolf).

The rapid social changes that occurred in Central Asia following the dissolution of the Soviet Union in 1991 are likely to have had a negative impact on human–wolf relationships. Economic and logistical frameworks implemented by the state to support wolf hunting ceased and, as a result, livestock husbandry became more vulnerable to wolf attacks (Lescureux and Linnell 2013). However, it would be a mistake to think that ancient positive beliefs and attitudes towards wolves have totally faded today. Villagers in Kyrgyzstan still oppose the eradication of wolves despite the fact that human–wolf conflict has increased in the post-Soviet era (ibid). Similarly, in Turkey, people's belief systems still include traces of animal cults, and the mythological character of the wolf still continues. There exists, for example, a legend dating from 1922 which mentions a wolf's guidance to a commander on the battlefield during the Turkish War of Independence (Boratav 2012). As a result, and despite ongoing human–wolf conflicts, the Turkish public does not support sport hunting or any kind of lethal wolf control in Turkey (Can 2004). The wolf, although generally referred to as 'beast' by rural people, has never been considered an evil animal in Turkish culture. This is a likely major contributing factor to the survival of relatively large wolf populations in Turkey and Central Asia. Today, Turkey has a wolf population of about 5000–7000 (Can 2004) and Central Asia is home to one-third of the world's wolf population. Arguably, for wolves, it is better to be a beast than to be evil.

CENTRAL ASIA: THE PRIORITY REGION FOR WOLF RESEARCH

In more than one-third of the countries where wolves existed, they have either been exterminated or are on the verge of extinction (IUCN Wolf Specialist Group 2013). The story of Romulus and Remus does not have much resonance with Europeans today. Although it has been reported that the European attitude towards wolves is relatively positive (Milheiras and Hodge 2011), there is evidence that levels of tolerance of wolves might have been overestimated (Karlsson and Sjöström 2007). Negative sentiment in countries such as Sweden and Germany cannot simply be attributed to livestock damage caused by wolves, or to insufficient compensation measures (Stöhr and Coimbra 2013). Negative perceptions of wolves, as incorporated in traditional representations and practices, still have resonance in the 21st century, leading to tensions between humans and wolves in Europe (ibid). As a result, the ongoing conflict between humans and large carnivores is a challenge for wolf management and is probably the biggest obstacle for recovering wolf populations in Europe. In the US, there is evidence that acceptability depends on reducing both the perceived and real threats posed by wolves (Treves and Bruskotter 2011). While science-based management plans can mitigate most of the problems that wolves cause to people (IUCN Wolf Specialist Group 2013), the future of wolves also depends on broader human tolerance, and this future is uncertain while humans still consider wolves evil (Stout 2001). Throughout human–wolf history, a perception as 'evil' or 'beast' has meant the difference between being exterminated or tolerated.

In summary, most of our scientific knowledge of wolves is based on studies conducted in North America and Europe, and our understanding of wolves and human–wolf interaction is incomplete. The contemporary negative perceptions of the wolf in Central Asia seem to be related to current socio-economic conditions rather than the persistence of negative beliefs (Lescureux and Linnell 2013), as is the case in Western culture. As wolves have been suggested as models for understanding early humans (Fritts *et al* 2003), a better understanding of wolves within Turkic-speaking people's culture will not only enable us to construct the complete picture of human–wolf interaction, but it may also enable us to better understand the minds of our ancestors, and through them, ourselves.

ACKNOWLEDGMENTS

I would like to thank Tom Moorhouse and Ian Convery for their helpful comments on this chapter.

BIBLIOGRAPHY AND REFERENCES

Armutak, A, 2002 Animal patterns in the Eastern and Western mythologies: I. Mammals, *Istanbul University Journal of Faculty of Veterinary Medicine* 28 (2), 411–27

Bayram, B, 2006 The holy wolf and cyclop in Oğuz Tales and Çuvaş Alp tales, *Journal of Turkish World Studies* VI (1), 19–27

Boitani, L, 2000 Action Plan for the conservation of the wolves (*Canis lupus*) in Europe, *Nature and environment, No 113*, Council of Europe Publishing

— 2003 Wolf Conservation and Ecology, in *Wolves: Behavior, Ecology, and Conservation* (eds L D Mech and L Boitani), University of Chicago Press, Chicago

Boratav, P N, 2012 *Turkish mythology*, BilgeSu Publication, Istanbul

Brenton, D M, 2004 The role that fiction plays in our cultural response to the animal kingdom: an analysis of attitudes toward wolves in American Literature, unpublished MSc thesis, California State University, Dominguez Hills

Can, Ö E, 2004 *Status, conservation and management of large carnivores in Turkey*, Council of Europe, Brussels

Coleman, J T, 2003 Wolves in American History, unpublished PhD thesis, Yale University

Esin, E, 2001 *Introduction to Turkish cosmology*, Kabalci Yayinevi, Istanbul

European Commission, 2012 *Status, management and distribution of large carnivores – bear, lynx, wolf & wolverine – in Europe: Part 2*, December 2012, Brussels

— 2013 *Status, management and distribution of large carnivores – bear, lynx, wolf & wolverine – in Europe: Part 1*, March 2013, Brussels

Fogleman, V M, 1989 Attitudes towards wolves: A history of misperception, *Environmental Review* 13 (1), 63–94

Fritts, S H, Stephenson, R O, Hayes, D, and Boitani, L, 2003 Wolves and humans, in *Wolves: Behavior, Ecology, and Conservation* (eds L D Mech and L Boitani), University of Chicago Press, Chicago

Fuller, T K, Mech, L D, and Cochrane, J F, 2003 Wolf population dynamics, in *Wolves: Behavior, Ecology, and Conservation* (eds L D Mech and L Boitani), University of Chicago Press, Chicago

Henrich, J, Heine, S J, and Norenzayan, A, 2010 The weirdest people in the world?, *Behavioral and Brain Sciences* 33 (2/3), June, 61–83

Hickey, K R, 2000 A geographical perspective on the decline and extermination of the Irish wolf *Canis lupus* – an initial assessment, *Irish Geography* 33 (2), 185–98

Hunt, D, 2008 The face of the wolf is blessed, or is it? Diverging perceptions of the wolf, *Folklore* 119 (3), 319–34

Hyo-Joung, K, 2005 On the motive of turning into animals in the Myth of the Old Korean Kingdom Goguryeo, *International Journal of Central Asian Studies* 10 (1), 119–31 [in Turkish]

International Wolf Center, 2013 *Gray wolf timeline for the contiguous United States* [online], available from: http://www.wolf.org/wow/united-states/gray-wolf-timeline/ [15 December 2013]

IUCN Wolf Specialist Group, 2013 *Manifesto on wolf conservation* [online], available from: http://www.slashdocs.com/nvnsku/iucn-wolf-specialist-group-manifesto-on-wolf-conservation-2000.html [29 April 2014]

Juraev, M, and Nurmuradova, B, 2010 People's beliefs about wolf cult in Uzbek folklore, *Journal of Turkish World Studies* (X-1), 145–50 [in Turkish]

Kalafat, Y, 2006 The belief of binding the mouth of wolves in Turks, *Sosyal Bilimler Enstitusu Dergisi* 20, 273–80 [in Turkish]

— 2009 Turkish Mythology, considering with the currently old Turkish belief and Prof Dr Bahaeddin Ögel, *The Journal of International Social Research* 2 (8), 212–20

Karlsson, J, and Sjöström, M, 2007 Human attitudes towards wolves, a matter of distance, *Biological Conservation* 137 (4), 610–16

Koca, S K, 2012 Symbols in Turkish culture, unpublished PhD thesis, Social Sciences Institute, Sakarya University

Large Carnivore Initiative for Europe, 2004 *Status and trends for large carnivores in Europe*, Large Carnivore Initiative for Europe, Poland

Lescureux, N, and Linnell, J C, 2013 The effect of rapid social changes during post-communist transition on perceptions of the human–wolf relationships in Macedonia and Kyrgyzstan, *Pastoralism: Research, Policy and Practice* 3 (4), 1–20

Linnell, J D C, Andersen, R, Andersone, Z, Balciauskas, L, Blanco, J C, Boitani, L, Brainerd, S, Breitenmoser, U, Kojola, I, Liberg, O, Loe, J, Okarma, H, Pedersen, H C, Promberger, C, Sand, H, Solberg, E J, Valdmann, H, and Wabakken, P, 2002 The fear of wolves: A review of wolf attacks on humans, *NINA Oppdragsmelding* 731, 1–65

Madonna, K J, 1995 The wolf in North America: Defining international ecosystems vs. defining international boundaries, *Journal of Land Use and Environmental Law* 10 (2), 2–38

Marie-Claire, L, 2005 The story of the wolf in France [online], available from: www.loup.org [19 December 2013] [Original document in French translated to English using Google Translate]

Marvin, G, 2012 *Wolf*, Reaktion Books, London

Mech, L D, and Boitani, L, 2003 Wolf social ecology, in *Wolves: Behavior, Ecology, and Conservation* (eds L D Mech and L Boitani), University of Chicago Press, Chicago

Mech, L D, and Boitani, L, (IUCN SSC Wolf Specialist Group), 2010 *Canis lupus*, in IUCN 2013, IUCN Red List of Threatened Species, version 2013.2 [online], available from: www.iucnredlist.org [12 December 2013]

Milheiras, S, and Hodge, I, 2011 Attitudes towards compensation for wolf damage to livestock in Viana do Casrelo, North of Portugal, *The European Journal of Social Science Research* 24 (3), 333–51

Muhly, T B, and Musiani, M, 2009 Livestock Depredation by Wolves and the Ranching Economy in the Northwestern US, *Ecological Economics* 68, 2439–50

Özkartal, M Ö, 2012 A general view over animal symbolism in Turkish epics (Samples from the Book of Dede Korkut), *Milli Folklor* 24 (94), 58–71 [in Turkish]

Roux, J, 2011 *Ancient Turk mythology*, BilgeSu Publishing, Istanbul

Smith, D M, 2007 What is a wolf; The construction of social, cultural and scientific knowledge in children's books, unpublished PhD thesis, University of Illinois at Urbana-Champaign

Stanmore, I, 2011 *The Disappearance of Wolves in the British Isles* [online], available from: http://www.wolfsong alaska.org/British_Isles [19 December 2013]

Steele, J R, Rashford, B S, Foulke, K T, Tanaka, J A, and Taylor, D T, 2013 Wolf (*Canis lupus*) predation impacts on livestock production: Direct effects, indirect effects and implications for compensation ratios, *Rangeland Ecology and Management* 66 (5), 539–44

Stöhr, C, and Coimbra, E, 2013 The governance of the wolf–human relationship in Europe, *Review of European Studies* 5 (4), 1–19

Stout, G R, 2001 Wolves and wilderness: A mythological and psychological approach, unpublished PhD thesis, Pacifica Graduate Institute

Treves, A, and Bruskotter, J T, 2011 Wolf conservation at a crossroads, *BioScience* 61 (8), 584–5

Treves, A, Naughton-Treves, L, and Shelley, V, 2013 Longitudinal analysis of attitudes toward wolves, *Conservation biology* 27 (2), 315–23

Zarcone, T, 2005 Stone People, Tree People and Animal People in Turkic Asia and Eastern Europe, *Diogenes* 52, 35–46

27

After *nanoq: flat out and bluesome: A Cultural Life of Polar Bears*: Displacement as a Colonial Trope and Strategy in Contemporary Art

Bryndís Snæbjörnsdóttir and Mark Wilson

nanoq: flat out and bluesome is a project by Snæbjörnsdóttir/Wilson that began as a survey of displaced taxidermic polar bears in the UK conceived with a view to restoring specific and discrete histories to relics whose purpose had hitherto been generic and symbolic.

Since completion of the project *nanoq: flat out and bluesome* (2006), the photographic archive from the survey has gone on continuous tour of a host of zoological, maritime and polar museums in northern Europe, including those within the Arctic region itself, such as in Longyearbyen, Svalbard and Tromsø, Norway. One of the prime ambitions of the project is to bring singularity to the remains of specimens whose individual, cultural purpose has been to act as representative for a species – and sometimes, even more generically, its environment. In addition there are those specimens in private hands which function as company mascot, conversation piece and inevitably, hunting trophy.

Beginning my discussion on the subject of animals and their presence in art from this very work [*nanoq*] is important, as [it] could, from multiple perspectives, be understood as a quintessential work thus far produced within the discourse of human–animal studies. […] Snæbjörnsdóttir and Wilson have developed an international reputation for the creation of conceptual works of art that simultaneously embody the core and transcend the boundaries of human–animal studies' strictly academic discourse in order to communicate to wider audiences. (Aloi 2012, s74)

For the artists, *nanoq* first and foremost concerns the issue of representation and how representation itself must always be a depletion and distortion of that which is represented. Historically, by removing the bears from the arctic and populating museums with these and similar colonial plunder, the will to construct self-congratulatory narratives through the display of the 'tamed wild' is clear. In 2004, when ten polar bear specimens were purposely displaced again for the eponymous installation at Spike Island in Bristol, their mutual effect upon each other was to simultaneously cancel out their representational and iconic currency and render visible as counterfeit the promises they had been called upon to convey. This chapter explores these contradictions and other readings of the project, of which Hunterian Museum director Dr Sam Alberti has stated led UK zoological museum curators to reappraise their approach to the collections in their charge.[1]

The significance of *nanoq: flat out and bluesome* hinges fundamentally on the concept of displacement – initially and specifically in the displacement of polar bears from their natural environment in various parts of the arctic region and into municipal and private collections. The initial acts of appropriation brought them either as living creatures into zoos, or as skins and, thereafter, into museums as taxidermic specimens. More generally, *nanoq* examines the simple act of displacement of 'material' from one place to another, in the form of trophies, souvenirs or objects of curiosity, as proof of a world beyond one's own, which act, depending both on intention and interpretation, transforms those who appropriate into colonial thieves, mercenaries, conquering heroes, philanthropists or cultural benefactors.[2] Through the second displacement both physically and photographically, in the service of art, the artists wished to test how prevailing taxonomies and hierarchies in Western thought, as presented in museum and private collections, could be productively disrupted as a way of imagining a more networked constitution of this specific material.

The intention was to intersperse the individualised images of bear specimens back into zoological and polar collections, in order to raise questions pertaining to singularity and individuation, between the layers of established and accepted knowledge-accretion, traditionally predicated on 'generic' representation. By redeploying prepared animal remains and re-presenting them, for instance in the light of their own specific provenances within museum contexts, the idea was to productively disrupt the symbolic role of such remains, shifting the emphasis of their meaning to one of discrete, networked singularity. How would the perspective allowed by this accumulation of singular accounting enable a new reading of colonial enterprise and polar exploration historically, and what bearing might this have on contemporary approaches, for instance, to the arctic as both habitat and environment? Through *nanoq: flat out and bluesome* the artists wanted to explore the relationship between taxidermy and photography and ask how the serial resituating and site-responsive presentations of *nanoq* would prompt a cumulative reappraisal of contemporary assumptions regarding taxonomy, polar history, wildness and environment – not to mention taxidermy.

After the completion of the project, in an uninterrupted series of showings of the *nanoq* photographic archive from 2006 to 2012, the artists continued to weigh the meanings of the project and, in the light of recent controversies surrounding contemporary environmental iconography, came to question through a series of further related exhibitions the cultural durability and perishability of symbols used.

The complete photographic archive from the artists' survey of the bears *in situ* comprises 34 framed, colour images taken in their respective public and private collections and settings. The provenance of each specimen is incorporated into the work, either as part of the image in the form of text at the bottom of the photograph's white margin or, in the case of the larger prints, engraved into a brass plate inserted into the bottom section of the hardwood frame. By

[1] Dr Alberti made this observation to the *Cultures of Preservation* conference, hosted by the Natural History Museum, London, in May 2011.

[2] The first polar bear (or 'white bear', as it was reported) displacement on record in England was a gift to Henry III in 1252 from Haakon IV of Norway.

dislodging the historical role of the taxidermic mount from its position as representative of a species and environment to one strategically less clearly inscribed and consistent, conventional expectations are disrupted, potentially to transform the meaning of the specimen. In a process that returned a specific history to each specific (ex)animal, a previous condition of generality was supplanted. By distinguishing the specimens as individuals and setting off a chain of new readings, ripples were sent through a set of historical and contemporary fields – museology, polar history, hunting, environmentalism, anthropology and so on. In this configuration, a constituency of polar bear specimens was assembled that hitherto, if ever, had existed only notionally.

A significant and noteworthy element of the *nanoq* photographs is the context within which each specimen had come to reside. From one perspective, when the complete archive is presented, owing to the repetition or constant of the bear's appearance it is possible to examine the images for this variance quite separately. A typical setting from which the specimens were sourced was a tableau of stolen 'exotica', the subtext of which is colonial conquest. It is rare in such a display for any direct allusion to the provenance of the specimens to be present. It is this characteristic cultural omission that prompted the artists to reintroduce the work back into zoological collections in order to identify and interrogate the presumptions and taxonomic behaviour traditionally at work in such institutions. As bastions of knowledge and public education, museums have been profoundly influential on collective readings of the world. The principle of the art intervention was as simple as it was quietly ambitious. Having attracted the viewer's attention and driven a wedge under the foundations of classification by using something as iconic as an ice bear, it would have the potential, at least, to destabilise everything within that context – or at least render open for review the taxonomic subject across multiple fields within the museum.

The juxtapositions of artefacts are often, as one might expect, bizarre, as the following examples will serve to testify. In Fyvie Castle in Aberdeenshire, for instance, the half-bear table specimen (together with a similarly truncated seal mount) is surrounded by an assortment of 15th-century armour, weaponry and zoological specimens, both indigenous and exotic. In Blair Atholl in Perthshire, the specific assortment of objects gathered at the bottom of the stairwell is striking. The ancient polar bear mount is flanked on one side by a curtain of Royal Stuart tartan and on the other by a suit of Samurai armour (see Fig 27.1).

In the case of the Halifax bear (see Fig 27.2), the contrivance of the museum's 'attic' installation functions conspicuously in more ways perhaps than intended. The effect is supposed to be one of neglect and nostalgia, a Victorian time capsule, suggesting the detritus of another age. In bringing a particular focus upon the polar bear, it became one of the most resonant discoveries of the survey. The 'forgotten' bear is starkly different in that rather than being representative, typically for instance of 'species', of an arctic environment or of a powerful or aggressive predator, this specimen is representative of redundancy, obsolescence and 'junk'. In the context of a project that among other things sets out to prompt a reappraisal of contemporary assumptions regarding taxonomy, polar history, wildness and environment, this singular find manages simultaneously to drill with significance into all these matters and more. And in relation to the general fusion of tragedy and comedy intrinsic to the photographic archive, the presence of this image assumes a particularly destabilising pathos.

Each new exhibiting context requires the archive to be curated and adjusted specifically in relation to the host collection. The venues vary from arctic and natural history collections (eg the Scott Polar Research Centre, Cambridge; the Fram Museum, Oslo; the Oxford University Museum of Natural History) to art galleries. In 2008 the artists were invited to develop

Fig 27.1.   Blair Atholl, 2004, Snæbjörnsdóttir/Wilson, Lambda Print.

Fig 27.2.   Halifax, 2004, Snæbjörnsdóttir/Wilson, Lambda Print.

the project in the context of the international exhibition *HEAT: Art and Climate Change* in Melbourne, Australia. There, with the help of the organisers of the show a survey of taxidermic polar bears in Australia was conducted leading to an artwork entitled *Polar Shift*.

nanoq: flat out and bluesome references and deploys a number of strategies drawn from a range of other models – from contemporary exhibitions to the methodologies of other specialist fields. Of particular influence was the exhibition *Private View* (1996), curated by Penelope Curtis and Veit Görner in which the Bowes Museum collection of French decorative arts had been sensitively and provocatively interspersed with contemporary artworks, including examples by Thomas Grünfeld, Damien Hirst and Anja Gallaccio. The artists had a rather special experience of the show, viewing it while the finishing touches were being put in place and before the exhibits were labelled. Within this distinctive and often bizarre historical collection this meant that there was a compelling uncertainty as to which exhibits were the temporary inquisitors and which the hosts.

The installation of ten polar bear specimens at Spike Island in Bristol (see Fig 27.3) depended on spectacle both for initial and lasting effect. A former light-industrial space, Spike Island is typical of many contemporary art spaces that have emerged over the last 20 years in its reimagining, as a place for art, a site with an entirely other history. In bringing the specimens together the plan was to supplant the often misleadingly simplistic, tidy and managed narrative displays discussed above, with something unexpected, surprising and much more difficult to assimilate. By stripping the specimens of any contextual supplementary information (diorama, props, texts, etc), the intention was that beyond the context of the gallery space and the vitrines in which they stood, the bears themselves would be the prevailing context for each other. Any clue that might, coincidentally or otherwise, serve to qualify this closed network of signs became eclipsed by the dominant spectacle of the specimens. The visitor was presented with something instantly recognisable and, owing to its plural constitution, simultaneously uncanny, prompting her/him

FIG 27.3.   *nanoq: flat out and bluesome*, 2004, SNÆBJÖRNSDÓTTIR/WILSON, SPIKE ISLAND, BRISTOL.

to reach (inadequately) for the references that would allow the image or experience to settle. What do I know of polar bears? What indeed is a polar bear? Are these polar bears? The specimens were anything but confirmatory of each other – the differences between them were striking, and lest the viewer was by this time coaxed into reading this as an exotic natural history display, this inconsistency served as a reminder of their profoundly constructed (and therefore representationally unreliable) nature. The immeasurable cultural identities, orientation, conditioning and confinement of the human visitor, complicit in the acts and practice of collection that for so long underpinned the colonial accrual and narratives of knowledge, served further to extend the work. Here the legacy tropes of enlightenment thinking were simultaneously summoned, destabilised and critiqued.

There were other objectives to be addressed in the installation. To one side of the warehouse space a wall was built, upon which an inverted, perspectival map of the British Isles was rendered in (vinyl) line. The reference to Tony Cragg's *Britain Seen from the North* (1981) was a conscious allusion but one that was instrumentalised to make a point specific to the erstwhile plight of displaced, slaughtered or captive bears arriving from the arctic as by-products of their respective expeditions. Alongside this was a column of text providing information on the ten bears included in the installation but providing no indication as to which provenance belonged with which mount. At the far end of the gallery space was a specially built seating podium, designed to serve the programme of lectures, seminars and a conference taking place each day during the period of the exhibition, as well as acting as a landscape reference to the original environment of the specimens. Discreetly back-projected on one side of this construction was a video work documenting the painstaking removal of bears from their respective collections, their journey to the exhibition and their installation in this space.

In addition to the programme of daily events, the artists organised a one-day conference (*White Out*) at which four invited speakers presented papers. Subsequently, together with the audience, issues were discussed around the many associated themes prompted by the project including museology and display, taxidermy, photography, the colonial impulse, arctic exploration, the whaling industry, subsistence and trophy hunting, and shifting attitudes to environment. A publication presenting all of the information gathered during the project – the provenances, the photographic archive, documentation of the installation – included essays from those speakers and writers who took part in the conference.

There is tacit acknowledgment that much contemporary art functions laterally across the specialist disciplines of others, drawing on their methodologies and putting them and their knowledge to other, new work, thereby simultaneously referencing, realigning and subverting such knowledge. This process, familiar in principle to artists and art historians as *détournement*,[3] mobilises such non-art practices and models, for example those in science or commerce, in the service of art and is at once both mindful and subversive of these borrowed methodologies (Debord 1967, 206). Similarly, during the development of *nanoq*, among the institutions and individuals with whom participation and to some extent collaboration was crucial, there numbered museums, keepers of collections, taxidermists, artists, hunters, anthropologists, art academics and the public.

---

3   *Détournement* reradicalises previous critical conclusions that have been petrified into respectable truths and thus transformed into lies.

For the conference and the publication, the selection of participants was based on their suitability to address certain aspects embedded in the project, such as contemporary uses of animals in art (Steve Baker), the relationship between photography and taxidermy (Michelle Henning), taxidermy as trophy (Garry Marvin) and subsistence hunting in the arctic (Ivars Sillis).

Since the completion of the *nanoq* project and through its numerous manifestations – the installation, the photographic archive, the conference and the book – the project has been interpreted and read in a variety of ways and continues to be analysed in papers, articles and chapters in museum journals, art publications, animal studies periodicals and academic publications.[4] Giovanni Aloi has written:

> Most importantly for the context of contemporary art in which the work operates, *nanoq* aims at adding a critical dimension to the resurfacing of taxidermy, a revival that brought pre-served animal bodies centre-stage of the mainstream scene through the work of Damien Hirst, and most recently many other artists (Tessa Farmer, Steven Bishop, Vim Delvoye, Oleg Kulik, and Polly Morgan, to name a few). The counterpoint *nanoq* offers to viewers is crucial in the problematisation of a phenomenon that may otherwise only be understood as a hollow fashion rather than the complex one it really is. In other words, *nanoq* functions as a 'problematiser' to any other contemporary work of art employing taxidermy as a medium. In doing so, it embraces human-animal studies arguments on the representation of the animal in art, simultaneously questioning the solidity of its very roots. (Aloi 2012, s76)

These interdisciplinary qualities and ploys are (perhaps not always consciously, but necessarily) staple ambitions within the field of animal studies. Further to this, a crucial aspect of appeal to academic researchers whose work is concerned with relationships between humans and other species is that contemporary art, while privileging non-linguistic methods and deploying, as it does, retinal, aural and other sensory means, finds other, less obviously partisan resources than text in order to test the world and carry its ideas. The dominance of language, its refinement, nuanced nature, flexibility and authority has been blinding, or perhaps deafening, in its effect: its tendency towards reductionism and absolutism has been significant in shaping the direction of our thought regarding the world, the environment and our relationships to it (Snæbjörnsdóttir and Wilson 2010, 211–26). Specialist knowledge, while adept at drilling down, has never intrinsically best served the relational – the matrix of connections between things, our knowledge of them and their potentialities – as an alternative and more holistic way of reading our position in the world and our relationship to those with whom we share it. Amongst the many aspects of its fascination, the field of animal studies has a broader implicit functional effect, as does in part, we believe, our practice; by pulling focus on significant historical oversights – the valuable otherness of others, the error of a perceived moral and evolutionary superiority in the world – we may move from an aggressively suffocating and ultimately both suicidal and zoo-cidal anthropocentric perspective

---

4 Most recent of these is Rachel Poliquin's 2012 volume *The Breathless Zoo: Taxidermy and the Cultures of Longing*, in which *nanoq: flat out and bluesome* is the subject of the book's Introduction (pp. 1–5). A new Thames & Hudson publication devoted to taxidermic representation is currently also being prepared which will feature this work from Snæbjörnsdóttir/Wilson.

to one where, within the ecosystem, we acknowledge our symbiotic potential. Expanding on her term 'geopathology', by way of the project *nanoq: flat out and bluesome* Una Chaudhuri observes:

> These photographs – displayed at various museums and printed in a beautiful book about the project – are exemplary documents of what might be called a zoo-geopathology: the infliction – by humans, on the other animals – of the vicissitudes of displacement. Leafing through these brilliant photographs is like journeying through the very definition of the uncanny – in its etymological sense of 'the unhomelike': the oddly estranged, the strangely out-of-place.
>
> (Chaudhuri 2012, 47)

The research that is positioned between disciplines, as is our own, must also negotiate and identify a role in managing not only the incongruities that may exist between the views from such discrete fields, but also those aspects altogether *disallowed* by the rational and analytical frameworks of those disciplines, eg anthropomorphism, sentiment, compassion – all attributes and tropes that will not readily submit to the calibration of scientific scrutiny. Pavel Büchler cautions that '…the more we overlap in our work with the practices of other fields – the more we trespass on others' territories – the clearer and more specific we need to be about our specialist identities and roles to get away with it…' (Büchler 1999, 44).

We would argue that it is in the juxtaposing of perspectives and the management of incongruity that art radicalises conventional givens. But in an article in which our work is discussed, Emily Brady references the shifting emphasis of environmental art away from its early tendencies of imposition towards more situated concerns of ecology and inter-relationality: 'It might seem that the criticisms raised against early forms of environmental art will have less relevance given that environmental art has been motivated by ecological concern more and more' (Brady 2010, 54–5).

This seemingly innocuous comment unwittingly illustrates the ways in which strategically pluralist artistic intentions may themselves be selectively cherry-picked. In this process, art projects too may be subject to hijacking by theoretical or representational contraction. Another case in point is when in 2006/7, two years after the first showings of *nanoq* and indeed six years after its conception, the national press (including *The Times*, *New Scientist*, the *Guardian*, *Time Out*, *Telegraph*, *Daily Mail* etc) seized on it as a work primarily concerned with arctic (and therefore global) environmental decline. In the six years between *nanoq*'s inception (2000) and exhibition at the Horniman Museum in London, which prompted the flurry of newspaper articles, the function of the polar bear in Western popular consciousness had dramatically changed, becoming as it did a motif or even (vulgarly) the 'poster child' for global warming. This environmental, ecological dimension was one, but only one, of many embedded in the project, which by the caricaturisation of this interpretation became (in those articles at least) eclipsed and overlooked. There is no question that the work we make is political and consciously so. Acceptance of or identification with our work by established groups or interested parties may well be well-intentioned, but such attention may privilege a specific dimension within the work to the exclusion of others, thereby unpicking the cohesion upon which its real integrity, as art, depends. And yet for as long as we foster, for instance, sympathies for ecological concerns ourselves, it seems churlish and even in itself misleading for us to rebuff such simplified and uncluttered readings.

Since that time, Una Chaudhuri (2012) has written of *nanoq*:

The artists themselves characterize [the polar bear specimens in] their project as 'a notional community', made up of 'animals that had shared a similar fate'. In the age of climate change, that shared fate includes that of the human animals wandering the gallery space, turning that space, suddenly, into a space of ecological consciousness and – possibly – a platform for action... [and] *nanoq*'s incarnation as gallery installation opens a space of performance that I call 'the theatre of species', naming an emergent performance practice of our times. Climate change, which turns familiar sites into landscapes of risk or disaster, also reminds us that we humans are one species among many, among multitudes, all equally contingent and threatened. [...] The theatre of species addresses what we could call a 'zoögeopathology' – the planetary health emergency that is challenging the anthropocentric geographies we have lived by for so long. (Chaudhuri 2012, 50)

Art deploys strategies essentially pluralist in their presentation of questions through material, context and situation. Art embraces complexity and uncertainty, and exploits these qualities in an invitation to reappraise what we think we know. Displacement and *détournement* are very much a part of this process. It is equally a fact that journalism, for instance, also has its own motivations and drivers, which so often demand that complexity is sacrificed on the altar of directness and public accessibility. When a geographer or an anthropologist or a writer steeped in the practices of performance cites an artwork, a new focus is brought to bear and potentially, something different is revealed. This possibility is not an accidental effect. As artists, with confidence in the processes of art research and practice, we believe that art, far from being autonomous or separatist, has work to do in relation to the world of the day. Anticipating the way in which art may function as an instrument to be picked up and used by others, it is our strategy and indeed characteristic of much socially engaged art practice to involve some of those other parties from an early stage. This involvement brings integrity and range to projects and allows them to be rooted in diverse fields of knowledge and to draw sustenance from them. In this way art is increasingly acknowledged as a sophisticated, dynamic and effective research tool, integral to which is its capacity to hold in balance disparate narrative and analytical threads. However, where specialist critics and reviewers of art might well seek to plumb the artwork on multiple levels, the farther it is displaced – that is, taken from the context of art – the more likely it is to be stripped of its more networked integrity. In such circumstances it is to be desired that such new contexts of the work will act, in the spirit of the art, to complicate the scrutiny of individual meaning-strands, rather than be mobilised towards the facile and symbolic. In this sense, displacement, rather than being an act of theft, is seen as an agent for cognitive and cultural adjustment – a process of realignment by which new thought is made possible.

BIBLIOGRAPHY AND REFERENCES

Aloi, G, 2012 Deconstructing the Animal in Search of the Real, *Anthrozoös* 25, s73–s90

Brady, E, 2010 Animals in environmental art: relationship and aesthetic regard, *Journal of Visual Art Practice* 9 (1), 47–58

Büchler, P, 1999 Other People's Culture, in *Curious: Artists' Research within Expert Culture* (ed S Brind), Visual Arts Projects, Glasgow

Chaudhuri, U, 2012 *Readings in Performance and Ecology*, Palgrave Macmillan, New York

Debord, G, 1967 *The Society of the Spectacle*, Buchet Chastel, Paris

Snæbjörnsdóttir, B, and Wilson, M, 2010 The Empty Wilderness, Seals and Animal Representation, in *Conversations with Landscape* (eds K Benediktsson and K Lund), University of Iceland, Reykjavik, 211–26

28

What Heritage? Whose Heritage? Debates Around Culling Badgers in the UK

Pat Caplan

This chapter concerns some of the debates centred on the recent proposals by both the English and Welsh governments to cull badgers in an attempt to lessen the incidence and prevalence of bovine TB (bTB) in cattle. It discusses some of the local meanings of 'heritage' and 'nature', and shows that both are highly contested terms in which different categories of people seek to appropriate ownership of them. It draws most particularly on fieldwork conducted in west Wales (one of the 'hot-spots' for bTB) with farmers, anti-cull activists and others, both Welsh- and English-speaking, and also makes use of local and national media, scientific reports and children's literature in order to seek to understand the varying positions taken on the proposed culls.

In two earlier articles, which focused on the proposed badger cull in west Wales between 2009 and 2012, I sought to discuss the complexities of the situation. The first paper (Caplan 2010) considered some local reactions to news of the cull, ranging from those who supported it wholeheartedly to those who argued against it, as well as a number who were somewhere in between. It also discussed some of the reasons why people adopted such differing standpoints, showing that a range of issues helped explain why people thought as they did. These included perceptions of risk, ethnicity, ideas about animals and wildlife, the selective uses of both scientific literature and emotional arguments, and finally some of the local and regional politics involved.

The second paper (Caplan 2012) continued the story of the proposed cull in Wales from mid-2010, when there were confrontations and arrests, to the effects of the changed political environment after the 2011 Welsh elections, when the policy was switched from one of culling badgers to vaccinating them. The paper argued that epidemics like bTB are a way of talking about a wide range of issues, including the state (at varying levels) and its policymaking, human–animal relations, and farming and conservation.

In this chapter, I consider some of the ways in which a disaster like bTB is dealt with at the level of individuals, communities and governments, as well as how the idea of 'our heritage', in the sense of landscape and the survival of both farmed and wild animals, is appropriated and contested, and how each side in the badger cull debate considers itself and 'its' animals to be the ones suffering in both social and individual terms. Again, the focus will be on north Pembrokeshire, the site of fieldwork, with occasional comparative references to the situation in England.

Farming and Bovine TB

For farmers, especially dairy farmers, bovine TB has in recent years become a major disaster in Wales and England, although some areas are more badly affected than others. Government policy is one of regular testing, and animals that are 'reactors' (ie test positive) are taken away for slaughter, although post-mortem examination reveals that not all are actually infected with bTB.[1] The farmers then receive compensation[2] but have to wait some time before being able to restock their herd and, during the period of restriction, they may not sell any animals either. Bio-security on farms has been considerably enhanced, with further restrictions on movement of animals and more frequent testing. Many farmers find this regime onerous, and few consider the compensation they receive in the event of a herd breakdown to be sufficient. Thus bTB increases the economic burdens on a farming community already stressed by factors such as the exceptionally bad weather of the past few years and the drop in the price of many farm products (especially milk).

There is much debate in the scientific community about how bTB is transmitted, but there is no doubt that badgers (*Meles meles*) can also be infected by the disease. Badgers have been a protected species in the UK since 1973,[3] but many are said to be killed illegally by farmers who consider their numbers to have increased too greatly since they were placed under legal protection, or who are concerned about the badgers possibly infecting their cattle. The main farming unions, the National Farmers' Union (NFU) and the Farmers' Union of Wales (FUW), both advocate the culling of badgers, whom they regularly describe as a 'reservoir of disease', and on the whole this policy has been supported by the main farming newspapers, the *Farmers' Guardian* and the *Farmers' Weekly*. Nonetheless, fieldwork in west Wales and scrutiny of the many media reports both there and in England suggest that not all farmers are in favour of this measure.

In the 1990s the then Labour government in Westminster instituted a large-scale set of trials (the Krebs trials) in England, and in 2007 a report was published by the Independent Scientific Group, which suggested that large-scale culling would have only a small impact on the prevalence of bTB in cattle. This was largely because of the so-called 'perturbation effect', which means that badgers, normally localised for long periods in their underground setts, will, when disturbed, move to other territories, thereby possibly spreading bTB to new areas. Shortly thereafter, however, the UK government's Chief Scientist, Sir David King, used both the Independent Scientific Group's (ISG) final report and a number of other studies to argue *for* a cull of badgers (King 2007), but the then Minister for the Environment (Hillary Benn) decided in the light of the scientific evidence available that there should be no culling of badgers.

Such diversity of views in the scientific community has continued, and the extent to which it is possible to use scientific data to argue for completely different policies is very striking, as I found from my fieldwork, and as other scholars have noted (Wilkinson 2007). In short, science is at present far from being able to provide clear-cut and unequivocal answers.

---

1 See: http://www.rethinkbtb.org/a_failure_of_policy.html [10 June 2013].
2 The cost of compensation to the Welsh Assembly Government (WAG) rose from £1.8m in 2000/01 to £15.9m in 2007/08. According to the WAG, it is reckoned that this cost could exceed £80m by 2014 (www.wales.gov. uk/bovineTB), although this latter figure has been hotly disputed.
3 The *Badgers Act 1973* made it an offence to attempt to kill, take, injure badgers or interfere with their setts without a licence. The *Protection of Badgers Act 1992 (c. 51)* consolidated the *Badgers Act 1973*, the *Badgers Act 1991* and the *Badgers (Further Protection) Act 1991*. See: www.opsi.gov.uk/ACTS/acts1992/ukpga_19920051_ en_1 [10 June 2013].

A Short History of Proposed Badger Culls in the UK

In parts of Wales meanwhile, the problem of bTB in cattle was also intensifying and, with increasing devolution of powers to the Welsh Assembly Government (WAG – later the Welsh Government), it was decided in 2009 that there would be a trial cull of badgers in one of the 'hot-spots' for bTB, north Pembrokeshire. It was the King Report that was used by the Welsh Assembly Government to justify the decision to cull all badgers in the 'intensive action' pilot area (IAPA) of North Pembrokeshire[4] – a prime example of the tendency of both sides to 'cherry-pick' their use of scientific arguments in this debate.

The proposed cull was supported by both Coalition parties in the Welsh Assembly Government, Plaid Cymru (the Welsh nationalist party) and Labour, and also by most of the Conservatives. Elin Jones, the Minister for the Environment, herself a farmer's daughter, maintained that a cull of badgers was the only way to begin to make some impact on the disease, and in this she was supported by her Chief Vet. It was argued frequently that 'all the tools in the toolbox' had to be used to bring bTB under control, and that the need for the cull was based on the 'fact' that badgers constituted a 'reservoir of disease'.

In August 2009 an opposition group, Pembrokeshire Against the Cull (PAC), was formed and soon began organising protests and set up email listings and a website. Around Christmas 2010, PAC held a 'badger dash' over the Preseli hills, which was filmed and put up on *YouTube*. Like the pro-cull side, PAC argued that its case was supported by science, making frequent use of the Independent Scientific Group's report. It also strongly advocated that a cull should be replaced by vaccination of badgers.

By 2010, the Welsh Assembly Government had passed a series of legal measures that would allow access to private land, whether landowners agreed or not, for the purposes of identifying badger setts and carrying out a cull in the Intensive Action Pilot Area. This was to be done by trapping badgers in cages and then shooting them. The sett surveys began in May of that year, and immediately led to major confrontations between the contractors carrying out the surveys, who all wore balaclavas and face masks and were accompanied by large numbers of police, on the one hand, and owners of farms and small-holdings who did not agree with the cull, on the other. Several people were arrested, and a film of one confrontation was put on *YouTube*. The badger cull had now also become an issue of civil liberties.

In the meantime, the Badger Trust, which has branches all over England and Wales, sought a judicial review of the Welsh cull. Leave to hear this review was granted, but the initial result in April 2010 was a disappointment for the Trust and its supporters since it was ruled that the cull was lawful. However, leave to appeal this decision was granted, and in June of the same year the cull was declared unlawful as constituted. The Welsh Assembly Government was thus forced to consider bringing in new legislation, and so a further public consultation was held. It was announced in March of 2011 that the cull was going ahead, but in the Welsh Assembly elections two months later, Plaid Cymru lost a number of seats and the new government was formed by Labour. Its leader announced the setting up of an expert panel to consider the whole issue afresh, and in March 2012 it was stated that there would

---

4   This decision in Wales was taken despite the fact that the former ISG members (see Bourne *et al* n.d.) challenged the King report. See: http://webarchive.nationalarchives.gov.uk/20130123162956/http:/www.defra.gov. uk/animalh/tb/isg/pdf/isg-responsetosirdking.pdf [28 January 2014].

be no cull after all, but that instead badgers would be trapped and vaccinated over a four-year period.

Paradoxically, in England the reverse process happened. When Labour lost power in 2010, the new Coalition government, dominated by the Conservatives, lost little time in announcing that there would be a public consultation on the DEFRA website around the issue of culling badgers. At the end of 2011 the then Minister Caroline Spelman announced that a cull would go ahead in England the following year. This was to take place in two pilot areas (Gloucestershire and Somerset) by licensing groups of farmers to employ marksmen to shoot badgers. In this way, the costs of the cull would be largely borne by farmers, not by the government. For a variety of reasons,[5] this cull was postponed for a year and meanwhile opposition mounted. However, in February of 2013, it was stated that the cull would begin in west Gloucestershire and west Somerset in June.

VIEWS OF BADGERS, VIEWS OF THE OTHER

In British culture, badgers have a peculiarly iconic status. To some extent such a view has been fostered by the character of the badger in the classic children's book *Wind in the Willows*, in which the badger is seen as mysterious but wise. In west Wales, the badger is also the emblem of the Wildlife Trust and 'badger viewings' feature in much of the tourist literature. Further, for the large incomer population of English-speaking 'alternatives' (Williams 2003) who have come from England to live in this area to practise sustainable farming or small-holding and to protect wildlife, the idea of a badger cull is abhorrent. In interviews people frequently spoke about 'my badgers' and 'loving my badgers'. In interviews and meetings they spoke of them as emblems of 'the wild', of nature, locality, and, by their continued existence in spite of the huge changes in the countryside with the industrialisation of farming, as symbolising resilience and resistance. They blamed the increase in bTB on the intensification of farming methods, including such measures as larger herds of cattle and the increase in keeping animals in sheds for much of the year, as well as the frequent movement of cattle.

There is a contrary view of badgers, one that is held by some farmers, whether English- or Welsh-speaking, who see them as a nuisance when they spoil the sward by digging it up to find worms, and it was mainly farmers who spoke of them as a 'reservoir of disease', although this term was also used regularly by the Welsh Assembly Government. Badgers are also subject to the rural 'sport' of badger-baiting, in which these animals are pitted against dogs, a practice that is illegal but still continues, albeit surreptitiously, including in north Pembrokeshire. Such opposed views have, according to the historian Angela Cassidy (2012), a long tradition in Britain, where the badger may be seen as both 'good' and 'bad'.

---

5   These reasons included the delay caused by the Badger Trust bringing another (unsuccessful) judicial review; the discovery that there were far more badgers in the areas concerned than had originally been thought; that the police presence at the 2012 Olympics and Paralympics meant that officers would not have been available to police the cull; and that the 'closed season' for shooting badgers was too close to allow time for any effective measures.

ANGER, FEAR AND SUFFERING

In interviews carried out in west Wales between 2009 and 2012, these contrary views aroused strong emotions, notably anger and fear. On the anti-cull side there was anger against those who would cull badgers or support such a cull, including the Welsh Assembly Government, its agents (such as police and contractors) carrying out sett surveys, farming unions and farmers who not only supported the cull but even killed badgers illegally. As a member of PAC explained in 2011:

> I used to be on the phone for hours listening to people who were under such stress and who had moved here to get away from stress. I heard from two women in their 70s who had had an unexpected visit from men in balaclavas. Yet the press talked mainly about the farmers' fears! (Interview with English-speaking member of PAC, 28 August 2012, North Pembs)

On the pro-cull side, there was anger against those who would prevent this, whether by legal or illegal means, and accusations that they lacked a full understanding of the situation and put wildlife and badgers before cattle and farmers. Farmers who were pro-cull claimed to be in danger from animal rights 'extremists', as did contractors (hence the balaclavas), while those opposed to the cull were afraid not only of the state agents, but even of their neighbours who were pro-cull. In short, there was a danger that the proposed cull would produce a process of 'othering' people who did not share one's views. A Welsh farming woman commented on the loss of a neighbour's herd: 'And yet those protestors don't want any badgers killed. They are to blame' (Conversation in North Pembs, 28 August 2012).

For many people, the only way of dealing with this situation and preserving some semblance of community harmony was through silence:

> We could have got to a very nasty place of incomers vs locals. So people didn't stick their necks out. Some couldn't even tell their own families what they thought. One farmer who was against the cull told me he could not speak out because he would have lost his livelihood since his father was so pro-cull. (Interview with English-speaking member of PAC, 28 August 2012, North Pembs)

> I wouldn't discuss it (the cull) with my neighbours because I have to live with them. Our closest neighbours are Welsh, they would turn out in the middle of the night to help us with an animal. So I wouldn't tie myself to a gate here [on her father's farm] for the badgers, although I might have done so at the farm in Cilgwyn where the first arrests took place. (Interview with English-speaking member of PAC (by telephone), 12 September 2012)

Those in favour of the cull argued strenuously that it was necessary for the preservation of farming as a viable form of livelihood. They also noted that it is farmed animals that give the landscape its particular character – sheep in the hills, and cattle in fields with trees and hedges. As one logo I saw on a farm stated: 'No cattle, no countryside'.

In this discourse, those who are suffering are the farmers who lose cattle and the governments (and taxpayers) that have to compensate them. As one Welsh-speaking farmer noted:

If you've been building up a good herd over the years, watching the genetics, and then you lose it, that is devastating. And those with pedigree herds feel that the compensation is not enough for them. (Conversation in North Pembs, 24 August 2011)

He went on to comment on the loss of animals from his nephew's herd:

They all had names, the girls (daughters) were crying when they were taken away. You shouldn't let yourself get that fond of animals.

Farmers bitterly resent the accusation that, since they are in any case breeding them for slaughter, they do not care about animals. On the contrary, some farmers grow so fond of their animals that their loss through diseases like bTB is keenly felt. 'Are badgers more important than cattle?' was a frequently posed question.

In Wales, there is a further issue: the Welsh-speaking farming community is often seen as the primary carrier of Welsh culture and language, and hence its survival is of primary importance to the nationalist project. Loss of farming means potentially loss of culture, identity and sense of self, possibly even physical displacement as farming ceases to be viable and people are forced to abandon it.

Landscape and the Wild

Many of those against the cull valued the 'wild' aspect of landscape, symbolised by the existence of badgers in woods and 'secret' valleys. In fact many of the incomers had gone to live in Pembrokeshire for precisely those reasons, often expressing this as 'being closer to nature'. They saw such areas as an important part of the heritage not only of Pembrokeshire, but also of Wales and the UK more generally, citing the large numbers of tourists who come to the area from outside each year. Such views of heritage are also espoused by the Pembrokeshire National Park, which encompasses both the coastline and the Preseli hills, and which has been designated as such precisely for its outstanding 'natural' beauty. Here there are strict controls on planning, although this has not prevented quite significant changes to landscape over the last several decades as farming has become more industrialised.

At the same time, people live and make a living in the Park, and there is often resentment on the part of locals that they have to put up with a stricter regime of controls than their neighbours outside the Park. For farmers, especially those in the uplands, there is a perpetual struggle against many elements: areas of land that are rocky and infertile, the often unpredictable weather, animal disease[6] and the increasing intervention of the state in the form of compliance with regimes of surveillance (see Caplan 2012). In this area of Wales a term often used by Welsh-speakers when speaking English is 'tidy'. This goes beyond its usual meaning in English and is used as a term of approbation signifying 'appropriate' or 'in good order'. For example, on one occasion I asked a group of men why they were (illegally) removing a hedge. 'To make the field tidy' was the response; in other words it would be more useful to the farmer in its enlarged state. In a nearby

---

6   North Pembrokeshire had also suffered from both BSE (bovine spongiform encephalopathy, also known as 'mad cow disease'; see Caplan 2000) and then Foot and Mouth Disease.

area, the owners were sometimes asked by neighbouring farmers why they did not drain their wetland fields so that they would be more economically productive, even though this would have meant a loss of biodiversity. People would talk of 'improved' as opposed to 'unimproved' land, the former often having larger fields, sometimes with drainage. The contrast was marked at certain times of year when the 'improved' fields, on which fertiliser was spread, were bright green whereas the 'unimproved' would be varying shades of brown and muted green.

It should not be thought, however, that in the farming community there is no appreciation of either the beauties of the landscape or the intrinsic value of wildlife. On the contrary – people talk frequently about these matters, with 'I couldn't live anywhere else' often heard. In the village horticultural 'shows' held each summer, there is almost invariably a photographic competition that includes categories of wildlife and landscape, which reveal, at least on the part of some rural dwellers, a keen appreciation of both. But farmers also recognise that the landscape has, at least to some extent, to be tamed if they are to wrest a living from it. They know well that the landscape is made by humans and their activities over many years. Many speak of their regret at the industrial farming methods they are now forced to adopt because of economic and other pressures, comparing their lot with that of their parents' and grandparents' generations. But at the same time, they resent pressures that oblige them, as they see it, to comply with restrictions that seek to preserve the landscape so as to make the area more appealing to tourists: 'like living in a bloody museum'. They want to preserve what they see as their own heritage: a vibrant, Welsh-speaking community, which can only be viable if their farms are also economically viable.

CONCLUSION

In this chapter I have briefly considered reasons why some people support a cull of badgers, while others are opposed, and why this debate has aroused such strong emotions. It can be seen that badgers sit on a fault line in worldviews. For those against the cull, what is paramount is the need for a system of farming that is sustainable in the long term and which allows for the existence of wildlife such as badgers. In this discourse, the ones who will suffer from a cull are not only the badgers but also the people who value them and everything they represent. The loss of heritage will be in the form of a barren countryside and landscape, a localised version of wider globalised scenarios.

For those who are pro-cull, badgers constitute just one more element among the many difficulties in making a living as a farmer. They fear that a continuing increase in bTB will mean that the farming of animals will cease to be a viable form of livelihood, and so are willing to try anything that might help stem the disease. A shift away from farming animals to arable farming is already happening, and paradoxically one of its effects is a change in the character of the landscape.

BIBLIOGRAPHY AND REFERENCES

Bourne, F J, Donnelly, C A, Cox, D R, Gettinby, G, Morrison, W I, and Woodroffe, R, n.d. Response to *Tuberculosis in Cattle and Badgers: a Report by the Chief Scientific Adviser* [online], available from: http://webarchive.nationalarchives.gov.uk/20130123162956/http://www.defra.gov.uk/animalh/tb/isg/pdf/isg-responsetosirdking.pdf [27 January 2014]

Caplan, P, 2000 Eating British Beef with Confidence: perceptions of the risk of BSE in London and West Wales, in *Risk Revisited* (ed P Caplan), Pluto Press, London and Sterling VA, 184–203

— 2010 Death on the Farm: culling badgers in North Pembrokeshire, *Anthropology Today* 26 (2), 14–18

— 2012 Cull or vaccinate? Badger politics in Wales, *Anthropology Today* 28 (6), 17–21

Cassidy, A, 2012 Vermin, Victims and Disease: UK Framings of Badgers in and Beyond the Bovine TB Controversy, *Sociologia Ruralis* 52 (2), 192–214

Independent Scientific Group (ISG – chaired by Sir John Bourne) on Cattle TB, 2007 *Bovine TB: The Scientific Evidence: Final Report of the Independent Scientific Group on Cattle TB*, published by the House of Commons Environment, Food and Rural Affairs Committee, June, available from: http://archive.defra.gov.uk/foodfarm/farmanimal/diseases/atoz/tb/isg/report/final_report.pdf [27 January 2014]

King, D (Sir), 2007 *Bovine Tuberculosis in Cattle and Badgers. A Report by the Chief Scientific Adviser, Sir David King* (submitted to the Secretary of State, DEFRA on 30 July 2007), available from: http://www.bis.gov.uk/assets/biscore/corporate/migratedD/ec_group/44-07-S_I_on [27 January 2014]

Wilkinson, K, 2007 *Evidence Based Policy and the Politics of Expertise: A Case Study of Bovine Tuberculosis*, Centre for Rural Economy Discussion Paper Series No 12, University of Newcastle upon Tyne, April, available from: http://www.ncl.ac.uk/cre/publish/discussionpapers/pdfs/dp12%20Wilkinson.pdf [27 January 2014]

Williams, J, 2003 Incomers and Alternatives in West Wales, in *Welsh Communities: New Ethnographic Perspectives* (eds C A Davies and S Jones), University of Wales Press, Cardiff

29

The Great Barrier Reef: Environment, Disaster and Heritage

Billy Sinclair

The Great Barrier Reef (GBR) is an iconic natural wonder, which represents one of the most biologically diverse ecosystems on the planet (GBRMPA 2009). The river systems that feed into the GBR lagoon are highly biodiverse and have a long cultural history for indigenous and settling peoples. As times, economies and climate have changed over the centuries, so has the impact and influence of these river ecosystems on the communities that thrive along their banks. This has been shown to dramatic effect in the last ten years with the repeated flooding of rivers in Queensland and the resulting social, economic and environmental devastation (Agnew 2011; GBRMPA 2011a). The human tragedy associated with these events is severe, financially and emotionally, but the resilience and determination of the affected population demonstrates the drive and motivation of the human spirit in extreme circumstances. The environmental impacts are just as severe, however, and in many cases the resilience demonstrated by the natural ecosystem is lower than that displayed by the human population and the consequences are longer lasting as the incidence of coral bleaching (Jones, Berkelmans *et al* 2008), fish mortality and invasive species has increased dramatically (GBRMPA 2009; Agnew 2011; GBRMPA 2011a; Moore *et al* 2013). As these natural events continue to impact on humans and the environment, the need for adaptation and change to deal with them and find sustainable solutions is crucial.

The need for solutions is evident when we consider the nature of these heritage resources: the GBR is the most extensive coral reef ecosystem on the planet, covering an area of 348,000 square kilometres and encompassing 14 degrees of latitude, from its northern tip near Papua New Guinea, south along the Queensland coast to Bundaberg, just north of Brisbane. The GBR ecosystem is comprised of 2900 individual reefs and contains over 900 individual islands, over 1500 species of fish, more than 400 coral species, 5000 species of mollusc, over 200 species of birds, innumerable species of sponge, worm and crustaceans, plus hundreds of species of algae, which all function as a vital part of the reef. The GBR is one of the most iconic and functionally relevant ecosystems on earth and was declared a UNESCO World Heritage Site in 1981 (GBRMPA 2013).

History and Culture

First human contact with the reef occurred a considerable time ago. Aboriginal people have occupied great parts of the Australian continent for 40,000 years, and Aboriginal and Torres Strait Islander peoples must have fished and hunted its waters and navigated between the islands of the reef, using outrigger canoes large enough to enable them to travel from the mainland to the islands and outer reefs, navigating using the stars, currents and the knowledge of generations (Bowen and Bowen 2002). Their settlements moved up and down the coast, following natural migration and prey utilisation patterns, for thousands of years. They certainly used the reef long before the arrival of its famous 'discoverer', Captain James Cook, whose ship *Endeavour* struck a reef near Cape Tribulation in 1770, but reached the shore at the settlement now called Cooktown on the Endeavour River (Bowen and Bowen 2002; Lawrence *et al* 2002).

The GBR is critically important in both the history and the culture of Australia's Aboriginal and Torres Strait Islander peoples. It contains many sites of significant archaeological importance of Aboriginal and Torres Strait Islander origin, including middens, rock quarries, fish traps and rock art, with notable examples on Lizard Island, Hinchinbrook Islands and Stanley, Cliff and Clack Islands (GBRMPA 2013). Resident and migratory animals such as dugongs, minke whales, sharks and turtles have long been part of Aboriginal dreaming, and there are many important cultural sites and cultural values associated with many islands and reefs in the GBR Region. Today more than 70 Aboriginal and Torres Strait Islander Traditional Owner clan groups maintain heritage values of the reef area. These values may be cultural, spiritual, economic, social or physical, and demonstrate continuing connections with the GBR and its natural resources. Their traditional cultural practices and knowledge of marine resource use is under increasing pressure from the activities of modern day society in both remote and urban areas, such as those found on Great Keppel Island in the Capricornia region, whose traditional owners, the Darumbal people (descendants of the Konami-Whoppaburra people) are still in residence and facing issues of modernisation and development on traditional land (Evans 2007; GBRMPA 2013).

There are more than 30 shipwreck sites of historic importance on the GBR, as well as a number of historically significant lighthouses and World War II sites. Chinese sea cucumber fisherman and Japanese pearl divers also frequented the waters of the GBR in the 19th and early 20th centuries, and it continues to support fishing activities within Queensland and beyond (Bowen and Bowen 2002; Lawrence *et al* 2002; GBRMPA 2013).

THE MODERN GREAT BARRIER REEF UNDER PRESSURE

The 21st-century world has continued to change apace and faces continual challenges, especially in maintaining the natural environment in a manner acceptable to current stakeholders, but also for future generations. Given the ongoing importance and utility of the GBR, in 1971 the Great Barrier Reef Marine Park Authority (GBRMPA) was set up to manage the resource using an ecosystem-based approach to maintain the multiple-use area that supports a range of stakeholder communities and industries that depend on the reef for recreation or their livelihoods. The Authority also deals with the identification, protection and maintenance of both indigenous and non-indigenous culturally and historically significant sites throughout the Great Barrier Reef World Heritage Area (GBRMPA 1981). In 2009, GBRMPA published the *Great Barrier Reef Outlook Report*, which highlighted the main challenges for the future, identifying climate change, continued declining water quality from catchment runoff (Fabricus 2005), the loss of coastal habitats from coastal development and the impact of lawful and illegal fishing and poaching as priority issues (GBRMPA 2009).

A key point in the 2009 report was that the next few years would be crucial in determining the long-term future of the GBR ecosystem – not only from a biological perspective, but also from a socio-economic and cultural standpoint (GBRMPA 2009). The GBR is one of Australia's biggest tourist icons, attracting over 2 million visitors a year. It is estimated that reef-based tourism supports over 54,000 jobs and annually accounts for more than AUD \$5 billion in tourism revenue. In addition to being a National Park and World Heritage Site, the reef also supports a full-time commercial fishing industry, employing over 2000 people and generates in excess of AUD \$1 billion in income per year (GBRMPA 1981; 2013).

However, the GBR is in a state of flux and under pressure from a number of distinct and complex threats. The impact on the overall health and productivity of the GBR has been growing and in some instances, devastating (Schmidt 2013; Pandolfi *et al* 2013). Australian Institute of Marine Science researchers reported that the amount of coral on the reef has halved since 1985, with two-thirds of this reduction occurring since 1998 when the reef (and the rest of the world's coral reefs) experienced unprecedented levels of coral bleaching events (Bruno and Selig 2007; Sweatman *et al* 2011).

A clear indicator of the connectivity of these threats was the increased prevalence of Crown of Thorns starfish (COTS) across the GBR from 2009 and 2011, following the incidence of extreme weather events (Timmers *et al* 2012). These coral-eating starfish (*Acanthaster placini*) are an integral part of the biodiversity of a reef ecosystem, where they feed on some of the faster growing coral species, preventing them from taking over a reef and allowing some of the slower growing coral species to become established. When COTS numbers are low, corals can easily recover from the damage they cause (Black 2013). However, if conditions change, allowing COTS to increase, coral resilience falls and they can no longer recover, endangering the overall health of the reef. Such outbreaks are a naturally occurring event, but the frequency of the outbreaks has increased, from once in 80 years to once every 10 years (Walbran *et al* 1989; Black 2013). Previous outbreaks have followed large, drought-breaking floods in 2009 and 2011, which released large amounts of nutrients into the coral reef ecosystem, coinciding with COTS spawning season (November to January). The elevated nutrient load led to increased levels of phytoplankton, the main food source for COTS larvae. Far greater numbers survive and develop into juveniles, at which stage they start feeding on coral (Walbran *et al* 1989; Brodie *et al* 2004; Timmers *et al* 2012).

CLIMATE CHANGE AND THE GREAT BARRIER REEF

The *Great Barrier Reef Outlook Report* (2009) identified climate change as one of the greatest threats to the long-term health of the reef. Climate change will impact the coral reef in a number of ways (Kerr 2010; Wernberg *et al* 2011; Schmidt 2013; Pandolfi *et al* 2013). In the long term, ocean acidification is likely to be the most significant impact of a changing climate on the GBR ecosystem (De'ath *et al* 2009; Webster *et al* 2013). The oceans are a CO_2 sink – they absorb CO_2 from the atmosphere and are estimated to have absorbed half the excess CO_2 released by human activities in the past 200 years (Albright and Mason 2013). Almost half of anthropogenic CO_2 is found in the oceans' epipelagic/mesopelagic zones (depths <1000m), due to the slow process of ocean mixing. This is calculated to decrease the pH of the oceans by 0.1 units, a process termed 'ocean acidification'. Increased levels of dissolved CO_2 in the ocean mean fewer free carbonate ions are available for making calcium carbonate. Even such small increases in the acidity of the ocean decreases the capacity of corals to build skeletons and to create habitats for marine life on the reef (Albright and Mason 2013; Webster *et al* 2013). The current pH of the ocean (~8.2) is predicted to fall to ~7.8 by 2100 (GBRMPA 2009; 2012; 2013).

Rising sea surface temperatures will affect every aspect of the GBR. Temperature is a key environmental determinant in the distribution and diversity of marine life and it is critical to reef building, controlling the rate of coral reef growth (Jones, Berkelmans *et al* 2008; Albright and Mason 2013). As temperatures increase beyond a species' optimum temperature, their natural processes may cease to function. Atmospheric temperatures, variable cloud cover and freshwater runoff all contribute to rising sea surface temperatures (Albright and Mason 2013). The summer of 2010 saw the highest recorded sea surface temperatures in Australia. Projected increases in average sea surface temperatures indicate that by 2020 it could be 0.5°C warmer and more than 1°C warmer by 2050, which will have significant impacts on biodiversity (Jones and Berkelmans 2010; GBRMPA 2012).

Cyclones and Floods and the Future

Rising sea levels will have a significant impact, as much of the GBR coastline is low-lying. Even small increases in sea levels will cause land inundation, resulting in significant changes in tidal habitats such as mangroves and saltwater intrusion into existing low-lying freshwater habitats (Wolanski 2002). Sea levels on the reef have risen by ~3mm per year since 1991. In geological terms, this current sea level change in the GBR is very small, but sea levels had been constant for the past 6000 years, resulting in a well-defined depth profile across virtually all reefs. As this is such a complex process, predictions of a future increase in sea levels are variable, ranging from 0.68m across the GBR region to a global increase of up to 0.9m by 2100 (GBRMPA 2012).

The coastal reaches of the GBR takes water runoff from 6 natural resource management regions and 35 catchments, draining 424,000km² of coastal Queensland. Water quality flowing onto the coral reef has deteriorated significantly over the past 100 years and, each wet season, floods bring large quantities of pollutants from adjacent catchments onto the reef. Pollutant loads entering the reef vary significantly, but are thought to be reducing overall; nitrogen by 4%, phosphorus by 2%, sediments by 2% and pesticides by 8% (GBRMPA 2012).

Increasing frequency of sustained rainfall and the occurrence of severe cyclones increases the risk to the GBR and decreases the time available for recovery between events. Significant rainfall led to high levels of freshwater entering the reef in 2009–10 and the subsequent localised coral bleaching events experienced by shallow inshore. Tropical Cyclone Ului (Category 5) passed through the Whitsunday region in early 2010, causing further significant localised damage to parts of the reef (GBRMPA 2011c). Summer 2010–11 was the second wettest on record for Australia – for South East Queensland rainfall was 300mm – 400% higher than normal (GBRMPA 2011c). This intense rainfall caused extensive flooding in many coastal areas of southern Queensland. Large areas of the inshore GBR region were exposed to persistent flood plumes from the Fitzroy, Burnett and Mary rivers, causing significant impact on the marine environment and adjacent coastal communities. The greatest impacts of the flooding were concentrated on the inshore reefs close to the mouths of major rivers such as the Fitzroy River. Floodwaters caused severe damage to shallow fringing reefs with 85–100% corals mortality. The inshore reefs of the Keppel Islands, in the Capricorn Bunker Group, have historically shown substantial resilience to floods and coral bleaching, but even here recovery at the severely damaged sites will take decades (Jones, Berkelmans *et al* 2008; Jones, Cantin *et al* 2008; GBRMPA 2011b; 2011c; Gilmour *et al* 2013).

The GBR has been exposed to cyclones and floods as part of its natural history, but the recent changes in frequency and severity are challenging this ecosystem in new ways (Cheal *et al* 2002) due to European habitation and development. This new paradigm is likely to increase as human impacts continue to utilise and develop the natural resource that is the GBR and its surrounding catchments (GBRMPA 2011b; 2011c). Most recently, there have been several large-scale developments for industrial growth along the coast of Queensland, which have far-reaching implications for the survival of both communities and the GBR. In June 2013 the UN World Heritage Committee, at its meeting in Cambodia, endorsed its draft report from May 2012, which noted 'concern' at the 'limited progress' in halting coastal development and other threats to the reef (UNESCO 2012). This effectively put the Australian government on notice that the Great Barrier Reef will join UNESCO's 'in danger' list at its next annual meeting if improvements are not made. The committee was concerned that the Queensland and federal governments have made 'no clear commitment toward limiting port development to existing port areas'. The report also urges Australia to ensure that the expansion of existing ports does not damage the 'outstanding universal value' of the reef. More than 150 Australian and international scientists, representing 33 institutions, signed a letter on the eve of the World Heritage meeting calling for urgent action to safeguard the reef. Critics have claimed proposed expansion of coal and gas export terminals, such as at Townsville, and major new export developments such as Abbot Point, will cause significant damage to coral, turtles, dugongs and other wildlife through increased shipping and waste from dredging (ABC News 2013). Ports on the GBR coast currently export 156 million tonnes of coal per annum, and there are plans to expand that to 953 mtpa within the next decade. By 2020, an estimated 7000 ships will traverse the GBR every year (up from 5000 in 2010). Federal and state governments suggest they have worked hard to address the concerns, but environmental groups say more needs to be done.

The Queensland city of Gladstone is at the core of the ongoing debate, with the fifth largest multi-commodity port in Australia and the world's fourth largest coal exporting terminal, all within the World Heritage Area of the GBR. However, from a regional perspective, the issue is how to strike the right balance between economic growth for the population and the need for environmental sustainability. The coal and gas industry in regions such as Gladstone is crucial to the long-term survival of communities (economically, socially and culturally) and cannot just be halted without some form of alternate strategy in place to deal with displacement and the levels of social depravation that would follow. Such developments and growth have to be balanced with protecting the GBR ecosystem and all that it provides for the local communities and the natural world itself.

When the GBR is subject to changing environmental pressures and responds to them, there will naturally be knock-on impacts and consequences for human communities and industries. Indigenous communities have lived with these dynamic changes throughout the centuries and evolved and adapted to continue the sustainable processes they have practised (GBRMPA 2011c). As human populations and industry adapt, they learn strategies for implementation that are crucial for effective conservation of the GBR. Changes in patterns of land use following extreme weather events can create new 'hotspots' of pressure on the ecosystem (Munday *et al* 2010; McLeod *et al* 2013).

Marine tourism makes a significant contribution to the regional economy in Australia, particularly in communities bordering the GBR. Such communities are especially vulnerable to transient declines and changes in the health of the reef, which affect their socio-economic condition (Gooch *et al* 2012). Cyclone Yasi, one of the largest and most powerful cyclones to affect Australia since records began, crossed the Queensland coast near Mission Beach in

February 2011 (GBRMPA 2011b). Minor damage was reported to fringing reefs up to 500km from the eye of Cyclone Yasi, but no significant impacts were incurred at the major tourist centres of Port Douglas, Cairns or Airlie Beach, and the majority of tourism operations were still actively providing reef experiences within days of the cyclone (GBRMPA 2011b; 2011c). Tourism operators did suffer impacts following the extreme weather, however, with many reporting lost operating days due to damaged infrastructure (berthing facilities, vessels etc). A 10% decline in visitor days to the Great Barrier Reef Marine Park in the 2010–2011 financial year was reported compared to the previous year, highlighting the direct relationship between sustainability and access to healthy reefs. For many such operators, declining visitor numbers to the region was the source of most economic hardship. The disruption to air, rail and road network routes across Queensland resulted in visitors changing travel plans to bypass or curtail their stay in the GBR region (GBRMPA 2011b; 2011c). Following the 2010–2011 season, tourism operators suggested that the high profile of Cyclone Yasi and the floods in social, local and international media presented an overall impression that the entire GBR was severely damaged and that this was a contributory factor causing tourists to postpone or cancel their plans for travel to the region. The consequences of this were that, even if operators were operational, visitors from outside the region (especially internationally) perceived the GBR as being unlikely to provide a good tourism experience (GBRMPA 2011b; 2011c).

The commercial fishing industry suffered a range of impacts following the extreme weather in 2010–2011, including the ability of fishermen to access fishing locations, compromised water quality and terrestrial debris that caused damage to fishing gear (NCCARF 2012). In the southern section of the GBR, inshore fishermen were particularly impacted, especially those targeting reef fish. The turbidity of the water from river outflows and runoff degraded the water quality, and many fishermen saw dramatic declines in catch rates of species such as coral trout in shallow reef areas. This was not a function of a change in the abundance of coral trout, but rather appeared to be explained by a decrease in the 'catchability' of the fish, a phenomenon consistent with previous weather events (Lawson 2011; GBRMPA 2011c). Conversely, changes in water quality had a positive impact on the mudcrab and barramundi fisheries (Heupel *et al* 2011) and catches increased in many locations due to favourable conditions caused by strong freshwater flows into estuarine systems (Lawson 2011; NCCARF 2012).

Whereas fishermen could diversify and survive by switching their target species, other marine-based industries, such as recreational scuba diving operators, had fewer options (Gooch *et al* 2012). Diving centres such as Cairns and the Whitsundays had enough in-built resilience in terms of access to a range of dive sites and high turnover in customers to cope with the weather in 2010–2011. However, river runoff from the Fitzroy River in Rockhampton spread more than 5 miles offshore at the height of the floods; sediment-rich, low-salinity water covered the majority of the Keppel Bay reef system (Brodie *et al* 2012). There was little or no scuba diving in the area for almost a year.

CONCLUSION

There is a plethora of ongoing research, frequent discussions at a variety of levels in stakeholder groups and continuing warnings about how the changing climate will impact on weather, rain, floods and the GBR. In the natural history of the GBR, such events are not new – it has evolved in tandem with such events and will continue to do so for millennia. The GBR is a dynamic, living ecosystem, which evolves and changes as the environment evolves and changes. The most pervasive problem faced by the GBR and the communities who live with it is the systematic pace of change to which they are subject. In a natural system, changes are either cataclysmic or gradual; however the current environment is evolving at an unnatural pace – a pace with which natural evolutionary processes are struggling to cope. It has been shown in the scientific literature that populations of harvested and hunted fish (commercially and recreationally) on average have a 20% smaller body size than previous generations and reproduce at a younger age – the impact of the harvesting and hunting forcing the species/populations to evolve their phenotype and the age at which they reproduce (Darimont *et al* 2009). Such forced evolution – where the natural process of change is artificially increased and causes potentially deleterious change in the demographic of the species – has been a common concern in relation to the extent of human impacts on natural ecosystems. The GBR is not immune to such forced impacts. The whole GBR ecosystem is facing unprecedented levels of pressure, from increasing sea surface temperatures from anthropomorphic sources (and the concomitant pressures on the reef itself and its associated biodiversity that develop from increased temperatures), pollution from agricultural runoff and from ever increasing coastal developments (such as the coal terminal at Hay Point, the resort and golf course development on Keppel Island, the coal terminal at Gladstone and the dumping of dredge spoils) and all the global issues surrounding marine biosecurity and shore-side development. Yes, without a doubt the GBR will change as the years progress; it always has done. How the management of the developmental and environmental issues progress and the remediation plans to mitigate the impact of these pressures develop, how stakeholders for indigenous communities engage and are engaged by these processes and how the natural plasticity of the reef itself can deal with the pressures of forced evolution will have a huge influence on the overall survival of the reef as an accessible natural resources for a huge number of stakeholders.

BIBLIOGRAPHY AND REFERENCES

ABC News, 2013 *150 marine scientists warn of port construction threat to Great Barrier Reef* [online], available from: http://www.abc.net.au/news/2013-06-05/scientists-warn-of-port-construction-barrier-reef-threat/4734182 [18 June 2013]

Agnew P & F Association Inc, 2011 *Flood: Horror and Tragedy*, Agnew P & F Association, Pan Macmillan, Australia

Albright, R, and Mason, B, 2013 Projected near-future levels of temperature and pCO_2 reduce coral fertilization success, *PLoS ONE* 8 (2), e56468

Black, K P, 2013 Dispersal and Recruitment in Crown-of-Thorns Starfish: Overview and Future Directions, in *The Bio-Physics of Marine Larval Dispersal* (eds P W Sammarco and M L Heron), American Geophysical Union, Washington DC

Bowen, J, and Bowen, M, 2002 *The Great Barrier Reef: History, Science, Heritage*, Cambridge University Press, Cambridge

Brodie, J, Fabricus, K, De'ath, A G, and Okaji, K, 2004 Are increased nutrient inputs responsible for more outbreaks of crown-of-thorns starfish? An appraisal of the evidence, *Marine Pollution Bulletin* 51, 266–78

Brodie, J, Schroeder, T, Rohde, K, Faithful, J, Masters, B, Dekker, A, Brando, V, and Maughan, M, 2012 Dispersal of suspended sediments and nutrients in the Great Barrier Reef lagoon during river-discharge events: conclusions from satellite remote sensing and concurrent flood-plume sampling, *Marine and Freshwater Research* 61 (6), 651–64

Bruno, J F, and Selig, E R, 2007 Regional Decline of Coral Cover in the Indo-Pacific: Timing, Extent and Subregional Comparisons, *PLoS ONE* 2 (8), e711.doi:10.1371/journal.pone.0000711

Cheal, A J, Coleman, G J, Delean, S, Miller, I R, Osborne, K, and Sweatman, H P A, 2002 Responses of coral and fish assemblages to a severe but short-lived tropical cyclone on the Great Barrier Reef, Australia, *Coral Reefs* 21, 131–42

Darimont, C, Carlson, S M, Kinnison, M T, Paquet, P C, Reimchen, T E, and Wilmers, C C, 2009 Human Predators outpace other agents of trait change in the wild, *Proceedings of the National Academy of Sciences of the United States of America* 106 (3), 952–4

De'ath, G, Lough, J M, and Fabricus, K E, 2009 Declining Coral Calcification on the Great Barrier Reef, *Science* 323 (5910), 116–119

Evans, R, 2007 *A History of Queensland*, Cambridge University Press, Cambridge

Fabricius, K E, 2005 Effects of terrestrial runoff on the ecology of corals and coral reefs: review and synthesis, *Marine Pollution Bulletin* 50, 125–46

GBRMPA, 1981 *Nomination of the Great Barrier Reef by the Commonwealth of Australia for inclusion in the World Heritage List*, Great Barrier Reef Marine Park Authority, Australia

— 2009 *Great Barrier Reef Outlook Report 2009*, Great Barrier Reef Marine Park Authority, Australia

— 2011a *Extreme Weather and the Great Barrier Reef*, Great Barrier Reef Marine Park Authority, Australia

— 2011b *Impacts of tropical cyclone Yasi on the Great Barrier Reef: a report on the findings of a rapid ecological impact assessment*, Great Barrier Reef Marine Park Authority, Australia

— 2011c *Extreme Weather and the Great Barrier Reef*, Great Barrier Reef Marine Park Authority, Australia

— 2012 *Great Barrier Reef Climate Change Adaptation Strategy and Action Plan 2012–2017*, Great Barrier Reef Marine Park Authority, Australia

— 2013 Great Barrier Reef Marine Park Authority [homepage], available from: http://www.gbrmpa.gov.au [30 May 2013]

Gilmour, J P, Smith, L D, Heyward, A J, Baird, A H, and Pratchett, M S, 2013 Recovery of an isolated coral reef system following severe disturbance, *Science* 340 (6128), 69–71

Gooch, M, Vella, K, Marshall, N, Tobin, R, and Pears, R, 2012 A rapid assessment of the effects of extreme weather on two Great Barrier Reef industries, *Australian Planner* 50 (3), 1–18

Heupel, M, Knip, D M, de Lestang, P, Allsop, Q A, and Grace, B S, 2011 Short-term movement of barramundi in a seasonally closed freshwater habitat, *Aquatic Biology* 12, 147–55

Jones, A, and Berkelmans, R, 2010 Potential costs of acclimatization to a warmer climate: Growth of a reef coral with heat tolerant vs sensitive symbiont types, *PLoS ONE* 5 (5), e10437

Jones, A M, Berkelmans, R, Mieog, J C, Van Oppen, M J H, and Sinclair, W, 2008 A community change in the symbionts of a scleractinian coral following a natural bleaching event: field evidence of acclimatization, *Proceedings of the Royal Society B: Biological Sciences* 275, 1359–65

Jones, A, Cantin, N, Berkelmans, R, Sinclair, B, and Negri, A, 2008 A 3D modeling method to calculate the surface areas of coral branches, *Coral Reefs* 27, 521–6

Kerr, R A, 2010 Ocean Acidification Unprecedented, Unsettling, *Science* 328 (5985), 1500–01

Lawrence, D, Kenchington, R, and Woodley, S, 2002 *The Great Barrier Reef: finding the right balance*, Melbourne University Press, Victoria

Lawson, A, 2011 *Trout turn tail in wake of Hamish*, FRDC Research Report 2010/65

McLeod, E, Anthony, K R N, Andersson, A, Beeden, R, Golbuu, Y, Kleypas, J A, Kroeker, K, Manzello, D, Salm, R V, Schuttenberg, H Z, and Smith, J E, 2013 Preparing to manage coral reefs for ocean acidification: lessons from coral bleaching, *Frontiers in Ecology and the Environment* 11, 20–7

Moore, J, Bellchambers, L M, Depczynski, M, Evans, R D, Evans, S N, Field, S N, Friedman, K J, Gilmour, J P, Holmes, T H, Middlebrook, R, Radford, B T, Ridgway, T, Shedrawi, G, Taylor, H, Thomson, D P, and Wilson, S K, 2013 Unprecedented Mass Bleaching and Loss of Coral across 12° of Latitude in Western Australia in 2010-11, *PLoS ONE* 7, e51807

Munday, P L, Dixson, D L, McCormick, M I, Meekan, M G, Ferrari, M C O, and Chivers, D P, 2010 Replenishment of fish populations is threatened by ocean acidification, *Proceedings of the National Academy of Sciences of the United States of America* 107, 12930–4

NCCARF, 2012 Extreme Weather: Adaptation Insights from the Great Barrier Reef, *Marine Adaptation Bulletin* 3 (4), 6–7

Pandolfi, J M, Connolly, S R, Marshall, D J, and Cohen, A L, 2013 Protecting coral reef futures under global warming and ocean acidification, *Science* 333 (6041), 418–22

Schmidt, C, 2013 As threats to corals grow, hints of resilience emerge, *Science* 339 (6127), 1517–19

Sweatman, H, Delean, S, and Syms, C, 2011 Assessing loss of coral cover on Australia's Great Barrier Reef over two decades, with implications for longer-term trends, *Coral Reefs* 30, 521–31

Timmers, M A, Bird, C E, Skillings, D J, Smouse, P E, and Toonen, R J, 2012 There's No Place Like Home: Crown-of-Thorns Outbreaks in the Central Pacific Are Regionally Derived and Independent Events, *PLoS ONE* 7 (2), e31159

UNESCO, 2012 WHC.12/36.COM/ Mission Report, Reactive Monitoring Mission to Great Barrier Reef, Australia, 6th to 14th March 2012

Walbran, P D, Henderson, R A, Jull, A J T, and Head, J M, 1989 Evidence from Sediments of Long-Term *Acanthaster planci* Predation on Corals of the Great Barrier Reef, *Science* 245 (4920), 847–50

Webster, N S, Uthicke, S, Botte, E, Flores, F, and Negri, A P, 2013 Ocean acidification reduces induction of coral settlement by crustose coralline algae, *Global Change Biology* 19, 303–15

Wernberg, T, Russell, B D, Moore, P J, Ling, S D, Smale, D A, Campbell, A, Coleman, M A, Steinberg, P D, Kendrick, G A, and Connell, S, 2011 Impacts of climate change in a global hotspot for temperate marine biodiversity and ocean warming, *Journal of Experimental Marine Biology and Ecology* 400, 7–16

Wolanski, E J, 2002 Natural resource system challenge. Oceans and aquatic ecosystems, in *Knowledge base for sustainable development: An insight into the encyclopedia of life support systems* (ed E J Wolanski), UNESCO Publishing/EOLSS Publishers, Oxford, UK, 3: 311–24

Endpiece

Phil O'Keefe

In ending this book I have been asked to reflect back on a life working in disasters – in effect, a personal heritage of collected memories, materials and cultures. This has been a difficult task. On the one hand there is disease and death, the data of morbidity and mortality; on the other hand, there are the successes in rebuilding livelihoods and properties, a geography that is, frequently, one of affect. Affect is observed firstly in the disaster-impacted community itself, which is usually the first responder, and then seen in the external parties who help deliver relief, rehabilitation and reconstruction. Some explanation is required of the political economy of global disaster response as it is practised in the 21st century, not least because it has essentially become an industry. As this is polemical, it is perhaps best done in an older form of argument, namely theses statements.

Thesis 1

For the last two centuries, with the rise of capitalism, we have gone beyond the mastery of nature to the production of nature.
Under globalised capitalism, nature is produced (Smith and O'Keefe 1980). In the production of nature, people change their relationship to nature from that of husbandry, an organic relationship with the environment, to one of mastery, even if that mastery is never fully achieved. People also change their relationship to each other as individuals, communities and classes. Importantly, these changes also change the nature of risk and thus disaster type (Beck 1992; 2009).

It is relatively easy to illustrate these arguments. Fracking increases earthquake risk; flooding is increased by land-use change; climate change heralds increasing extremes of temperature and precipitation beyond what people accept as 'normal' weather. It is not simply the physical world but the biological one where fundamental changes are occurring, for example HIV/Aids, SARS epidemics and BSE outbreaks. Technology itself generates new risks, not least with nuclear engineering and biotechnologies. Science can only bring a limited solution to these problems, not least because the modern positivist science project comes encumbered with a dual vision of nature. This dual vision holds that nature is both 'external' to humans while simultaneously nature is 'universalist', incorporating humans. But a global capitalism necessarily produces a humanly nature in the realm of exchange, or market, values. Talk of a post-normal science to address complex problems simply reinforces the contradictory duality of the scientific approach to nature (O'Brien and O'Keefe 2014). Under globalised capitalism, new risk is produced that science cannot control.

Thesis 2

Understanding vulnerability is the key.
To understand risk, including new risk, requires an understanding not only of the hazard event, including its frequency, but more importantly, the conditions of vulnerability. Vulnerability is a

people process where the generation of vulnerability itself lies in the production of poverty. Poverty, in turn, can be specified along gender, ethnic and religious lines, but it is essentially explained, at a global level, along class lines. Particularly in those formerly pre-capitalist modes of production that are now incorporated under a global capitalism, the loss of traditional coping mechanisms by which common wealth was shared exacerbates problems of recovery after disaster (O'Keefe *et al* 1976).

It should come as no surprise that, if disasters are mapped, most occur in the least developed countries. And if local wars are mapped (for those 'complex emergencies' are treated as disasters by the humanitarian community), the results are much the same. Disasters happen to poor people in poor places (Middleton and O'Keefe 1997).

THESIS 3

Disaster planning and response is an industry.
Humanitarian aid itself has grown as a global industry since 1989, a year that marks the collapse of the former Soviet Union and its allies (Rose *et al* 2013). It has grown from around US $500,000 million a year in 1990 to almost US $20 billion by 2013 (Global Humanitarian Assistance 2013).

In 2012, over 75 billion people received some level of humanitarian aid. These include some 10 million refugees and 20 million Internally Displaced People. Most of this movement is a direct result of local wars; frequently wars in which the developed world is involved. Kosovo, Iraq, Afghanistan and Somalia together account for over 80% of current humanitarian assistance. Wars in which the West was directly engaged are the major focus of humanitarian assistance. They are largely in Muslim countries where many of the international non-governmental organisations (NGOs) working are Christian. Unsurprisingly, the military are increasingly involved in humanitarian delivery, not just abroad but at home, not least following Hurricane Katrina in the United States in 2005 (Adams 2013).

It is difficult for humanitarian organisations to hold to their mission of humanitarianism, neutrality, impartiality and advocacy, not least because most of their monies come from Western governments. The last decade has seen a profusion of technical and professional improvements, such as Sphere (see: www.sphereproject.org), that serve to divert attention away from the core issue of accountability.

In both the United States and the United Kingdom, there has been significant revision to emergency planning. The emphasis, not least post 9/11, has been on building resilience. And there is evidence that resilience has been built, largely in the 'blue light' emergency services, not in the affected communities (O'Brien 2008).

ANTITHESIS 1

Towards an Economy of Affect
People frequently do things without monetary compensation, such as bringing up a family or caring for the sick and the old. In disasters, that is what happens all the time. The local communities, rather than the 'blue light' services, dominate first responses. It is in the community that relief, rehabilitation and reconstruction take place. It is essentially a local giving and receiving economy, an economy of affect. That affect also includes memories of trauma and lessons learnt. As the editors indicate in the Introduction, collecting and curating these materials and memories

has the potential to result in understanding, commemoration and deterrence, but care must be taken to consult, listen and respond to local communities.

The editors have highlighted much of this economy of affect and consider how it might be captured, remembered and commemorated, against a background of global natural disaster, biological threat and technological failure. The studies of disaster and heritage contained in this volume speak of global landscapes of memory, trauma and loss. Their inclusion here allow the voices of the vulnerable to heard. These are voices less of pain and anguish but of a contained anger. Listen carefully.

Phil O'Keefe
Emeritus Professor of Geography
Northumbria University, UK

BIBLIOGRAPHY AND REFERENCES

Adams, V, 2013 *Markets of Sorrows, Labors of Faith: New Orleans in the Wake of Katrina*, Duke University Press, Durham NC

Beck, U, 1992 *Risk Society: Towards a New Modernity*, Sage, London

— 2009 *World at Risk*, Polity Press, Cambridge

Global Humanitarian Assistance, 2013 *Global Humanitarian Assistance: Report 2013* [online], Development Initiatives, Bristol, UK, available from: http://www.globalhumanitarianassistance.org/wp-content/uploads/2013/07/GHA-Report-2013.pdf [23 April 2014]

Middleton, N, and O'Keefe, P, 1997 *Disaster and Development: the Politics of Humanitarian Aid*, Pluto Press, London

O'Brien, G, 2008 UK emergency preparedness: a holistic local response?, *Disaster Prevention and Management* 17 (2), 232–43

O'Brien, G, and O'Keefe, P, 2014 *Managing Adaptation to Climate Risk*, Routledge, London

O'Keefe, P, Westgate, K, and Wisner, B, 1976 Taking the Naturalness out of Natural Disaster, *Nature* 260 (5552), 566–7

Rose, J, O'Keefe, P, Jayawickrama, J, and O'Brien, G, 2013 The Challenge of Humanitarian Aid: an overview, *Environmental Hazards* 12 (1), 74–92

Smith, N, and O'Keefe, P, 1980 Geography, Marx and the Concept of Nature, *Antipode* 12 (2), 30–9

Contributors

Marc Ancrenaz is the Scientific Director of the French NGO Hutan and co-director of the Kinabatangan Orangutan Conservation Programme, a community-based programme active in wildlife research, conservation and community development located in Sabah (Malaysian Borneo) since 1996. He is a scientific adviser for the Sabah Wildlife Department. He has an extensive background in wildlife research, medicine and population management as well as in wildlife management policy formulation, with 25 years' experience of working in wildlife range countries (Gabon, Congo, Saudi Arabia, Indonesia and Malaysia).

Rupert Ashmore is a Lecturer in Art and Design History at Northumbria University. He completed his doctoral thesis in 2011, examining the representation of communal trauma through landscape imagery in photography and film. His current research investigates how internet social media platforms, photographic archives and amateur video production combine to facilitate communal memory and place-making in northern British communities.

Josephine Baxter farmed in South Cumbria for 26 years. She was a support worker on the Farmers' Health Project (http://www.lancs.ac.uk/fass/ihr/publications/farmershealth.pdf) and for three years was a Research Associate on Lancaster University's study of the health and social consequences of the 2001 Foot and Mouth epidemic in Cumbria, and a co-author of 'Animal Disease & Human Trauma'. She has written both fiction and non-fiction about Cumbria and was a contributor to an earlier volume of this series, *Making Sense of Place* (2012).

Jo Besley is a doctoral candidate in the Museum Studies programme at the University of Queensland. She was formerly Senior Curator at both the Queensland Museum and Museum of Brisbane and has published in academic, sector and popular publications.

Özgün Emre Can is a Conservation Biologist at the University of Oxford's Wildlife Conservation Research Unit, Department of Zoology. He completed pioneering field-based studies on wolves and other carnivore species in Turkey where his efforts have been instrumental in changing the official paradigm for carnivore conservation. He has worked with remote rural communities, colleagues from universities and NGOs, and with decision and policymakers in Europe and Caucasus in various efforts to conserve carnivores in those regions since the late 1990s. His current research focuses on clouded leopard and human–carnivore conflict in the Himalayas. He is a member of the council of the International Association for Bear Research and Management and a member of several IUCN Specialist Groups.

Pat Caplan is Emeritus Professor of Social Anthropology at Goldsmiths, University of London. She has carried out fieldwork on Mafia Island, Tanzania, since 1965, in Chennai, India, since 1974 and in west Wales since 1994. Her interests have included social inequality; risk and trust; ethics; personal narratives and biography; and local perceptions of modernities. Most recently she has been researching the effects of animal diseases on farming communities in west Wales. She has published a number of monographs and co-edited books, articles and book chapters, and has also been involved in the making of two films and the establishment of two websites. She has been a Chair of the Association of Social Anthropologists of the UK and a Trustee of ActionAid, UK.

Chia-Li Chen is Director of the Graduate Institute of Museum Studies and Project Manager at the Centre for Museum and Cultural Heritage at Taipei National University of the Arts. She is the author of *Museums and Cultural Identities: Learning and Recollection in Local Museums in Taiwan* (VDM Verlag) and *Wound on Exhibition: Notes on Memory and Trauma* (Artco Publisher, Taiwan). Her research interests focus on three main areas: museums and contemporary social issues (especially the engagement and representation of disabled and minority groups), museums and disease, and the history of community and literature museums.

Ian Convery is Reader in Conservation and Forestry in the National School of Forestry, University of Cumbria. His main research interests are related to community resource use and place studies, with a particular focus on protected areas and conservation management.

Gerard Corsane is a Senior Lecturer in Heritage, Museum & Gallery Studies in the International Centre for Cultural & Heritage Studies (ICCHS) at Newcastle University. He is also Dean for International Business Development & Student Recruitment in the Faculty of Humanities & Social Sciences. He teaches across the Heritage and Museum Studies postgraduate programmes in ICCHS and is working on a new e-learning suite of courses in Heritage Management and Tourism. His research interests are in community participation in sustainable heritage tourism, ecomuseology, integrated heritage management and the safeguarding of intangible and tangible cultural heritage resources.

Peter Davis is Emeritus Professor of Museology at Newcastle University. His research interests include the history of museums; the history of natural history and environmentalism; the interaction between heritage and concepts of place; and ecomuseums. He is the author of several books including *Museums and the Natural Environment* (1996), *Ecomuseums: a sense of place* (1999; 2nd edition 2011) and (with Christine Jackson) *Sir William Jardine: a life in natural history* (2001).

Hugh Deeming is a Senior Research Assistant at Northumbria University and the scientific technical officer for the EU FP7 emBRACE project (www.embrace-eu.org). His principal research interest lies in the investigation of 'community resilience'. In addition to the Hull Floods Project, where he worked extensively with the UK Cabinet Office to integrate the research findings into the National Recovery Guidance, he has worked on a number of hazard-related projects funded by the EU (ARMONIA, SCENARIO, MICRODIS and now emBRACE), the Environment Agency (eg SC060019: 'Improving Institutional and Social Responses to Flooding') and the Cabinet Office/DSTL ('Community Resilience Research').

Susannah Eckersley is a Lecturer in Museum, Gallery and Heritage Studies at Newcastle University. Her research interests are interdisciplinary and span the full breadth of what is understood as 'heritage' (in other words, museums, galleries, tangible and intangible heritage). She has particular knowledge and interests in: cultural policy; museum architecture; museology; economic and social regeneration; built heritage; difficult histories (in particular in relation to German history); memory and identity; migration, diversity and representation. One commonality for her between these diverse interests is that they all encompass the wider issue of the relationship between heritage, culture, history and the state.

Esther Edwards is a technical specialist in the use of satellite remote sensing, geographical information systems and environmental management. She has extensive expertise in airborne remote sensing for clients in Africa, Europe and South America. Her current research focuses on the use of mapping tools to understand societal behaviours in the context of environmental hazards, and also pedagogy of fieldwork for Geography undergraduates on international field trips. She is an active member of the Changing Landscapes Research Group at Bath Spa University, and also co-leads the collaborative research and teaching activities to the Indian Himalayas.

Sarah Elliott is an independent scholar with research interests in ecomuseology and the theories of new museology, recently positioning both within Turkish Area Studies. The emergence and significance of postmodern approaches in contemporary Turkish museology is the focus of current British Academy-funded work; previous AHRC-funded PhD research at Newcastle University examined the impact of large dams on the cultural heritage of south-east Turkey, attempting to address the issues through an ecomuseum-centred methodology. Hasankeyf, a *sui generis* medieval town threatened by the Ilisu Dam, was the case study for the latter.

Kai Erikson is Professor Emeritus of Sociology and American Studies at Yale University. He has been studying the effects of disasters on human social life for a number of years. He is the author of, among other titles, *Wayward Puritans: A Study in the Sociology of Deviance; Everything In Its Path: Destruction of Community in the Buffalo Creek Flood*; and *A New Species of Trouble*.

James Gardner (BSc (Alberta), MSc, PhD (McGill)) is Professor Emeritus, Natural Resources Institute, University of Manitoba, and Adjunct Professor, Department of Geography, University of Victoria in Canada. Formerly Provost and Professor at the University of Manitoba and Professor and Dean of Graduate Studies at the University of Waterloo, he has pursued research and teaching in geomorphology, hydrology, glaciology and resources and hazards management with field studies in mountain environments in Canada, Europe, India, Pakistan and China. He has published widely on alpine geomorphology and resources and hazards.

Takashi Harada is Professor of Sociology at Konan Women's University, Kobe, Japan. He studied the 'Actor Network Theory' during a year spent in England in 1993–94. Shortly afterwards, he worked as a volunteer in a public shelter following a major earthquake in Kobe in 1995. The experience of that earthquake led him to study voluntary activities and the ways in which people experience major earthquakes in Japan.

Richard Johnson (CGeog (Geomorph) CGeol) is a mountain Geomorphologist with extensive expertise in both academic research and commercial applied geomorphology for UK and international clients. His research has focused on the mountainous English Lake District, exploring relationships between hydro-meteorological and sediment system dynamics, particularly in relation to fluvial floods, debris flows and shallow landslides. He was part of a research team that investigated the disastrous 1995 and 2005 floods and landslides impacting the Cumbrian region. He is Chair of the Changing Landscapes Research Group at Bath Spa University, and also co-leads collaborative research and teaching activities to the Indian Himalayas.

Isabelle Lackman is President of the French NGO Hutan. She obtained her PhD from the National Museum of Natural History, Paris, France, in 1998 for her research on the socio-ecology of Hamadryas baboons in Saudi Arabia. Since 1998 she has been co-director of the Kinabatangan Orang-Utan Conservation Programme (KOCP). For the past 16 years, Hutan-KOCP and the Sabah Wildlife Department have been working to develop and implement innovative solutions to conserve the orang-utan in Sabah. The main components of the programme include: long-term research on wild orang-utans and habitat; orang-utan conservation policy development and implementation in Sabah; mitigation of human–wildlife conflicts; capacity building for Malaysian conservation professionals and local communities; environmental education and awareness; involvement of local communities in the management of protected areas; and community development projects compatible with wildlife and habitat conservation.

Ellie Land is an international award-winning film-maker, researcher and educator. Her films have attracted a number of awards, commendations and special mentions from a variety of prestigious international film festivals. Her most recent film *Centrefold* has won numerous awards, including Best Non-broadcast Factual at the Royal Television Society awards. *Centrefold*, commissioned by the Wellcome Trust, was released in July 2012. It attracted international media attention and is a key player in the international debate on the ethics of labia surgery. While making films she continues to conduct practice-based research in the area of animation, documentary and participatory film-making. She is also the founder and regular contributor to the blog animated-documentary.com. She holds an MA in Animation from the Royal College of Art and is currently Senior Lecturer in Animation at Northumbria University.

Andrew Law is currently a Lecturer in Town Planning and Degree Programme Director of the BA Degree in Architecture and Urban Planning at Newcastle University; broadly speaking, over the last ten years he has been mainly concerned with the sociology of history, memory, nostalgia and heritage. His early research was concerned with built conservation and the history of Mock-Tudor architecture. With respect to the latter, along with co-author Professor Andrew Ballantyne, he published *Tudoresque: In Pursuit of the Ideal Home* in 2011. However, in recent years his work has become more concerned with the sociology and politics of history, memory, nostalgia and heritage in contemporary China; at present he is conducting research on place-making, built heritage and the Hanfu movement in Beijing, Wuhan and Xi'an.

Peter Lurz lives in Scotland and is an internationally recognised tree squirrel expert. He has studied and carried out research on squirrel ecology, behaviour, competition and conservation in the UK, Italy and the USA for over 20 years. Although he has worked and published on hedgehogs, badgers, bats, arctic foxes, ducks and black grouse, his great passion has been and continues to be squirrels.

Aron Mazel joined Newcastle University in 2002 after a 25-year career in archaeological research and heritage and museum management in South Africa. Between 2002 and 2004, he managed the Beckensall Northumberland Rock Art Website Project, which won the Channel 4 ICT British Archaeological Award in 2006. Aron is also a Research Associate in the School of Geography, Archaeology and Environmental Studies at the University of the Witwatersrand (South Africa). His research interests include the management and interpretation of heritage; museum history; the construction of the hunter-gatherer past; the dating of rock art; and Northumberland rock art. Book publications include *Tracks in a Mountain Range: exploring of the history of the* *uKhahlamba-Drakensberg* (2007, with John Wright); *Art as Metaphor: The Prehistoric Rock-Art of Britain* (2007, co-edited with George Nash and Clive Waddington); and *uKhahlamba: Umlando wezintaba zoKhahlamba / History of the uKhahlamba* (2012, with John Wright).

Arthur McIvor is Professor of Social History and Director of the Scottish Oral History Centre at the University of Strathclyde, Glasgow, Scotland. His research interests lie in the history of work and the body, and he has published widely in this area, including (with Ronnie Johnston) *Lethal Work* (2000) and *Miners' Lung* (2007). His most recent book, *Working Lives: Work in Britain since 1945*, was published in 2013. He is currently working on the wartime 'Reserved Occupations' and on the history of disability and impact of deindustrialisation in the Scottish coalfields since 1945.

Will Medd was Principal Investigator on the Hull Floods Project, when he was a Lecturer in Lancaster Environment Centre, Lancaster University. He left his Lectureship in December 2013 to develop his coaching work, specialising in offering coaching for academics. He is co-author of *Your PhD Coach* (McGraw Hill/Open University Press, 2013) and *Get Sorted: How to make the most of your student experience* (Palgrave Macmillan, 2015). www.willmedd.com

Stephen Miles is an Affiliate Research Fellow at the University of Glasgow's Crichton Campus in Dumfries, Scotland. He completed his PhD in Heritage, Tourism and Development in 2012, the subject of which was *Battlefield Tourism: Meanings and Interpretations*. His research interests concern heritage tourism at sites of conflict and interpretation and visitor experience at such sites. He also has an interest in the construction and meanings of place, particularly at 'dark' sites. He has spoken at several conferences, and his work has been published in articles and as a book chapter.

Brij Mohan, a Geographer by training, is a graduate of both the Punjabi University in Patiala and the University of Jammu in Jammu and Kashmir, India. He is Principal of the Harvard Convent School in Baghapurana (Punjab), where he continues to teach Geography to secondary school children. He has co-authored multiple geography exam textbooks for the ICSE Board, New Delhi.

Rob Morley is an Ecologist and Natural Resource Manager. After gaining his BSc/MSc he joined ETCUK and was posted to Gorongosa in 1997 as an Ecologist. Following his PhD in elephant population dynamics, he opened a consultancy in Mozambique. He joined Sustainable Forestry Management (Africa) in 2008 and moved to South Africa. His areas of interest include landscape management, payment for ecological services, environmental planning and rural development.

Maggie Mort is Professor of Sociology of Science, Technology and Medicine at Lancaster University, UK. She works primarily with participative ethnographic approaches in projects examining the response to perceived 'crises' in health and medical practice (such as the introduction of new technical systems) and in the response to disasters both from citizen and policy perspectives. She coordinated the Health and Social Consequences of the 2001 Foot and Mouth Disease Epidemic in Cumbria study (http://www.footandmouthstudy.org.uk/) and is co-author of *Animal Disease and Human Trauma* (Palgrave). She is currently coordinating the ESRC-funded study: 'Children, Young People and Disasters: recovery and resilience', in collaboration with Save the Children. Recent publications include 'Ageing with telecare: care or coercion in austerity?' in *Sociology of Health & Illness* (2013).

Phil O'Keefe undertook his undergraduate work at Newcastle University and his doctorate at the School of Oriental and African Studies, University of London. After teaching in Africa, he joined Bradford University's Disaster Research Unit where he led work on vulnerability analysis. From 1976 to 1980, he taught and researched at Clark University, USA. In 1980, he joined the Royal Swedish Academy of Sciences, providing leadership for a range of energy, environment and development programmes through the Beijer Institute. For the last 30 years, he has been based at Northumbria University, specialising in humanitarian work. His last major programme was climate change adaptation in 14 developing countries. He has written over 30 books and published over 250 articles in the scientific press.

Bryony Onciul is a Lecturer in Public History at the University of Exeter. Her main research interests include Indigenous history, community engagement, public history, Indigenising and decolonising museology, (post)colonial narratives, identity and performance, understanding place, and the power and politics of representation. Her research considers these issues in an international context; she has worked extensively with the Blackfoot Nations in Canada and also explores these issues in the UK, America, Australia and New Zealand.

Tim Padley has a BA in Archaeology and Geography from the University of Exeter and an MA in Scientific Methods in Archaeology from the University of Bradford. He has worked at Tullie House since 1997, and carried out most of the research and provided much of the curatorial input for the Roman Frontier Gallery that opened at Tullie House in 2011.

Catherine Roberts, MA, is an Associate Member of the Institute for Dark Tourism Research (iDTR) at the University of Central Lancashire (UK), where she is currently completing a PhD thesis on Dark Tourism and Other Death-Mediating Relationships in Contemporary Society. With a professional background in museum interpretation and education management, she is currently a freelance consultant for UK and European heritage and education projects and has a particular interest in how society manages conflict, remembrance and identity.

Shalini Sharma is an Assistant Professor at the Tata Institute of Social Sciences (Guwahati), India, where she teaches courses in social movement, environmental and memory studies. She was a Felix PhD Scholar at SOAS, University of London, 2009–13. Previously she worked with the International Campaign for Justice in Bhopal. Besides teaching, she coordinates the Remember Bhopal Trust and is actively engaged in the Bhopal Memorial and Oral History Project, led by survivors of the 1984 Gas Disaster. She is also currently translating poetry from north-east India into Hindi in order to bring public attention of this much-marginalised region to mainland India.

Billy Sinclair is Reader in Genetics and Conservation in the Centre for Wildlife Conservation at the University of Cumbria. He started out researching the evolutionary genetics of trees, but eventually found kelp forests to be more interesting than pine forests. He is interested in inter-acting with marine life and trying to understand its biology and behaviour. His main research interests are in deep-sea cephalopods, evolutionary biogeography of fish and the communities that interact (catch or dive) with them.

Jonathan Skinner is Senior Lecturer in Social Anthropology at the University of Roehampton. His interests are in tourism and tour guiding, arts health, disaster recovery, dance and physical performance. He has worked in the US, Northern Ireland and on the island of Montserrat in the

Eastern Caribbean. He is author of *Before the Volcano* (Arawak, 2004), editor of *The Interview* (Berg, 2012), *Writing the Dark Side of Travel* (Berghahn Books, 2011), and co-editor of *Great Expectations* (Berghahn Books, 2011) and *Dancing Cultures* (Berghahn Books, 2012).

Bryndís Snæbjörnsdóttir and **Mark Wilson** are a collaborative art partnership. Their interdisciplinary art practice is research-based and socially engaged, exploring issues of history, culture and environment in relation to both humans and non-human animals. Through their practice they set out to challenge and deconstruct notions and degrees of 'wildness' and culture. Underpinning much of what they do are issues of psychological and physical displacement or realignment in relation to land and environment and the effect of these positions on cultural perspectives. Their artworks have been exhibited throughout the UK and internationally. They are frequent speakers at international conferences on issues related to their practice. Their works have been widely discussed in texts across many disciplinary fields and regularly cited as contributive to knowledge in the expanded field of research-based art practice. They conduct their collaborative practice from bases in Iceland, the north of England and Sweden. The artists are 2014/15 Research Fellows at the Centre for Art and Environment based at the Nevada Art Museum, Nevada, USA. Dr Bryndís Snæbjörnsdóttir is Professor in Fine Art at the University of Gothenburg, Sweden. Dr Mark Wilson is Reader in Fine Art at the University of Cumbria, UK.

Philip Stone is Executive Director of the Institute for Dark Tourism Research (iDTR) at the University of Central Lancashire (UK). He holds a PhD in Thanatology and has published extensively in the area of dark tourism, including as co-author/editor of books such as *The Darker Side of Travel: The Theory and Practice of Dark Tourism* (Channel View Publications, 2009); *Tourist Experience: Contemporary Perspectives* (Routledge, 2011); and *The Contemporary Tourist Experience: Concepts and Consequences* (Routledge, 2012).

Clare Twigger-Ross is a technical director at Collingwood Environmental Planning, undertaking and managing consultancy and research in environmental and social assessment, sustainability policy and decision-making. She leads the practice's work in social research and social appraisal, and has particular expertise in understanding vulnerability and differential impacts from flooding. She has a specific interest in the impact of environmental risks (eg floods) on place and identities.

Gordon Walker is Professor at the Lancaster Environment Centre, Lancaster University, UK. He has a profile of research on the social and spatial dimensions of environment and sustainability issues. This includes work on environmental justice and the social patterning of environmental goods and bads; social practice, sociotechnical transitions and energy demand; community innovation and renewable energy technologies; and the concepts of vulnerability, resilience and governance in relation to forms of 'natural' and technological risk. His latest book is *Environmental Justice: concepts, evidence and politics* (Routledge, 2012).

Marion Walker is a Senior Research Associate in the Department of Sociology, Lancaster University, UK. She has research interests in the geography of education, research with children and young people and working with innovative methodologies. She was the lead researcher for the 'sister' Hull Flood Project: 'Children, Flood and Urban Resilience: understanding children and young people's experience and agency in the flood recovery process' and has strong links with the Lancaster Environment Centre working for the Catchment Change Management Hub, a knowledge exchange programme where people who are interested in the well-being of their local rivers can share understanding across river catchments. She is currently working in collaboration with Save the Children and is the lead researcher on the ESRC-funded study: 'Children, Young People and Disasters: recovery and resilience'.

Diana Walters works as an international museum and heritage consultant, specialising in access, participation, intercultural dialogue, education, management and professional development. Originally from the UK, she has worked in over 20 countries as a project manager, facilitator, researcher and lecturer. Currently based in Sweden, she works part-time for the Cultural Heritage without Borders NGO, overseeing museum and interpretation development in the western Balkans and in other countries in transition.

Nigel Watson is a Lecturer in Environmental Management at the Lancaster Environment Centre, Lancaster University, UK. He was a co-investigator on the Hull Floods Project and works primarily in the field of water and environmental governance, conducting original research and advising governments and catchment groups on institutional arrangements and collaborative decision-making. He has held appointments as a Visiting Researcher at the University of Oregon in the USA and as a Royal Bank of Canada Visiting Professor at the Water Institute, University of Waterloo, Canada.

John Welshman was educated at the universities of York and Oxford, and is Senior Lecturer in the History Department, Lancaster University, UK. His research interests are in the history of public policy in 20th-century Britain, on which he has published widely. His books include *Titanic: The Last Night of a Small Town* (Oxford University Press, 2012) and *From Transmitted Deprivation to Social Exclusion: Policy, Poverty, and Parenting* (Bristol, Policy Press, 2007; second edition 2012). His most recent book is *Underclass: A History of the Excluded Since 1880* (Continuum, 2006; second edition Bloomsbury, 2013).

Graeme Were has a PhD in Anthropology and is Convenor of the Museum Studies Postgraduate Programme at the University of Queensland and Director of Postgraduate Studies in the School of English, Media Studies and Art History. His recent work includes the monograph *Lines that Connect: Rethinking Pattern and Mind in the Pacific* (University of Hawaii Press, 2010) and the co-edited (with J C H King) volume *Extreme Collecting* (Berghahn Books, 2012).

Rebecca Whittle (formerly Sims) has research interests that centre on the sustainability of community–environment relations. Her current focus is on researching and developing local and alternative food systems that combine environmental sustainability with social and community benefits. She has also applied her environmental research interests to other fields, including sustainable energy use in the workplace and social aspects of hazards management – most notably during the Hull Floods Project, on which she worked as the researcher.

Index

HERITAGE MATTERS

Volume 1: The Destruction of Cultural Heritage in Iraq
Edited by Peter G. Stone and Joanne Farchakh Bajjaly

Volume 2: Metal Detecting and Archaeology
Edited by Suzie Thomas and Peter G. Stone

Volume 3: Archaeology, Cultural Property, and the Military
Edited by Laurie Rush

Volume 4: Cultural Heritage, Ethics, and the Military
Edited by Peter G. Stone

Volume 5: Pinning down the Past: Archaeology, Heritage, and Education Today
Mike Corbishley

Volume 6: Heritage, Ideology, and Identity in Central and Eastern Europe: Contested Pasts, Contested Presents
Edited by Matthew Rampley

Volume 7: Making Sense of Place: Multidisciplinary Perspectives
Edited by Ian Convery, Gerard Corsane, and Peter Davis

Volume 8: Safeguarding Intangible Cultural Heritage
Edited by Michelle L. Stefano, Peter Davis, and Gerard Corsane

Volume 9: Museums and Biographies: Stories, Objects, Identities
Edited by Kate Hill

Volume 10: Sport, History, and Heritage: Studies in Public Representation
Edited by Jeffrey Hill, Kevin Moore, and Jason Wood

Volume 11: Curating Human Remains: Caring for the Dead in the United Kingdom
Edited by Myra Giesen

Volume 12: Presenting the Romans: Interpreting the Frontiers of the Roman Empire World Heritage Site
Edited by Nigel Mills

Volume 13: Museums in China: The Politics of Representation after Mao
Marzia Varutti

Volume 14: Conserving and Managing Ancient Monuments: Heritage, Democracy, and Inclusion
Keith Emerick

Volume 15: Public Participation in Archaeology
Edited by Suzie Thomas and Joanne Lea